Telemetry Phenomenology Commonwealth

Corporate Surveillance and the Colonization of Personality

Michael Roth

Printed in the United States of America

First Printing, 2019

Library of Congress Control Number: 2018914819
Paperback ISBN: 978-0-578-43488-9
Hardcover ISBN: 979-8-484-36445-9

Lensgrinder, Ltd.
Kirkland, WA 98033

www.lensgrinder.com

Studies in Phenomenology & Social Philosophy

Lensgrinder, Ltd

Kirkland, WA

For Georgia

Contents

Acknowledgements

A very special thanks goes out to Bill Vaughan who read the entire manuscript and provided valuable insights on some of its content.

I would also like to mention Eric Jaeger, Laura Haber, Bryan Belmont, Eray Ozkural, Chip Hanauer, Steven J. Wagner, Pete Resnick, Lynette Reid, Bill Martin, and Babette Babich for the occasional exchanges, public and private, that contributed to some of the content of this book.

There is always a corporation at work in every publication. Shout out to Lensgrinder, Ltd. and Big Brother Amazon, Inc.

Epigraph: "...broken clock..."

Introduction

> Where everyone is the other and no one themselves, humans are cheated of their ecstasy, their loneliness, their own decisions, and their own direct connection to the absolute outside, namely death. Mass culture, humanism and biologism are the cheerful masks that, according to the insights of philosophers, conceal the profound boredom of an existence devoid of challenge. The task of philosophy would then be to shatter the glass roof over one's own head and directly make the individual the monstrous once again. (Sloterdijk, 2017 p. 172, riffing on Heidegger)

Human existence is both rooted in a deep relationship with others and an equally fundamental relationship to one's own potential. As the rule dictates, everything follows from a contradiction. More than an abstract tension between being with others over and against one's unique individual self, there is a real concrete contradiction enabling this 'cheating' to take place in the act of making public a line of thought that offers a material resolution to this formal conflict. "Beware the urge to be seen from far away", the author writes to readers he cannot see and who cannot see him. This sense in which the book separates us surfaces in both space and time. I am here now, and you are wherever you are in the future reading this. Even if we are friends and have some tangible past connection, you are not here now, and I am not there then. I tell you to beware this urge to be seen from far away while being seen from far away, while seeing you from far away. I fail to heed my own warning as I make it and you will see from way out there that everything follows from this. Not only does the urge itself deny both the deep connection to others *and* to one's own potential, but the warning realizes a further aspect of that same urge. In the desire to be known and to be seen, we run the risk of being transformed into performers for an audience where both the emergency and the warning are announced by actors on the stage. It may turn out that calling the subject of this contradiction 'privacy' and then publishing a book about it is a fool's errand.

When I write I am alone, in a sense, but not at all. Even if no one ever reads this, I sense the possibility of the future readers in every sentence. I am now, at this minute, revising with you in mind. Even if I am just talking to myself, spinning human yarns, to pass the time or enjoy the day or working through some set of menacing thoughts that keep coming back again and again. There are other people here in every emission that I set onto the page. They may be future people or distant people, but they force me to elaborate my meaning, to make clear what I am getting at, and to express the facets of it in a way that I would never have thought to do were I just meditating or –if it is even possible– sitting alone with myself. I do not think it is possible to develop a line of thinking without presenting it somehow, as a piece of writing, a performance, as a work of art like a painting or a motion picture, as a recording, or some other kind of artifact. If your memory is significantly better than mine, even an inner monologue could function in this way. The process of presenting or publishing the line of thought self-develops, it forces the thinker down pathways they did not see before the ideas accumulated in a sustained piece of work. The persistence of the thinking in object form presents the thinker with a constantly growing stream for analysis and feedback. As the thinker becomes heavier and heavier with each passing sentence or image or note, the burden of the thought process grows, but so do its possibilities.

What compelled me to write out this warning in a presentable form was this very sense that the urge to write this book was itself a sign of the malady the book tries to unearth. I am here now, and I desire to be seen from far away. What follows is not an elaborate attempt to avoid this conclusion. The tension is real, and it is effective throughout the project, haunting it at every turn. It is the reason why I have something to contribute to the discussion. I am not asserting that this is a universal problem, that it is the essence of human beings to be caught in this trap, to be threatened by this danger, and in need of this warning. I am a living human being and I have observed this urge in my peers and I have observed it in myself. I see it at work in many areas of our lives: in our politics, in our entertainment, in our consumer habits, in our education, and in countless other places. And when I say 'we' or 'our', I am talking about you and me. It is just us here and even if that is not the end of it, it is the beginning. Because our engagement only spans that far and has no onlookers or commentators, no audience to speak of, you will frequently find yourself in a position to confirm or reject the applicability of any such statement. Publication may make a work available to the public but reading and writing are always a one on one relationship. It may very well be the case that you bring an entire clan, tribe, or nation with you to this engagement, but as far as the genre is concerned, you and I are speaking on behalf of those tribes, we are the tribes-becoming, making their positions known here and now or there and then with the authority of our individual being. Even if the era of the book is dying out and hardly anyone reads any more, the exchange between us is proof that it has not yet died and the possibility of issuing such a warning still stands. The fact that you must privately soliloquize your objections is not contrary to my point, it is my point.

Given these claims that I have made without providing any data, how can I possibly explain the fact that amidst this condition where people have become bound by the urge to be seen from far away, to be the center of attention, and to have all eyes cast upon them (or to at least act as though that were the case), there is nonetheless a loud and persistent demand for the protection of privacy? The hypothesis that drives the current work is that privacy is a shorthand reference to our concern for the associations that we make, the relationships we have, and the groupings that we identify with. Our concern for such privacy amounts to a basic desire to maintain control over these areas of our lives. The reaction that comes from a threat to it is a fear of a coercive and disciplinary power that will do something to us that we do not want. The conditions we suppose are that we do not want this invasion and that we are able to foresee and forestall it. I have my doubts, and they arise from my sense that these invasive means are so deeply affective in our desires and orientations that we cannot see them. In what follows, I explore some of the common notions of privacy and demonstrate the ways in which they become confused when confronted by modern machinery specifically designed to co-opt those associations, relationships, and groupings that are the bedrock of any human being's lived experience.[1] I hope to provide an alternative demonstration of how experience — rationalized as privacy, even in the hands of many who would claim to protect it— is itself a socially crafted device that can be used for remote manipulation of the agent. It is the cultivated inner sanctum that becomes the subject of all telemetry, emitting events that self-describe a person for the sake of building an aggregated understanding of associated, related, and grouped behaviors that can be pushed and pulled via intervention.[2]

---

[1] See the appendix for additional context.

[2] The entire first part of this book will provide a detailed definition of "Telemetry." Briefly, in modern-day computing systems, the term refers to the logging information that the software emits to describe its operations as it performs them.

Telemetry

"The act of describing the detailed workings of a distributed application when it is triggered into operation by an interiorizing exterior force is already an engagement in deconstruction: aka sniffing or philosophizing with my nose."
–Further Investigations of a Dog

Chapter I:
What is Telemetry?

Collective and wide-scale monitoring is crucial to the advent of any instance of rationalization within human civilizations. Telemetry, remote measurement, is the pattern by which this collective monitoring is performed. The word itself is both contemporary and ancient. The Greek origin accounts for the ready-made elaboration of it as remote measurement, but the term has come into common use with the rise of networked computing and the need to monitor processes running headless on remote desktop computers. 'Headless' indicates that the process is running without user involvement. Instead a computer may be hosting a daemon process or service that is executing the operation in the background with no interaction between user and software. Although this machine-based execution of the process is crucial to any complete understanding of the current state of telemetry, it is not crucial to the notion in its origin.

As an example, take the case of workplace operations performed by manual labor producing goods on a factory floor. As part of an effort to increase profitability, executive management may use an efficiency expert to examine and evaluate production for the sake of fine-tuning it. We can say that this agent acting on behalf of the enterprise is performing remote measurements of the practices and procedures. The evaluation is in one sense close to the work being measured, since without some point of contact –via observations and measurements– it will not reflect its true status; but in another sense, the evaluation takes place far away since the observer is not directly performing the action but is gathering data and reporting on it. The observers are not decision makers, they merely gather information and produce analysis. Even if the analysis includes recommendations for improvements to the process, the observer does not make the final decision.

In cases where the observer is replaced by computers and the measurements are performed by a digital mechanism relaying data across a network, there is need for a local agent that is close to the process, close enough to gain a detailed view of its operations, but not identical to it. Whether that agent is a software library or component embedded in the executing application or a human observer carefully watching the operator's movements and actions, the results of the measuring process are noted and compiled for the sake of later deliberation. The company leadership will study the results of the report and make decisions or changes to the production process based on it. Telemetry refers to the specific measurements made on remotely executing procedures, and it also refers to the collection of that data. The data source might be multiple operations happening in sequence or multiple instances of the operations happening in different parts of the factory or a combination of the two. Since the data comes from multiple such observations, it must be brought together, fed into sets of analysis producing operations (analytics), and aggregated. The telemetric can fuel any form of decision making or rationalization of processes, in collection and analysis, to carry out its task through the vehicle of a multi-faceted system with multiple moving parts that are both orchestrated and integrated. The purpose of this chapter is to begin showing those moving parts that make up the end to end system for carrying out detailed and highly optimized remote measurement. We will focus on the details of a specific kind of modern computing environment, but the intent is to uncover a set of patterns integrating observation and performance to formulate a logic of procedural governance.

We will later see that the telemetric patterns offer specific innovations for introducing feedback loops into the various kinds of rationalizing measurement activities enabling the consumption of those measurements directly within the process being measured, but first we will focus explicitly on the measuring itself. The examined procedures and processes are part of a larger scale operation of some kind like, for example, the production of pins. Observations of individuals engaged in the production of pins may be performed by multiple observers taking account of how different people accomplish the task in cases where there is no division of labor. The actions of multiple observers carrying out multiple observations of many different pin producers are brought together in a reporting mechanism that may both capture some of the details of the various observations as well as some of the aggregate statistical information. On average, how long does it take to produce a pin, how many can the average worker produce in a day? The observations may capture details on these same statistical aggregates with the intent of trying to understand the implications of the measurements. Among the slower producers of pins, what sorts of practices were used, how were the actions broken down into parts? Likewise, among the faster producers. From this, analysts may be able to formulate a theory about the practices that slow down manufacturing and those that speed it up. Depending on the focus in each of the observations, many different factors may become relevant through the reporting. The general attitude of the workers carrying out the process, the workers' physical characteristics, the environment within which the work is done, the amount of help and mentoring that goes on between junior and senior workers, and how the content of that interaction impacts both the quality and the rate at which the pins are produced. This analysis informs the decision makers about the details of the process and offers them raw material for optimizing the process of production. Gathering data in this way so that it can factor back into the design of those very same observed practices constitutes a feedback loop. The data gathered by the observers can be used to instruct and train the pin producers so that they use more uniform methods approximating the actions of the most efficient among those observed. Telemetry and the telemetric are crucial to the practice of using feedback loops to boost the effectiveness of any collaborative activity. This, in turn, amounts to the further rationalization of the activities. Why measure the data unless the results of those measurements are actionable? And one form of action is to loop the findings from the measurements back into the process that produced them in the first place.

This does not mean that optimization amounts to laying off those observed workers who were less efficient than others. The observations come from a quarter that is not personally invested in the outcome, rather they are merely displaying the process of production in its detailed properties as though it were a machine to be tuned. Suppose the analysis identifies three sets of procedural types: 1) those who are highly productive and using efficient methods to do their jobs, 2) those who are trying to be highly productive but using inefficient methods, and 3) those who are not productive and lack any sense that their actions could be improved. The analysis may formulate an approach where the second group is optimized to be more like the first group thus making the third group redundant. The standard worker response to the efficiency expert was that layoffs were coming. Although that is a possible outcome, it is not the only one. In such investigations, we describe the human involvement in measured processes as though it were merely a mechanical part. This is an essential characteristic of the telemetric approach. If the observations determine that the third group in the analysis are burdened by personal concerns effecting their productivity, the outcome may be that investment should be made to help solve those personal problems. The decision will be based on data and cost and not likely to have anything to do with specifically human concerns, unless all other factors are equal or nearly equal. Or they will give them some weight based on the impact to morale layoffs might have. This factor, however, could swing the result either way as it might have the impact of making the remaining workers more motivated from fear or it might depress the remaining workers and make them less productive. From the perspective of the enterprise, these matters are all just math applied to the calibration of a machine.

As the production process undergoes greater rationalization –for instance, through the division of labor that Smith argues made the process of producing pins so much more efficient– the relative impact of these measurements will increase. The telemetry articulates the invisible hand, making it visible. Once the task has been broken down into parts, the actions themselves yield more information for analysis, further helping to streamline the process. The micro actions of each aspect of the production line can be observed for a more fine-grained insight into their specific efficiency, and then measured again for comparison with other workers using the same or different methodologies. This allows the process to undergo a still more fine-grained testing of actions and operations with each iteration providing greater precision in the results. The observation is now more finely focused on the smaller component parts that go into the overall production of the item. The decision making becomes better informed in the design of the production process and better informed about the impact that minor adjustments might have on it. At some point, the efficiency experts and decision makers developed the ideas and the know-how to fashion a mechanical device to perform the process. As the actions making up the production process are broken down into component parts, the specification for such a machine becomes more and more clear. Rationalization already mechanized the process and provides the critical first step in emulating it in the machine's design. The machine can then be geared to just these micro actions with the most efficient movement in mind. And once a machine is introduced into the process, the observations continue to make improvements to its design. When we gained the ability to digitize the machine that makes the pin, we gained the ability to digitize observations of it.

The machine not only contains features that allow it to construct a pin this way or that, it contains features that include measurement capabilities. And the more and more fine-grained the actions of that machine, the more and more fine-grained are the measurements that can be made on the process of production. The data from these measurements can be streamed into collection services that do the work of bringing the data from all the machines together, allowing for statistical aggregations in a fraction of the time the human observers required. Even the creation of reports can be automated so that one can gain something of a constant and all-pervasive view into the entire process of pin production in near real time. Not only does the production process admit of automation, but so does the measuring process. To the extent that machine behaviors can be reduced to software driven techniques rather than hardware or mechanically based, the automated observations can be used to directly feed back into the software engineering process. If the software is 'intelligent' enough to adjust operations in real-time based on feedback received from various monitoring systems, the software can be further enhanced to make automated adjustments to its operations based on telemetrically generated feedback. By 'intelligent' we need not be depending on something exotic or from science fiction. A configuration-based rules engine in the form of a simple state machine is sophisticated enough to be able to adjust its behavior using data coming from a telemetry-based feed. The executing code need not change, in this condition, rather the configuration contains the description of the operations that need to be run, and the configuration can be updated – without recompiling the code– from the telemetry data and whatever intermediate analytical processing it goes through.

It is not just that a more micro-engineered process enables greater remote measurement of the processes of production, but that it *requires* greater remote measurement as well. The burden of constantly watching the machines defeats the purpose of automation since the need for direct human observation would add cost and decrease reliability. The machines can run 24 hours a day, 7 days a week, all year long. Capital's dream of the true work week has been achieved. Are the machines running okay? Are they doing what they are supposed to do? If a person is needed on site to ensure this, they might miss something, they might get tired, they might require time off. Automating the process of observing the machines and then attaching that automated measurement directly to the process is the logical consequence and fullest realization of the dream. Now the machines are watching the machines and the machines watching the machines are also watching themselves while providing constant monitoring of their up time, of their efficiency, and of their maintenance requirements.

There may still arise circumstances, however, where the machines are not able to respond automatically to some conditions. In those cases, the telemetry system can raise an alert and notify a human being. The human beings brought into the equation through notification can be monitored through processes used to improve the automation, improve the monitoring, and –if possible– automate the response to the problem. Software based ticketing systems, for example, can be used to contain all user provided data describing the human being's intervention in the system. The human being is alerted to the condition at such and such a time, acknowledges the alert so many seconds or minutes later, declares the issue mitigated after some further measure of time has passed, and so on. The human being inputs the problem condition and the steps taken for mitigation, leaving a record of what happened and what was done to address it. Such software to manage the process of human involvement in an otherwise automated set of procedures makes it possible to measure and optimize even that portion of the operation that transcends the automation, thus giving the whole process a more predictable and measurable set of operative parameters. A manual process can thus work as a training ground that takes feedback from human intervention and uses it to augment the automated process.

Telemetry systems integrating with other telemetry systems providing a total end to end experience that aims at automation through and through. The people who intervene in the production process based on machine notification will report on their actions and that process will be observed and made more efficient. Reports will be generated, metrics captured and used to understand where the inefficiency of the system lies. Ideally, the people engaged to intervene in the problem situation would be the same people capable of improving the automation. That is, if the technicians were experts in both the measuring and in the processes being measured, they would also know how to modify the stream of operations so that the troubleshooting and problem-solving steps, to the extent that they can be formulated in terms of simple state machines, can be further rationalized as additional programmable steps and operations to augment the process. The process of responding to notifications and acting on the system makes the respondents experts in the efficiency of the automated system and thus gives them the perfect viewpoint from which to optimize that system.

In the world of software engineering, this last phenomenon is called 'dev ops' where the developers of the software system are also the maintainers of the operations. If the machine breaks down in the middle of the night or goes off the rails in the middle of normal production, the software developer is the person notified. Since the software developer is generally tasked with making improvements to the system, these alerts would constitute an inconvenience to the same people who are able to reduce the likelihood of the notification, giving them a clear incentive to streamline the process and improve the automation and management capabilities. Being woken up in the middle of the night to solve a computing problem is the best incentive to automate a response to the identified problem. These sorts of changes to the production process are the conditions that have made it possible for some of the largest corporations in the world –in terms of capitalized value– to become lean and mean in terms of their number of employees. The highest valued corporations used to be the corporations with the most employees –like the railroads, automobile manufacturers, and oil companies, but under this new model, there can be companies with an extremely high ratio of revenue to employee. It may be the case that some of these employees are higher skilled and associated with a more competitive labor market, forcing the companies to pay them more, but the better ratio justifies the impact based on productivity per capita making these modern giants the most profitable enterprises the world has ever seen. In the last section of this book, we will look more closely at the confluence of corporate organizations and the telemetric patterns of action enabling this extreme efficiency. At this point, I want to remain focused on the patterns without introducing larger scale interests and associations. This way, we will be able to see these patterns at work not only in the collective associations relevant to commonwealth, but in human experience and personal association when we turn our attention to phenomenology in the second section.

My experience with telemetry began when I went to work as a software engineer on an internet-based service. Previously, I had been working on products that were shrink-wrapped and sold from a shelf and in a box. The customer for those products purchased the box and the license to run it, installed it on servers located in their data centers, and then had to manage and maintain that product themselves. The product, in my case, was a management software platform that provided the capability to monitor other software services and systems and introduce automated responses to that monitoring, based on user-defined rules and workflows. The purpose of this product was to make it easier and less error prone for human beings to oversee the growing machine population that their companies required to provide various basic functions like email and document management. As the companies that were the typical customers for such products began outsourcing these functions to larger cloud-based services, software engineers started migrating into organizations focused on cloud-based products. These are products that provide comparable services, but which do not ship in a box nor require the customer to build out their own infrastructure and skillset for running the services. That first service I ever worked on was a cloud-based management platform that tried to provide the same functionality as the on premises, shrink-wrapped solution had provided. You could start outsourcing some of the management automation overhead to a cloud-based service thus reducing the footprint of locally owned and operated machines required to do that same work. My personal migration matched with the trend in the industry at that time with hundreds of thousands of software engineers moving from locally run solutions to remotely run solutions where the migration was euphemistically called a "move to the cloud". The advent of large-scale data collection to monitor such systems was essential for enabling this migration. It would have been impossible for large corporations to take on the infrastructure management functions for billions of users without a proper telemetry subsystem to monitor and manage all the resources required for the task.

The management service started out as a product for managing desktop computers and servers that were being used for various business objectives: car dealerships, dentist offices, banks, pharmacies, and so on. Over time and integral with this large scale move to the cloud, as the industry was moving into various kinds of computerized devices, the management service became more and more device centric. The common paradigm of the device may have been the smart phone, but the Internet of Things (IoT) made it possible to envision all sorts of common household items as devices that must be remotely observed, remotely monitored, and remotely managed for proper functioning and efficiency. Your refrigerator, your car, the burglar alarm system in your house, these are all devices requiring headless operations and management under an IoT platform. And the notion of service-based management is where the implementation of telemetry really hits its stride. To the extent that devices of all kinds start coming with their own local 'agents' that are capable of measuring the ongoing operations of the device, to the extent that the data describing these measurements can be sent to central stores for comparison and statistical analysis, all processes of production and operation become visible to the "decision makers" working to make things run more smoothly whatever that might mean: more profitability, more productivity, or even with greater impact and value to users.

Seeing so much from far away ruins our eyesight for what is up close. The telemeters can find their way into a process even when they are rejected by that process and made redundant by it. The view from far away saturates all those things that the human beings in the equation can see, think, and do. The agents collect the data, the agents stream the data to the central collection store, and the processing components munge and land the data when it arrives there. Then the human observers step in and analyze it to better understand the trends and tendencies through their evaluation. The human role in the understanding of the processes and procedures becomes the position of the data scientists, with their preferred toolbox being statistics. It may be the case that this data-oriented point of view, this analysis, reveals itself as the new meaning of being up close, the new way in which our "vision" works. The world as we come to know it is the world of the statistical aggregate, of data driven decision making, of the trend, of the line chart and the various tactics and techniques that can be used or codified to improve the performance and increase the quality of service. Telemetry "in and of itself had to reshape human consciousness." (Gleick, 2011 p. 47).

Quality of Service

When measuring the quality of service of a cloud based, high volume, data collection and distribution service, we speak of the case where the monitoring fails us as 'flying blind.' When the monitoring goes down or no longer provides the correct perspective on the system's operations, we are sightless, unable to see. And yet, human beings do not constantly look at the monitoring output, instead they rely on alerts and notifications to trigger when something has gone wrong. It is the monitoring that has stopped seeing and those who would be notified in the event of a failure will not be notified. The monitoring system is a manner of seeing, and based on this seeing, the operators of the system are notified by phone of problems or issues requiring their attention. This is not a simple form of seeing, it amounts to keeping watch. Under conditions of high-volume data transmissions, however, what counts as a matter requiring the operators' attention? The service itself must be described or defined in such a way that the monitoring output captures its notion of health, its proper working condition. When that health has been captured and programmed into the monitor, it is poised to ensure that the health of the system is maintained and to send an alert when it is not.

Consider a service that is a system of relays. A message is received at one point using a standardized protocol like http and then the message is processed. A high volume of requests and the limitations of the point of entry require that the message be dequeued from the service in rapid cadence. Such a point of entry service is called a listener indicating that the service waits for external callers on the network to initiate calls to it or make requests. Waiting for requests is called 'listening'. Queue size on a listening service is a finite resource and must be managed carefully if the service is to relay large numbers of requests coming in from multiple callers. Some requests, actions instructing the server to do something, can be pulled from the service listener and then dispatched to an internal handler that examines the message for various conditions before handing it off to another component that will persist the content of the message in some durable store or relay the message to another service. The hand off may include a validation step and the recipient service may respond with a success code once it validates the inbound request and estimates that it has all it needs to properly process that request. This response clears that request from the listener service's queue, making room for more inbound requests while continuing to process it internally and taking extreme caution to ensure the request gets –eventually– to where the service has promised to deliver it. All this depending on characteristics of the message, what kind of a message it is, what resource it targets, what system it is relevant to, all of which will tell the internal handlers what needs to be done and where the message needs to be relayed. One common handler operation is a simple partitioning of the messages for the sake of storage. The relay system, let's say, ultimately wants to persist the contents of the message in a store, but a single store could not handle the volume of data received. The system might invoke partitioning logic to determine what types of messages go into one store rather than another. This may be an arbitrary piece of the message's content or it might be something explicit in the message's metadata or characteristics that qualifies it for this or that storage partition. All this introspection and examination of messages requires organization, rationalized and properly factored code that moves the requests through the system to their proper destination. When we say that we are 'blind' relative to such a system, what exactly is it that the monitoring no longer sees? Let's look more closely.

In the early days of service-based processing of operations, the primary 'monitor' of operations was the simple service log. This was a log maintained by the service listener describing the operations from its point of view, meaning that the internal logic and operations of the service were not captured, but the listener's operations were. The logs would indicate that a 'Get' request was made at this time and resulted in such and such a payload size being successfully dispatched in response. Or an internal error took place or maybe the resource itself was not located. The error surfaced by the service code to the listener service might be written to the log in those cases where there is something further to note. These logs could then be individually inspected, by a person trying to respond to a user complaint, or the logs could be processed by additional application code that parses them and looks for common patterns describing what kinds of responses were made to what types of requests. In this case, we might recognize that not all observed service behaviors were simple in their alignment with each other. That is, it may be the case that not all user requests targeting a resource got the same reply. It may be that other conditions occurring on the system were impacting the kind of response that the caller received. In this case, we might observe transient failures or unexpected behaviors specific to one single request. In some of those transient cases, the system may recover from the issue and the recovery may be sufficiently swift that the caller, who might make use of some retry logic, may have hidden the problem completely from its user. Simple listener service logs, in this case, soon proved insufficient for providing the level of detail required to understand the nature of problems and issues. Some errors, for example, might be due to a bad user request, a typo in the address or resource locator provided by the caller.

From the listener service's point of view the address seems fine, but the internal service code was unable to locate the targeted resource and so is in a better position to supply details on what makes that address incorrect. There are other types of errors that might succeed when the caller tries again, where whatever condition that caused it to fail the first time will have cleared up during the second try. A simple line item logging approach might fail to correctly capture the impact that event has on service availability, since the web log shows 1 failed request and 1 successful request. A simple computation might render this condition as 50% availability whereas a more complex approach that takes the user's point of view into account might show it as 100%.

In the case of internal errors, where something went awry in the internal handling of the operations behind the request, there could be any of a million different reasons for the problem. Since the amount of code and the conditions under which the code will have been processed can vary greatly from one request to another and since there might be some security concern preventing the system from surfacing the exact error message and call sequence triggered by the request, the log alone may be insufficient for understanding what happened. Maybe the problem occurred while dequeuing the message from the service listener or while routing it to the storage partition? Maybe it happened while sending the data over the network or because of memory pressure on the listener service machine? Or again, the caller timed out because the message sat in the internal queue for too long? There are just too many moving parts internal to the system and opaque to the listener service for simple logging to provide enough visibility into operations, what makes them succeed and what makes them fail. To get greater detail on the error, the system developers would have to return the call stack or sequence of procedural steps taken when the error was raised. This act of surfacing the call stack, however, leaks critical information about how the service functions and informs a savvy user of ways they might take advantage of it, ways they might leverage errors to hack the system's operations, take control of its processing, and gain access to resources they should not be able to access. The log clearly had limitations because it could only show what the listening service itself was doing and seeing. It could not surface the internals of that system without leaking the internal details of the system to an external caller or user. The only data that passes through the listener service is the data in the response returned from the internal service to the caller. Seeing is tactile in this case, the listener service can only log what it has touched, and a monitor of that log can only see what the listener or point of entry service has made public in the response. Telemetry makes it possible to avoid exposing these internal details of the system's operation to the general public and to the users of the system. Telemetry makes it possible to keep the system's internal details private.

Quality of service-based telemetry does not rely on the listening service for observation data on the performance of the system. Instead, the developers of the system can design a set of events to describe the internal operations and those events can be fired as part of internal processing. When the system dequeues a message from the service listener, it fires an event indicating that it has done this. When the routing subsystem takes the message from the queue, it can fire an event. When it determines the partitioning and routing mechanism to associate with the inbound message, it can fire another event. When it sends the data off to the external storage mechanism to persist the data until it needs to be retrieved again, it can fire yet another event. These events can then tell the complete story of the request arriving in the system and going through the various procedures and operations that comprise the processing pipeline. The events can be correlated so that when each of these events is fired an identifier is included associating it with other events fired during the processing of that same request from that same caller. The correlation vector or identifier can be passed to any other internal system, like the storage handler, so that their operations –to the extent that they are complex actions– can fire events associated with the common identifier used by each of the different systems that touch the request during its travels from machine to machine or from service to service. The identifier can be returned to the caller in the final response to assist in debugging something that has gone wrong during end to end processing. The caller can log the outcome and the identifier on their end and surface this information back to the system providers when evaluating any issues that might have transpired, or the caller might resend this identifier back to the service indicating that it is retrying an action.

The events that are being fired are forgotten by the system as it is firing them. By 'fire' what we mean is that a call has been made to a telemetry component or library and the calling code does not have any insight into what might happen to the event once that component gets ahold of it. This is not just 'fire' it is 'fire and forget'. The processing code is not going to worry about the success or failure of the fire operation, rather, the system is just going to dispatch the operational status to the telemetry mechanism and go on about its business handling the request and responding to the caller. What the telemetry component is doing with the event is –from the point of view of the processing code– an internal implementation detail of no concern to it. The component might be a wrapper over a log file or it might be a local aggregator of some kind. In the latter case, it might be a background process that receives the event content and handles it independently of the code that fired it. This separate process might wake up periodically and examine the events in its queue, it might perform a summary operation to associate the various events of the same kind or coming from the same source to build an aggregate picture of the various operations and outcomes. An aggregated time dimension might be introduced by this local processor. Rather than dispatching each event as it is fired, it might store up batches of events over a 1-, 2-, or 5-minute period. Any events that are otherwise equal except for the time at which they were fired would be pressed together into a single aggregate event thus reducing the size of the data coming from the machine. A system that processes 100 successful 'Get' operations against a resource would compress those 100 events into 1 single aggregate event containing all the common attributes plus the time stamp describing the moment of aggregation and a new value describing the count of events compressed together. All this complexity is opaque to the processing code which has forgotten the event immediately after firing it and moved on with its operations.

This kind of summary reporting spanning a time dimension and providing statistical aggregates, providing an average of any computed measurements, may be hidden inside the telemetry mechanism but it still poses a great risk to the system. On the one hand, if the local aggregation is minimalist, the system may be overwhelmed with a flood of data. There may be many machines running the same service and handling the same types of requests. If all those machines send their data to the same processing system in raw format, there may be too many events to handle. Pipelines for processing event output have limited capacity. Even if that capacity is very large with thousands and thousands of machines dedicated to the processing of received events, the proliferation of data sources for these events will always manage to challenge the capacity of the receiver logic and topology. The cost of this might be paid for by aggressive sampling on the receiver side such that only some percentage of the events are processed, and the rest are ignored or dropped. Or the design may call for additional aggregators in the receiver system resembling those on service machines back at the event source. All these mitigations to the scale problem pose dilemmas for the functional purpose of the telemetry system. That is, if the source servers aggregate for too long, then the latency of information transmitted to the data processing pipeline will increase and, as such, will reduce the timeliness of whatever actions are taken in response to that data. If the purpose of the system is to provide a real-time monitor for the behavior of the primary mechanisms (like the mail service or file service that is being monitored), then the introduction of a five-minute delay could be catastrophic. In the case of an outage impacting an entire data center, for example, waiting five minutes before notifying someone to begin investigation could have dire consequences, depending on the purposes that the system is serving. If those five minutes constitute five minutes without air traffic controller service or five minutes without emergency responder services, then the latency might be a matter of life and death.

The system design decision will need to take the time dimension into account in the most essential manner. It will be critical in terms of both the acceptable latency in the system and the acceptable scale of the processing pipeline. As such, the time dimension is crucial for connecting the requirements of the system monitored with the requirements of the system that monitors. The greater the latency that the system can withstand, the less pressure there will be on the system to address scale issues. The greater the scale that the system can withstand, the less pressure there will be on the system to address latency issues. In the end, cost is the most significant factor. If the enterprise has deep pockets and can afford a state-of-the-art telemetry system to go with its state-of-the-art service, then the scale requirements and the latency requirements can both be optimized. Regardless, the time dimension shows itself at the heart of this matter and the way the events are rationalized in terms of the time dimension will lie at the center of the design considerations.

The aggregate output of the intermittently processed events might then be written to a persistent location: either some local log that accumulates the ongoing operations of the machine or the aggregating and observing agent sends the report of operations off the machine to some other service that collects the data. Even the former approach is likely a first step in performing the latter approach. Under conditions of high scale, there might be a thousand or more servers handling requests for the same application. This means that all the user traffic to extract or write resources to the system, are spread out across thousands of different servers all doing the same thing: all taking requests from the listening service, routing them via partition logic, and then sending them off to some storage service. It is not feasible to manually go through thousands of different servers to check on the behavior of the service, so regardless of whether the observer writes its observations locally or immediately sends them off to some remote system, there must be some automated process for collecting the aggregated observations together and taking them to some common place where they can be integrated into a comprehensive view of the system across all its instances or nodes.

Once the events from the various servers are collected together in aggregated form, the quality of the service can now be determined. The events carry information about the state of the system, they describe the request that was received, for example, they describe the partition that needs to be targeted with the data, and they describe the data center where the payload must ultimately land. These attributes of the request, embedded into the events as if they were synthetic qualia, constitute the dimensionality of the event. The dimensions describe aspects of the processing pipeline and system at large to classify and categorize the operations based on functionality easily recognized by any user or contributor to the system. All would recognize that when they use the system, they make calls to it that target specific resources. If they discover resource or resource address as one of the key attributes in the events, they would understand that the data associated under this attribute constitutes the resource dimension or domain of possible values. To the extent that the events also capture the size of the posted or returned payload and the amount of time it took to process the monitored portion of the operation, we can speak of aggregating across this dimension in terms of specific measurements: measurements of size and duration. The introduction of dimensionality into the event allows the events to describe, in aggregated form, the behavior of the system across any and all its various attributes. Furthermore, multiple attributes or dimensions can be used to describe complex aggregates and associations that hold across the events as if they were distinct persona expressing themselves in the individual activities of the system.

In addition to considering resource or resource address as a dimension, we might consider the caller to be a specific dimension. Let's say for the purposes of our example that different calling identifiers like logins or IP addresses can be associated with each of the requests made to the service. Also, we might capture the server information itself in the event where the servers have been broken out into a topological hierarchy. Suppose then that the thousands of servers are spread around the world with some placed in the US, some in the UK, some in Malaysia, and still more in India. Within each of these geographic regions there might be further associations between the different servers. They may be connected to a common networking device or even reside within the same physical rack inside a data center. An identifier for these "scale units" within the data center may be captured in the events fired by each of the components processing the request making it possible to determine that callers are properly affinitized to a processing data center based on geographical proximity. In addition, the components that process the request may themselves be broken down into finer and finer detail. That is, it may be that each component is made up of multiple classes and each of these classes is made up of multiple methods or functions. The function that is currently executing might be associated with the firing event and multiple events may be fired by a component. The software processing, the topological design, the calling patterns, the resource targeting, all these different aspects of the operation may become dimensions in the events being fired.

These dimensions, and the ability to aggregate the captured measurements in terms of them, are the ingredients for providing the quality of service of the system being monitored. This is a fine-grained rationalization of the system. At the top level, the most coarsely understood aggregate of the dimensions, it might be possible to say something like 99% of the requests are being handled successfully. This may sound like a good number unless of course you happen to be in the 1% of requests that are being dropped due to an error unrelated to anything the user has done. And that is the key here, what kind of error? How can we zero in on it? It may be that the percentage goes from 99 to 100% when focusing on those requests handled by the US data center. It may also turn out that all the data centers except Malaysia are at 100%, but within Malaysia the number is more like 90% and the overall skew to 99% was because significantly more requests are being handled in the other data centers during a focused stretch of time. Going deeper into the Malaysian data center, we may discover that one scale unit is showing 50% success rate whereas the other scale units are all at 100%. Scale units are a small collection of machines chain linked together in some fashion to redundantly provide the service in case of a problem forcing a failover from one machine in the scale unit to another. It may be considered the smallest functional unit in the machine topology and might be made up of several machines with different roles that work together to provide the end to end service functionality to callers. As we dig into the one suspect scale unit, we discover that there is an API or function in the code that has a 0% success rate. And this may be due to some problem with the configuration for routing requests that come into that scale unit and which is required for the proper functioning of that method. By drilling into the dimensionality of the data, we went from thinking things are going well, functioning at 99% reliability, to thinking that there is a place, a specific place, in the system where a catastrophic failure has occurred and needs immediate attention. Having telemetry that carefully segments and then aggregates the operations was critical to being able to properly monitor the system's quality of service.

Once monitoring and measurements have followed this bifold process of segmenting and aggregating, it becomes possible to trigger notifications and alerts based on this monitoring. If the common store of event measurements is the source data for all monitoring, then the data can be constantly queried and evaluated for anomalies along the various dimensions being captured. Furthermore, the conditions themselves can be variously configured so that the notifications and alerts may be directed at people with different areas of expertise depending on the type of problem detected. Problems in configuration might need to be routed to one set of people, problems in networking connectivity may need to be routed to a different set of people, and so on. The better the process has been understood, the more fine-grained the details on the behavior and execution of the system, the more accurate the alerts can be at getting the right people to pay attention to the right problem at the right time. The outage we described above might appear to be a small matter, but after drilling into the details based on the fine-grained dimensionality of the data, we discovered that the problem effects an entire region. We might do further investigation to discover that because of the way the request stream works from the various callers of our system, the problem in Malaysia might impact every user sending requests that are handled by that data center. If the Malaysian data center is the only data center that handles traffic for far eastern operations, it may be that every single user of the system located in the Far East is suffering from this outage and is currently unable to properly use the system. This could be a catastrophe to a business or government operation providing a critical service or it could be an economic catastrophe for businesses losing sales and customer loyalty due to the outage.

Quality of Service thus not only provides numerical understanding of the system's behavior regarding its errors and successes, it also provides a detailed model of the system itself. Properly modeling the system is imperative to getting the telemetry. And modeling amounts to accurately defining all the dimensions at work in the system's operations. If there is an authentication mechanism and if the callers are known to the system, then one pertinent dimension will be the caller's login information along with their IP address which indicates their physical location on Earth. We already discussed the various dimensions that might be part of the system topology and the system architecture itself. Proper factoring of the code would likely result in hundreds if not thousands of methods making up any part of the system execution, but the designers must recognize that including each of these functions in the product model would create too much noise. They must be selective in how they come to understand the application, its major components, and the critical parts that make up a highly rational portrait of the system. This need not be determined once and for all and probably should not be. Throughout the life of the system holes will be discovered, blind spots, the model of the application itself must be sufficiently structured so that it allows for adaptations and changes over time in accordance with changes to the operating patterns and usage patterns of the system. Problems will be discovered for which answers cannot be provided. And this will cause revisions to the model, revisions to how the application's structure is understood and revisions to how the events are designed to capture that structure. There must be a regular and constantly evolving project of modeling the system's execution and behavior and that model must be used to design the type structure of the events and the dimensions that are captured in each event fired.

The model may suggest that the events themselves will be of different types and different structures. Some dimensions may be important at one point in the execution flow and other dimensions may be deemed irrelevant. Different types of events may be designed to correlate with different parts of the system's execution and those events may reflect the variable sets of dimensions that accompany that portion of the system. Furthermore, there may be normalization requirements in the dimensions that allow for a reduction in duplication. If each machine sends a single event at start up time, for example, that captures details about its current role, current place in the topology, the current operating system installed, and the current version of the application software, this event may fire only once per session on each machine and sessions may last for days weeks or months. Then each event fired during that session may omit all those related details and include only a single machine identifier. Later in the process, this machine identifier and the time stamps on the various events that it is attached to can be associated with the topology or machine description event that fired at process initialization. The events are part of a logic that aims at describing the system that is executing, the machine that it is executing on, the networking infrastructure that it is part of, the data that is passing through it, and the processing operations that are being performed. All of it, the entire world of the machine/system's behavior and life cycle will have to become part of the model that the events articulate. The model describes the system and its execution in context and the events realize that model at points in time throughout the system's life.

The processing pipeline that receives the events may execute additional logic against them as they arrive in a central store: cleaning them, sampling them, integrating them with some data that is of interest to the system but did not need to be stored on the source machine. Because the processing is taking place in a pipeline, it means there is a possibility that multiple disparate operations will be performed on the data as it arrives and flows through that pipeline. Different groups with different responsibilities to the system at large will have a separate set of operations that they each need to perform on the data as it passes through. And the operations may align with the specific functions of the groups providing the operations. Some of those groups may not be interested in a real-time look at the data and so their actions performed on receipt may take hours or days to execute. Other groups may be interested in real-time visibility and so their actions may have to fit within just a few seconds. The aim of all these different processes is to reproduce a picture or representation of the system according to a variety of well-established attributes and properties and serving a defined set of ends. The predicates are the dimensions associated with the data at the time the event was fired, and these attributes vary from one event type to another constructing distinct representations of the system with a purpose or set of purposes in mind. In quality of service, we have been concentrating on real time service monitoring, but later in the chapter we will look more closely at some other ends that might be served by the data. Regardless of which perspective we are taking, the pattern indicates that the emission of data is not pure or raw, but is already metric in nature, already taking shape from those system properties attached to the event in the moment the data was produced. The data identifies the system, describes it, in the very act of monitoring it. Whether we identify the system's products for the sake of monitoring its production of them or whether the two modes are different aspects of the same substance, the event itself surfaces their interplay.

The outcome of this modeling process is a set of data that can then be targeted with monitoring rules where those monitoring rules are measuring the health of the system at work and sending out notifications and alerts based on a comparison of the system's actual behavior to its supposed normative aims. This is crucial. A large system with billions and billions of events coming into its processing pipeline may decide that there are thresholds for each detected misfire. The rules themselves must use semantics that are complex enough to provide these kinds of caveats. We do not want to wake up a human being, or an expensive automated process, for just a few seconds of problematic performance. If a couple of requests are dropped in Malaysia, we may determine that there is no immediately actionable need to invoke some response to that condition. We may require, for example, that the quality bar dip below 99% on a very clearly defined set of cross sections to the dimensions and that the dip must persist below that level for a specified period. The misfiring method from our Malaysia example at 0% may clearly require intervention but should that process dip to 99% for a minute on a Tuesday, there is no need to invoke a recovery process. It might be desirable to trap the time and nature of the condition and store it in some notification with a lower priority or severity, indicating that it can be investigated at a future time with a less aggressive service level agreement. The notification and alerting process must be able to set a priority on the various conditions that it detects and configure a response that is determined by that priority, triggering a high latency response for lower priority alerts and a real-time response for higher priority alerts.

The model covers all the moving parts of the system, all its facets and dimensions, and provides a formal organizational representation that the various events are meant to realize. This realization of the model is its materialization in events that carry data of the pre-defined types with the pre-defined dimensionality. The telemetry system, to this effect, becomes a total rationalization of the primary system and is directly coded into that system. The system itself may have been well designed and may have been carrying out its purpose as best it could, but only once the telemetry system is added to it and fully fleshed out to capture the quality of service of the system, only then has the system entered upon a pathway to thorough rationalization. Only when the system's behavior begins to materialize in its events, in its telemetry, is the system realized as properly functioning, visible, manageable, and supportable. The process of fully rationalizing a system through its materialization in a model-based set or swarm of events is what I call the telemetric state of the system. And by this, I mean to invoke not only all the quality of service extensions and apparatuses that can be added to it, but the entire process of making the system viewable from afar, visible to observers that are far away, and manageable by decision makers who receive the inputs from those observers. The telemetric mechanisms are all those mechanisms that rationalize the system to fully realize a feedback loop allowing for greater flexibility in addressing, manipulating, and managing the system.

Metrics

While discussing quality of service, we described how the dimensionality of the data produced by the system, and hence its model as it materialized in these events, might be centrally processed according to multiple interests or points of view. And examining these different points of view may reveal what the event data is capturing and what the telemetry is measuring. When we articulate the different interests that exist relative to the telemetry, we articulate the set of metrics that the variously structured events provide. Metric, in this sense, is just the measurement along its various dimensions combined with the interests of the party evaluating that event within the central store. Suppose for example that the events are fired alongside two distinct types of operations: waiting for something to happen, on the one hand, and making something happen, on the other. In the former case, the measurement will describe the amount of time spent waiting and the dimensionality of the event will describe the context of the waiting as thoroughly as possible. In the latter case, the measurement will describe the amount of time spent performing the operation and the dimensionality of the event will likewise attempt to describe that context as best as it can. Two distinct metric types emerge here: dependent and independent. In both cases, we capture error conditions, aggregate conditions within specified time slices, and enable pivoting the data by each of the dimensions included in the event. The intent or meaning of the event will be different. In the case of the wait operation, we will measure the relationship the system has with what it is waiting for. To the extent that what it is waiting for are things outside its control, things performed by external systems integrated with it, the metrics that capture these operations will describe the dependent quality of service. In the case of the independent quality of service, only the system's internal reliability will be measured. Control and action boundaries are central notions in the determination of 'interest' relative to the parties involved and the measurements taken.

This higher order distinction between types of metrics would help in properly identifying and addressing problems found in the system. If the problem is detected on the internal independent operations, it makes sense to notify an agent, automated or human, to address the problem. If the problem lies in external operations that the system is waiting for, those same actions may be fruitless. Instead, the system needs to identify the correct external body with a controlling interest in the operation and notify them. If the problem is isolated to some specific functionality our system is using when it makes these calls, it may be that the external system has not yet detected the condition. This demonstrates why the type distinction in the metrics definition is so important and how it can be used to drive behavior. The external system only knows about some problems if notified by its partner because it cannot see the world from the point of view of external interests and so cannot detect problems specific to them. Our system, however, can notify the external system only if it recognizes that the monitoring operation that it is currently performing is dependent on an external system and able to integrate its notification and alerts with that system's notifications through standard alerting protocols: there must be a common language to express their interests and ask for assistance in realizing them. The type distinction between the different kinds of metrics is thus a crucial precondition for properly monitoring the problem. And by properly, I mean monitor in such a way that it results in improved system reliability and efficiency where interests can be expressed to the party that controls behavior in those areas of operation.

Monitoring that is done purely for the sake of describing the system's behavior is not worth the cost of building it. Rather, the monitoring must be able to effectively feed its observations back into the system that it is monitoring. If those problems originate outside the system, then the monitoring mechanism must be able to communicate with that system. This explains why large-scale systems like this tend to take precautions whenever external dependencies are introduced into the design. The architects and developers want to make sure that there are a set of protocols with detailed accommodations describing the communication. Only if such practices can be settled in a service level agreement, would it be feasible for the one system to take the dependency on the other. The negotiation and establishment of these service level agreements can be time consuming and spirited but are essential conditions to the introduction of the features that have the dependency.

The pattern that we see here is part and parcel of the notion of a metric. The metric is a set of delineations and categories, higher order typing of the events, to increase the intelligence of the system, increase the meaning implicit there, and identify the interplay of interests through visible artifacts produced during operations. The metrics make the system smarter by adding new distinctions and differentiations to the activities and projecting clear boundaries between them. The only difference between a waiting metric and a doing metric might be the presence of a dimension indicating what system it is waiting on. The domain values of the dimensions in the various cases might be different, but the type difference does not lie in the values of the dimensions, rather it is in the list of dimensions included in the type definition. The presence of the dependency system identifier in the one type of metric delineates the metrics of that type as different in form from metrics without that identifier. This kind of type differentiation articulates the improved intelligence of the system in identifying problems by yielding more meaning in the various exposed facets. There is more that the system can do because of these differences, more interests that can be expressed, and more relationships that can be surfaced. Behaviors are conditioned by these differences and more subtle actions can be taken because of the increased meaning these types bring to operations. Being able to notify an external agent vs an internal agent is a non-trivial power that the system achieves through such type distinctions. Since any code that is written against a metric must be able to drive its logic off differentiations and since the whole point of gathering data is to write code that does something with it, metrics are the source of power within the system. It is as if the dimensional qualia or metric emissions are being harvested for the sake of a higher order set of thrown projections serving ends that go beyond any of its component parts.

Some metrics only have meaning in the machine's context whereas other metrics have meaning in a user context. When remotely monitoring activity, some operations are focused on machines serving up the resources in the request, whereas other operations are focused on the requests themselves and the user context. In the former case, we might not even know who the user is since the events are based off performance counters that track the CPU or memory usage on the machine. In the user request context, a complex distributed architecture might mean that the request is bouncing around from machine to machine as the system computes the correct response to the request. In the former case, the dimensions available to the event firing mechanism containing measurements of machine state will not include information about requests and their pathways through the system. In the latter case, the events may be associated with one or more specific machines and the request, bound to an authenticated user, goes through the call stack until the appropriate response to that request has been determined and relayed. Once these events land back in the telemetry data processing pipeline, they are run through different processing operations corresponding to the different dimensions and measurements that the various types contain. Metrics accumulate meaning through attribution and accumulated meaning enables more interests to be expressed in the operations that will be performed against metric shapes of the data. Every attribute enables a new aggregate orientation, renders greater rationality in the collected process view, and permits richer responses to the data.

We have seen three distinct types of events. 1) The dependent API event where the operation that fires is waiting on some functionality provided by an external system. 2) The standard API event where the operation is processing the caller's request in terms of any of the various components that make up the system's service or application layer processing. 3) Performance counter-based events that describe the state of the fabric that underlies the application or system operations that are handling the caller's requests. Within these type distinctions, we can describe all sorts of variation as previously discussed. In the first case, we can differentiate between the various external systems or stores that the monitored system is in communication with. In the second case, we can make distinctions between the various component parts of the processing system, and in the third case we can distinguish between the various counter or machine attributes that are being captured: memory usage details or processing cycles by the CPU. The type distinction is drawn from the different attributes that each of the measurements has associated with it and the token distinctions are drawn from the different values assigned to those attributes. We may speak, therefore, of the dimensionality of the type on the one hand and the cardinality of the dimension on the other. Types and the interests associated with them are understood in terms of the various dimensions attributed to the type whereas dimensions are understood in terms of the various token values associated with the dimension as its domain. When modeling the system, we must capture the different types of events, but not just the types, we must account for the variety of domain values associated with each of them. This model thus provides us with a holistic differentiator of the intricate parts and pieces of the system. Holistic because it attempts to capture every facet of the system's execution and construct a complete story from those parts. A differentiator because it tries to separate the parts in as much detail and with as thorough a description as possible. The more closely associated the event generation is to each and every part of the system, the more exhaustive the model of the system will be. Like Foucault's suggestion that power is everywhere because it comes from everywhere, the model emerges from the entire system as it executes. The system emits the model in such a way that the differences in the system's operations are implicit in the events that are fired. More than just a system of relays, the metrics and the model show it to be a system of interests captured through an emergent process that replicates the immanent order into the monitoring output.

This constitutes the rationalization of the system. The engineers who construct the components and string them together in the processing of the requests to respond properly to the caller are not necessarily able to see the fully rationalized system. It is only when they begin to fire the events to fully and completely monitor and support it that this view materializes. Later we will see that it is only once the engineers begin to design and implement the feedback loop needed to keep the system operational that they get a full view of the system in all its nuances. Thus, telemetry provides the raw basis of this point of view and the telemetric is the process of gathering this perspective not once and for all, but continually as part of the ongoing operations. A model is not a static representation: telemetry provides constant visibility into the system's operations and interests, a constant surfacing of the inner workings from minute to minute so that the service can be maintained with maximum availability and reliability. This ongoing visibility via implicit replication into monitoring operations is the rationalization of the system and its maintenance. Rationalization is just the process of making the monitored system visible and keeping it visible throughout its ongoing operations. And essentially, this rationality is achieved through mechanisms of surveillance integral to the system. We are not describing observations performed by an external agent that is separate and independent. Rather the telemetry is emitted by the system itself. Every function that operates on the user request or that gathers performance feedback from the machine that hosts the application, fires events intrinsic to the operations that are performing the work. All the fits and starts of the machine emerge in such monitoring. The events are fired by one and the same process handling the caller's request and by machine-based operations that are reading the counters directly off the box. The telemetry is providing a remote measurement of the process, it is executing remotely for the sake of the decision center and on behalf of the specific area of analysis and deliberation calibrated to the metric design. In effect, however, the events are being fired by the system itself: the efficiency expert is a "company man".

This interplay between an observer that is up close and intrinsic to the system operation and the dispatch of the telemetry to some far-off place for deliberation and decision-making is, of course, essential to telemetry. The measurements must be made locally and up close to the processing if they are to be accurate and properly reflect its operations, but the singular occurrence is insufficient for fully grasping the meaning of the event. Only in the aggregate, only when compiled over time and across heaps of processed requests, can the meaning of the system's operations be properly surfaced. The example we began with, the efficiency expert, relied too much on the boundaries between individual human beings to fully capture this aspect of the relationship between telemetry and the system that is emitting it. A common design pattern in telemetry-heavy code is to "delegate" all code execution through a telemetry mechanism. This means that instead of directly calling some function that you need to get output from, you send a delegate representation of the method to a telemetry operation which encapsulates both the call to the method and the firing of the event. This is a nice design heuristic, but it reflects the essential nature of the emission of telemetry as an action performed by one function on behalf of another. The 'on behalf of' relationship between what is being done and the measurement of it, closely intertwines the actions, binds them together, such that the one is inseparable in practice from the other. The worker wears two hats: efficiency expert and pin producer, company men all the way down. There is no uncertainty principle at work here any longer since the role of observation has been absorbed into the intrinsic functionality of the system. Visibility is everywhere because it comes from everywhere.

The metric is this intertwining, it is the carrier of the local intelligence that has been captured by the fired event. The metric is what brings together the dimensionality and the cardinality of the operation and captures it in a neat little package dispatched to the remote processing pipeline responsible for aggregating and persisting the data for future analysis. The metric is the event that brings together the rational differentiation of the system's facets in each of its moments, in each of its operations, and in various of its aspects. All for the sake of carrying that data away from the locality where the operation took place and dispatching it to the store where its meaning can be brought to fruition by decision makers and deliberators studying the system to determine whether alarms need to be raised or notifications sent. It is the system expressing itself, the system expresses itself in metrics, emitted measurements originating in the very actions being measured. It does not just do the work that it does, it broadcasts about it, it sends out signals describing itself so that others somewhere else can pay attention to it. We might say that it desires to be seen at a distance and that its action has been structured to facilitate this.

The model, or ongoing modeling, of the system amounts to a collection of metrics. Say that we have a metric that describes each of the system's major operations: 1) the processing time required to accept the request and queue it up for routing, 2) the amount of time the request sits in the queue, 3) the amount of time that is spent determining where and how to route the request to some external storage mechanism, 4) the time spent waiting for that external store to respond to the request, and 5) the time spent preparing and packaging the data from the store for its ultimate return back to the caller. In addition to these application metrics, we might also have a set of completely orthogonal metrics that capture the memory usage on the machine and the CPU processing time used by it. The first set of metrics may be associated with each other via correlation identifiers that allow us to see the pathway traversed by the request through the system. We know that the 5 metrics capturing the work done by the processing can be associated together through this common identifier. The machine counters will relate to the machine context in a broader fashion. Performance counters generally work in specified intervals, let's say by capturing readings once per minute. In that minute, the same machine may handle many thousands of requests correlated by identifier. This means that we cannot capture one single correlation identifier to go along with the various machine performance readings that are captured during the processing of the request. Rather, in this case, we will have to rely on machine or topology identifiers that accompany both the performance counters and the system events. The topology identifiers in both types of metrics will make it possible to correlate any events that have been captured along each of these axes at a specific point in time. And this correlation will make it possible to support all kinds of decisions about how the system is behaving and whether there are any conditions that need attention. Specifically, we might discover that the machines in some location of the world are overburdened, processing too many requests at a given time. This tells us that we need to add more machines to a specific data center to handle the greater traffic that we see during peak hours in that time zone. The variable interests can be correlated to increase the power of the system, to increase the range of things that it can do.

The metrics must each capture some element of the system's operations, but they must do so in such a way as to be integrated, to be able to form a complete picture. The correlation identifier serves this purpose in the case of pathways through the system's execution and operation, but time and location (topology identifiers) can serve this same purpose. Spacetime geometry is crucial for surfacing the attributes of the component parts and the work they do. Again, the key here is that the process of rationalizing the system, of making it visible, amounts to a thorough modelling of the system in accordance with the complete set of its differentiable attributes and facets. And for that model to be complete and functional it must account for the requirement of integration and association. The dimensionality of the metrics must be designed in such a way as to allow for joining together operations of different kinds into a common narrative or picture of operations and performance while allowing for the elimination of those facets that are contrary to integrated interests and classified as problems to be monitored and resolved. The metric is the essential machine for bridging the distance between the conceptual model and its time-space sensitive execution. The materialization of the system is its realization as the fullness of its design comes to fruition in its metrics. The 7 metrics that we have seen in the example system reveal what the model for that system looks like, we have a high-level picture of what the system does when we examine the metrics it produces. And then the individual events, the swarm of billions of events, realize these metrics as the metrics realize the model. The metrics are a designed intersection of conceptual attributes and characteristics embodying interests and expressing the moments within which they are captured. Each of the instances of the measurement qualifies as an event that is emitted to self-describe what the system component is doing, what it is, in detail. And not just what it is, but what it ought to be and what counts as the desirable set of working conditions to be maintained over and against any problems.

Later in the book, we will see the way this pattern is realized in the phenomenology of the agents and then again still later in their common wealth of collective associations. Here we have shown that the exposure of attributes amounts to a rendering of the identity of the system components, a rendering of what they are and of how they are. The emergence and maintenance of those qualities and identities constitute an interest where the collective interests are the 'health' of the system or, as we will call it, its commonwealth. Furthermore, it has become clear that these renderings are meant to function as knobs that can be turned by supplemental procedures whose behavior is based on the properties and the property values assigned to those 'knobs': visibility increases power by increasing what can be seen and what can be done with it. Rather than immediately trying to situate these micro manipulations into large scale maneuvers like 'capitalism' or 'oligarchy', we defer that until later iterations of the pattern and, for now, turn toward the initial appearance of analytics: a demonstration that will complete the effort to reveal telemetry as the acorn that grows into an oak throughout the remainder of the book.

Analytics

This is not meant to be a general-purpose discussion of analytics. That comes later in chapter two. The metrics have been collected together into a common store that accumulates all events fired by the system at work. The metric store is the source data for the monitoring operations and rules that examine the system in its aggregated context for alerting and notification. This is the first type of analysis applied to the metric data store. We suggested earlier that alerting raises a condition to an operator who could act on it. The system needs attention, needs someone or some process to do something in response to a detected condition expressing interests contrary to the system's interests. Recall the case of the Malaysian data center in which some small sub-system has gone offline. Let's say, for example, that some file share location has become inaccessible. And let's say that in some conditions, the caller's request requires that data be loaded from that file share. This is a conditional operation, meaning that much of the time the data comes from some other source. Let's say that if the user has logged on to the system within the last 7 days, the data needed for the operation will come from a global cache. If the user has not accessed the system in 7 days, the data must be retrieved from the file share. If the file share is offline or inaccessible, these users will receive error responses when making requests. The events that are fired provide a metric into this condition. As the events land in the metric storage area, this information lies dormant. It is unknown, it is in an inert form. Say there are monitors, components of some application, that run every minute looking for specific patterns in the collected metrics. And these monitors discover errors and identify the decrease in quality of service for the Malaysian data center under whatever conditions it can detect in the events. And by whatever conditions, I mean the set of dimensions that are included in the data. When the monitor detects the dip in quality of service based on its inspection of the metrics, it raises an alarm. The analysis at this point is only partial. The monitor is a broader net than is required to zero in on the root cause of this outage. The rationale behind this partial state is simple. The number of monitors that would be required to hunt for each condition would be astronomical. As a result, the strategic approach is to use the automated monitoring to flag drops in QoS without attempting to get to the bottom of the problem. That process is left to the subject of the notification, the agent that is notified when the condition is detected. The analysis begins with a broad inspection of error conditions, and then the alarm is fired to bring in another agent to do more analysis, to perform a more fine-grained investigation. From the point of view of the monitor, it does not matter whether the invoked agent is human or another computerized process.

Suppose that a human being receives the notification that there is a dip in quality of service in the Malaysian data center. The operator has been notified that the dip in reliability is associated with a specific API that makes up the user identification system: GetUserData. This dimension, the name of the high-level function that is performing its work when the error occurs, already shows the operator where in the code the error lies. The operator uses that data to refine the query operations and discovers that the line in the code where the stack trace indicates the source of the problem is the line where the code tries to read a settings file from a network hard drive. The operator has been spoon-fed this information by the monitoring system because it was able to provide specific configuration information extracted at the time the monitor condition was met and the alarm raised. Upon seeing this information, the operator attempts to navigate to the network location identified in the alarm. If the network location is unavailable, the operator has a clear understanding of what is going wrong. At that point, the operator might need to contact a network administrator who is responsible for maintaining the network location. This process, if tracked in a good ticketing system, can be easily converted into a state machine enabling an automatic response that takes the human beings out of the equation.

This form of analysis, both the partial segment performed by the automated monitor and the partial segment performed by the human operator notified by the monitor, is the be all and end all of the telemetry processing. The reason that the system fires events that produce measurements of its operations, the reason that the system organizes these events into a rational format of modeled metrics, is so that monitoring can be coherently structured to execute its state machine: detect conditions and then produce alerts and notifications to enable corrective action to maintain any service level agreements for reliability and availability. This form of analysis is short term in nature. It is, in a sense, an immediate feedback loop meant to keep the system in a healthy state. This is the easiest form of analysis to automate. Conceptualizing both the human segment and the machine segment as state machines is part of the telemetric process of extending the reach of the automated procedures. If the monitor can identify that the misfiring line of code which it has already captured in its alert is related to an external system, then the monitor can be modified to notify the external system directly. The system operator for our system need not be bothered and the external team can be notified directly and then their responding agents can go through whatever workflow they use to identify the cause of the problem and rectify it. The division of interests between the different parts of the system is a separation of concerns appropriate to the division of labor at work in the design of the system such that different services provide different functions for addressing specific capabilities. The more structure present in the action design, the more progressively automatic is the response: first human action is made machinic, then the machines are programmed to take on the action context in its entirety.

It may be, however, that the data for this modification is not immediately present in the metric data store. It may be, for example, that the line of code that is misfiring is only visible in the call stack that accompanies the error and parsing such text is not feasible for the monitoring system. In that case, it may be that the events themselves must be modified to enable this automation of functionality. We might have to switch the metric type of the event so that it no longer looks at the coarse-grained operation being performed by the system (gathering user settings), but is focused on the more finely grained operations, discriminating between parsing the settings and the external call to retrieve them from the dependency system. In that case, we might need to start firing another event that provides a metric into the wait operation performed when we retrieve the user settings from the file share. This event will now show the dependency system as part of the metric that describes the error and the monitor will be able to use that dimension value when triggering an alert. The compilation of a decision tree for addressing the problem, as suggested before, is the perfect starting point for determining whether the metric has enough structure to support the interests expressed in the conditions being monitored. We would need to ensure that the metric contains enough intelligence to be able to drive decisions contained in the necessary state machine. Germinal analytics, therefore, cannot inject this intelligence ex nihilo, the metric emissions must have enough intelligence to instigate the translation into analysis aimed at action.

> Thus, the fictional internal life of the organism gets passed off as identity, as something about which is unique or private. The evolutionary biologist speaks to pain-qualia as being a temporary mechanism to aid the organism in harm-avoidance. Metric emissions and state machines posit qualia-like instances for the same reasons – temporary place-holders for aggregations of data. Thus privacy-qualia are micro-constructs of subjectivity by and for the telemetric systems.... (Vaughan, 2018)

This iterative process is typical of the analysis phases. Often when performing analysis, we discover that the data we are using does not bear out the distinctions and differences that we need when determining the correct response. In that case, it is necessary to go back to the source of the data and modify it so that it contains a richer description of what is happening. And this is how analysis progresses, it becomes a higher order feedback loop on the feedback loop itself. The operator who was awakened in the middle of the night is likely to complain about that and point out that it was not necessary since the monitor already gathered sufficient information to route the alert to the correct team. In discussion with the other members of their team, it would become clear that this is true, and that insufficient data was present on the metric to enable this action automatically. They would have seen that the event that was fired in the case of the error would have had to contain the information about the wait operation that the system was performing as it accessed the data on the networking file location. They would propose a change to the telemetry design so that this information could be added to the events being monitored thus giving the monitor enough information to make the relay to the external team on its own. This kind of higher order analysis goes beyond the direct troubleshooting of the problem and it begins to analyze the system's efficiency in so far as it captures problems and responds to them. It is important to note that the lower order form of analysis is aimed at correcting the system execution itself whereas the higher order form is aimed at correcting the feedback to automate the correct action. The first I have been calling alerting and notification. The second I have been calling automated management. Both are revealed as forms of analysis and serve the purpose of completing the feedback loop that is the primary reason for the self-surveillance and rationalization that drives the telemetric observation of the system.

The forms of analysis so far discussed require real-time access to the metric store since they are meant to provide real-time feedback into the proper execution of the system. There are additional forms of non-real-time analysis that the metric store must be able to support: root cause analysis, deployment analysis, and historical analysis. Often non-real-time analysis can make use of more dimensions with richer descriptive power over events. The added dimensionality may bring increased processing costs mitigated by directing these richer events to a lower priority pipeline that works more slowly in bringing the data together, aggregating it, and cleaning the dimensional values. The amount of time this pipeline requires increases processing time and makes it inadequate for monitoring, but these richer dimensions provide information that is useful to the longer-term maintenance of the system. The richer dimensions might include information about the calling and usage patterns helping to better understand who is using the application and how they are using it. This latter data can provide feedback for project and program managers to understand what sorts of features are missing from the product and what kinds of investments need to be made to increase the product's value to its users. In addition to these longer-term concerns, it might serve more medium-term concerns as well. The monitoring may be done not only to determine whether the current execution is in a healthy state, but it might be leveraged to indicate whether the general state is good. In the case where the metrics are gathered as part of the gradual rollout of new releases of the product, this kind of data can help the engineers determine whether there are quality or usability issues in the code they are changing. Deployment is usually done in stages where the new version of the code is slowly published to larger and larger audiences of the application. Monitoring the issues that these various audiences encounter as the code is being deployed can help determine whether the code is ready to be presented to larger audiences. In the case of the long-term analysis for features and the future makeup of the product, the time span of the projected action might be anywhere from a month to a year. In the case of the middle term analysis of rollout to assist in deployment, the analysis may be projected onto actions that will be taking place in the coming week or weeks. In the short-term analysis of maintenance and operations of currently executing code, the analysis may be projected onto work to be done in the next few minutes or hours.

The time factor associated with analysis, as described, is directly proportional to the complexity and richness of the dimensions that are used. Metrics with richer dimensions or more dimensionality per se, take longer to prepare for analytic consumption. As such, the various models of metrics need to be prepared with this time/complexity set of constraints in mind. Questions that drive the short term are: what is failing, where is it failing, and how is it failing. Questions that drive the middle term might go into greater nuances associated with the new features that are being deployed or into testing regressions on old features. No doubt the short-term analytics will play a role in the deployment analysis, but those metrics will just make up a subset of the relevant concerns. Additionally, new vs. existing features will have to be captured, where a feature is new, specialized metrics aimed at capturing usage will be in place to determine rate of uptake. In the last category, the analytics that attempts to discover missing features, far more creativity is required in the analytic procedures. Analysts must learn to pose questions of the data that trigger data compilation meant to show indications that something is being done because something else is not there, inconvenient workflows or wait times that users must exercise to work around the fact that something is missing. And these types of questions will often require long running queries with complex conditions that span many different dimensions. The more complex the dimensions, the richer the meaning of the events, the more complex the questions that can be posed to it. All this is based on increasing the intelligence embodied in the metric emissions facilitating its analysis once the data has arrived in the central store.

In general, analytics provides the basis for decision making and surfaces higher order questions and answers concerning the system's operations. It is not just a matter of determining whether the system is healthy or whether it is running as it is supposed to run. Rather, the analyst is meant to look for patterns that describe deficiencies in the product relative to the market place of similar products and patterns that describe optimal operations and that provide a competitive advantage for the product. Things of this nature go beyond the enclosed walls of the system as it is and start exploring what the system is in a larger context, how it looks when understood within the world at large, as part of a larger set of concerns and interests. Moving beyond the questions of whether it does what it has been built to do, the analytics would start to ask whether it should be built to do different things, more things, by providing features that are not currently there but which could be there and, if so, would provide greater value to the set of callers or a larger set of potential callers. This realm of analytics is the realm where the immediacy of the system is no longer taken for granted and the system itself becomes available for a more critical point of view. Because analytics that surface these aspects of the system require greater creativity and understanding, this has been an area where significant investment has been made and even more significant results have been achieved. Human beings are slow and limited in the set of patterns and questions they can raise. As the amount of data increases and the number of dimensions and metrics increases, the complexity of possible insights grows proportionately. Analysis has moved beyond the realm of the system operator and become a science all its own. In the next chapter, we will look more deeply at the growing science of analytics and how it has taken shape to provide greater self-understanding for the system's execution.

At this point, we should take note of the pattern of realization that the telemetric has introduced into the system's execution and how it brings that execution to fruition as a more thoroughly rationalized apparatus for carrying out its purposes. The system is executing its function, handling user requests, retrieving data, running operations from external systems, and preparing responses to send back to the caller. The system is initially blind to its quality of service, it does not yet have any self-understanding of its reliability and availability. The quality of service of the system is surfaced in a set of events that are fired by the system's execution. That is, close-up and intrinsic to the operation of the system, events are emitted that describe its inner workings. This emission begins to make the system visible to remote observers who can then tend to its operation from within it, as a part of it. The emitted events can be finely grained to introduce greater differentiation and distinctions in the system's operations. These differences –through the merging together of the relevant attributes describing the various moments of execution as the events are emitted– is the system's metric dimensionality. This metric dimensionality provides a more expansive model representing how the system goes about its business of responding to requests and sending back data meeting the caller's expectations. Once these metrics land in separate metric stores where data comes to rest and is directly consumed by monitors actively impacting the system's behavior, there is analysis taking place. The analytics on the system takes the metrics that have been gathered and puts them to work, uses those metrics to drive responsible actions, actions that aim at short-, middle-, or long-term changes. The system emitted these metrics for the sake of some action, they are part and parcel of a feedback loop that is meant to change the system execution in some projected span of time.

Visibility has been a means to an end and the end is the modification of the behavior of the system according to the types of action and interests fitting within the various time frames: the meeting of service level agreements, the introduction of changes in a predictable and reliable fashion, and the projection of new features and new capabilities that will improve the system's value proposition making it more useful to its callers and more profitable to its owners. The system's goals may be understood through a purely internal dynamic of continual and gradual increase in scope and capability. The purpose of this making visible to remote analytics and observation, and the purpose of the analytics to increase the system's capabilities and possibilities, may be understood as the system's will to power. All this rationalization is meant to increase the system's capabilities, to increase the fine-tuned details of its world and its ability to respond to that world in a meaningful way. All the differentiation and distinctions that the system surfaces can be increased and proliferated for the sake of making the system *be* more and for growing its powers of execution and operation. The fact that it may increase profit or well-being for callers is an important connotation, but it is not essential to understanding the workings of the system's telemetric realization as an increasingly rational mechanism. The workings of such a system do, in a sense, fulfill some set of larger goals that can be cast in human terms, but they also have a logic all their own that makes it possible to examine them as pure abstractions, as operations that perpetuate their own existence regardless of the impact they may have on any higher order contexts containing the system: a livelihood, a business venture, a public service. The power of the metric emissions triggering the analytical movement of telemetry lies in their ability to appropriate exteriority into the system, making it possible for the system's internal operations to run on purely immanent logic within which problems appear as external conditions cast into a language the system understands fluently.

Chapter II: Data Science: The New Paradigm

When placing 'science' in the subject of a statement is it immediately obvious what such a generic or universal usage might mean? Specific sciences are associated with specific domains: physical, chemical, biological, psychological, geological, anthropological, etc. 'Domain' is puzzling here. There are reductionist and emergentist theories that suggest, each in distinct ways, that there are boundaries between the domains and that the nature of those boundaries, and what the sciences on either side of them are concerned with, must be articulated in a complex theory covering the objective scope of the domain and how the objects relate to each other across those boundaries. It seems that some of the more radical events in the history of various sciences have challenged the nature of that science's domain or the nature of the boundary between that science and other sciences. None of this gets us much closer to understanding what we might mean by 'domain' which seems to be circularly used as both the target object area of the science and what organizes it into the specific science that it is. What are physicists? People who study the physical world. What is the physical world? The area of the world that physicists are concerned with.

Despite this difficulty with the details of the word 'domain', it seems plain as day that the physical world can be examined, poked at, or observed with different emphases. Over time these emphases have been solidified and extended by many different contributors who learned from each other, argued with each other, and established norms and standards leading up to the creation of disciplines. In and throughout this process, a specific domain of objects and object states became firmly established as their customary focus. There may be resistance to such an historical determination of a domain but without it, changes in paradigmatic notions are difficult to explain while maintaining a unified history of 'physics' or 'biology', etc. over and against the creation of a new science with a distinct domain. Once the domain notion has been established, however, there is the matter of the practices and behaviors of scientists. There is much talk of a 'way of knowing' that is the scientific method where a hypothesis is articulated and then experiments performed to test the hypothesis. Surely that plays some role in what 'science' is? But, digging deeper, there is also talk that scientific praxis does not always fit cleanly into that little box. There is grunt work, just data gathering. There are creative flights of fancy and thought experiments. There is the influence of math and how new strategies must be developed for crunching the various experimental data that have been gathered. Scientists do lots of different things and the range of typically scientific behavior likely has evolved over time through learning and collaboration, through information sharing, professional communities, and standards of publication. Scientists read about the strange imagination games other scientists used to come up with a monumental discovery and this affects what they think is fair game in carrying out their own pursuits. The sciences (not 'science') each have their own history and community and the practices appropriate to those sciences are the result of long-lived efforts and the sharing of outcomes and methods followed by the inevitable review process that helps establish and vet the feasibility of anything new that comes along.

This is a decidedly circular understanding of the sciences: to practice a science is to be a member of a scientific community where that amounts to partaking in the common norms and practices of that community, focusing on the issues that it tends to focus on, using the methods and approaches that it has sanctioned, and targeting the objects that the community tends to target. This circularity explains why originators of a new science may be initially outcast and that the process of normalizing the science is the process of growing connections between the innovator and others who are informed in the ways of the disruptive approach. Newton may be credited with the insights, but Newtonians must be credited with the established science. As the innovations gain broader acceptance among practitioners sympathetic to the original work, a community forms and standards enforcing the innovations come to be. Even scientific revolutions are not random and wild contributions, rather they must gain traction and build communities behind them to promote the new ideas and orientations that are their content and aim.

A philosopher like Popper, for example, cannot help but define scientific practice in terms of forms of communication and community contribution. The notion of falsifiability, whether it provides a legitimate demarcation for science or not, is essentially a community standard that relies on others to poke at one's conjectures, trying to refute them and – ultimately– holding onto those that manage to survive the process. Even the lone scientist striking out on their own to overthrow all previous standards, still developed with some influence from others, with some sense of domain and area that comes from an alternative approach to the matter at hand. Copernicus learned the Aristotelian view of the world and began the process of showing its limits and flaws while producing an alternative. He developed his initial insights within one community, orienting his sight toward a specific domain commonly studied by that community, and then revolutionized it with alternative insights and measurements. The process of "proving him right" is not something that happened within his life time. Rather, others who came to know of his insights had to come along, test them, and show that they withstood critical scrutiny. In short, the community that developed over the centuries following his death established the new science brought on by the revolution with his name on it.

Clearly, I do not know what 'science' is. And maybe that is because there is no 'science', just sciences. Even with this caveat, a circular definition does not inspire confidence. It seems that each of the sciences consists of contributors trying to produce, test, and reject or confirm insights into the meaning and existence of patterns in the observable universe. Not the universe in some open-ended sense, rather the universe as it is constrained by the domain of objects. That domain is effectively the scope of the expertise of the participants focusing on those patterns and what we might expect given the conditions of the objects and object states in that domain. Although it is a mistake to reduce the sciences to predictions, it is worth pointing out that predictions in the form of expected outcomes are often an important indicator of whether scientific insights are worth pursuing and failed predictions are a good indicator that something is missing from the presumed principles. When predictions are made, they are made in terms of the objects and object states that are specific to the domain of the science. Given the proposed nature of light or DNA or molecules, we would expect to see the following set of conditions at work on related objects and object states resulting in specifiable outcomes. Should those conditions occur, the expectation is either met or not. Should a prediction fail or succeed there is still a great deal of work to be done to make sure that the conditions were isolated and that there were not other factors playing a part. The predictions are made for understanding how the insight applies and whether it provides a proper picture or representation of the phenomenon in question, of the object states described in the outcome. The sciences aim at insights meant to describe or explain the physical world spun in terms of some domain, and they evolve those insights through data collection, analysis, thought puzzles, and tests against hypotheses, all aimed at producing results floated throughout the community of scientists helping the others with their work and benefiting from critical insights. They develop tools and instruments for measurement that are adapted to the specific objects and object states in their domain and their body of knowledge grows as a factor of the history of measurements and operations performed by such tools. It is possible that objects and object states outside the domain of the community might be relevant to some set of explanations or expected outcomes, but in that case the typical community members will be hard pressed to contribute to the critical discussion since they are not well-versed in such objects and their states. And it may be the case that new tools and gauges must be invented to perform the sorts of measurements that these objects and their states require. In those cases, other communities might need to participate in the process and the result might be a merging of concerns between two communities and –ultimately– the formation of a new science that sits between the two contributing ones.

This is a very squishy understanding of science, but it is within this context that I introduce the notion of data science. In terms laid out in the previous chapter, I am poised to show how 'data science' evolved out of the need for long term analytics aimed at perpetuating the reliable activity of the distributed system. Data scientists, I claim, are scientists. They do everything that scientists do. They only lack the domain. And I do not mean that sarcastically. A data scientist examines data collected by machine-based systems like the sample system described earlier. And they examine that data for the sake of coming up with insights for the sake of the system capturing the data. Initially, that might be a set of insights to improve the reliability of the system, or its efficiency, or insights aimed specifically at the growth or extension of the system into new areas of operation, new features, new properties that can provide greater usefulness to the users or potential users of that system. The feedback loop that the telemetric needs to grow the system's functionality is based off the massive amount of data that the system itself is emitting to describe its operations. The explosion of data that this entails becomes too complicated for the simplistic means that might be carried out by the typical operator or developer who is not devoted to the process of poring through the data to discover patterns for use in the future design or functioning of the system. This is a complicated process that requires study and investigation, there must be people who dedicate themselves to this project full time. The data scientist is a scientist proper and their role in the use of long-term analytics is to improve understanding of the system, a practice that lines up with any other scientific field.

Of course, the explanation of the data scientist as a scientist whose domain is the functioning of the same machine that produced the data in the first place might already suggest a key difference. The data scientist is essentially connected to the engineering project that lies at the origin of the data creation. Of course, it is not the scientist who makes those changes to the machine, rather the data scientist just discovers the patterns and provides a language for articulating them clearly in some form of algorithm. Other sciences have traditionally had a similar relationship to engineering. Insights in areas of physics can lead to the creation of weapons, insights in biochemistry can lead to the creation of pharmaceuticals, insights in biology can lead to advances in genetic engineering. These technical applications do not just describe or explain the world within which they act, they contribute to the engineering of solutions that modify the world, modify the kinds of things that exist within the world, or change the interactions between things that already exist. The fact that the data scientist's results are used to improve the execution of the data producing system that is the source of its analysis does not differentiate data science from the other sciences, it associates it with them. And I would like to credit the data scientist with an important discovery about the machine systems they started out by analyzing. What the long-term analysts saw when they looked at the telemetry output of the executing system was that there was not just data there about the system and how it was operating, how well it was performing the tasks for which it had been built. Rather, the analysts saw that there was data about the data that was being carried back and forth between callers and respondents. That is, the system was not just a thing to be monitored because we had to make sure that it was available 24 hours per day without error, it was also a carrier of important information. People were accessing data and the data they were accessing was indicative of what they were in the middle of doing. The transported data was not initially significant to the telemetry, the measurements were just there to make sure that the system was up and running. The long-term analysts started to see that the actual data in transport was equally interesting and that an understanding of it and how it was being accessed and used would be interesting to the system at work in executing the exchange between caller requests and system responses. Metadata entered common usage.

It might be more correct to say that the data scientist is a scientist whose domain is not known prior to the patterns or insights that are discovered in the data, but that the insights themselves might reveal the domain. That is, the data scientist's orientation is toward the patterns in the data regardless of what subject matter they may contain, regardless of what area of the physical world they pertain to. If the application is providing users with food delivery, then the insights would likely concern eating habits and food intake. If the application is providing devices with policy-based software installations, then the insights would likely concern the application usage and the standards of control that need to be placed on them given various conditions and various intended outcomes. The data scientist is not bound to a pre-determined domain, rather they are experts in statistical analysis of data patterns for the sake of learning what those patterns can tell us about the areas of the world from which the data has been gathered. As such, data science is the becoming self-conscious of the feedback loop that the telemetry system introduced for the sake of making the system present to itself. Through data science, the emitted data from the system's execution can be shaped and modeled into questions and answers that no one ever thought to ask about the system, about the way it is working, the way it is being used, and the way it could adjust to accentuate aspects of its execution to the greater advantage of its users and providers.

This may describe the early state of data science, but over time we would expect to see a greater division of labor between its practitioners. The expertise of individuals would increase in the specific machine domains discovered while doing their work and this expertise would segment them into distinct communities within the larger community of data science per se. There are already data scientists who are experts in specific areas of focus and specific areas of patterns and domains of data: medical, astronomical, psychological, etc. We should expect to see these communities grow and develop over time and, at some point, I would expect that "data science' would become just as unclear as the term 'science' is now. This may have been true of all sciences in their early days where there was a single community with a high-level domain that brought them all together, and it was only over time that biology and chemistry, et al took on the details and complexity that we currently associate with them.

Generalizing from these common origins to a common logic, however, is not valid. It would be fallacious to suggest that data science as such is paradigmatic of 'science' and the way that the sciences have been conducted in practice just because it may have a common origin story. The role that data science plays as critical analytics for feedback into the domain of study might resemble some elements of scientific analysis that have previously worked this way (like ballistics, clinical psychology, or nuclear physics), but there are plenty of counter examples as well (theoretical physics, biological anthropology, or linguistics). Not all sciences are oriented toward the modification of the world that its practitioners are studying. To the extent that Data Science is oriented in this way, it merely participates in an attribute of some sciences rather than embodying the archetype of all science. Nonetheless, it might be possible to suggest that data science, in its role as long-term analysis for the sake of completing the operational system's feedback loop, is the new archetype for science. It emerges as the new template that says that when a civilization invests in scientific activity, it expects results and results amount to a better understanding of some area of operations such that innovation and improvements can be introduced into that area. In other words, data science shows the telemetric pattern that scientific activity can have relative to the data that it uses to project results and, having shown this, produces a precedent to guide the expectations of investors. And data science may not be the first example of this, but it is the self-realization of it: the data scientist is trained with this specific end in mind, with the express intent of occupying a component part of the telemetric system driving data back into the system as knowledge for the sake of improving its operations.

It is easy to imagine that there might be scientists who are purist in their attitudes and have deep feelings about science as an activity in the service of the understanding. That is, who think that the sciences are communities of investigators assisting human beings in the project of getting the joke, of learning for the sake of understanding the universe with no ulterior motives or extenuating side effects. And this is likely an inspiring story to tell young would-be scientists who are just starting out in their education and want to help contribute to those endeavors and be a part of the various communities of investigators. Even these scientists understand that the practice of such sciences has costs associated with them: the livelihood of the scientists and assistants, the cost of offices or laboratories, the expense of funding studies or purchasing materials to be used in the studies. Where does this money come from? Such idealists would have to project their own purist notions onto their sources of funding to preserve the integrity of their vision. Even when there is some partial evidence that the funding is coming from sources that fit with these projections, one would have to ignore the part of human history where unintended discoveries have turned out to be extremely valuable. Great contributions have been made to the profitability of some enterprise without anyone expecting such an outcome. And this, in the long run, has taught investors that sometimes betting on the purists will pay off big. The increase to and love of understanding should suggest to these hypothetically pure figures that their point of view might be a useful fiction to keep things moving, as one of many useful ingredients in the recipe for profitable outcomes and innovations in the sciences. Like the philosophers in their university departments, the purists are part of something larger that justifies their cost.

In the previous chapter, we allowed 'analytics' to cover both automated and manual operations that happen in response to both automated and manual monitoring operations. When considering data science as long-term analytics and the completion of the feedback loop for effecting the behavior of the operational system, we likewise should accommodate both an automated and a manual understanding. The components of data science such as statistical analysis, algorithm design, model projection, etc. could be done by both machine and human alike. Various sub-areas of data science have aimed at automating automation, which is to say applying machine learning algorithms that, once trained on sufficiently sized data pools, can work out responses to new circumstances complying with patterns and practices already identified. The ideal condition for our sample system would not only be the use of data science to better understand future optimizations, but to plug the data feeds back into the system itself so that its actions can benefit from learned optimal responses captured in the analysis of previously emitted data. This kind of concern will make the content of the system significant to the system's operation. That is, to properly optimize the system's execution, to keep the system available and growing in use and value, the feedback mechanisms must consider the detailed and actual work of the system to suggest changes that optimize its response to real world usage.

Long-Term Analytics or Projecting Future Operations from Past Emissions

We have already seen that the data collected by the system as it emits events describing its operational procedures and responses to requests can, once collected in the data processing pipeline remote from the system's operations, be used to produce analysis for the sake of future modifications to the system. At a purely functional level, it may be that we discover that the system tends to have problems every Monday morning and we may note that this is due to a rapid increase in traffic around the start of the business day in some region. In response to this kind of long-term analysis, we may determine that we need to enact temporary and automatic resource expansion operations to account for 'rush hour' type events. We may determine the schedule for these events based on region and –to remain flexible– we might configure these schedules in a data structure that is readily available to aspects of the operational system throughout its execution. The system can then adjust its resource allocation based on this configuration. The findings were determined offsite through reporting and analysis of the system's operations and then fed back into the system through the specification of a new feature, a module that looks at the configured schedule and makes changes to the operational environment based on it. This is simple stuff, a few simple line charts based off the metrics would show this pattern to any curious engineer and suggest a simple feature that might be introduced to address it. Data science has significantly more tools at its disposal than this example shows.

The initial finding is that traffic increases in general on Monday morning as the work day is beginning. Where is that traffic coming from and where is it going? What kind of data is it accessing? People need access to the system all the time, they access it throughout the week and throughout the day, often they access it on the weekend. Why specifically would Monday morning be a time for a spike in use? Does the use reflect some specific activity performed at the start of the work week? Suppose that deeper analysis shows that the spike in requests is broken out into two distinct kinds of requests: 1) some very small segment of the user population is making a set of long-running calls that each extract a very large data structure. 2) a short period of time after these large extract queries are executed, we notice the same small set of users making lots of post operations sending data back to the service in smaller chunks. Each request is small but there are many of them, so the total content of the post operations rivals the size of data that was initially extracted. Then 3) we notice that many different users run queries to access the data that has been uploaded by the users who did the work in items 1 and 2. 4) A few hours after the beginning of the work week, this large flood in traffic ceases and the numbers go back to the normal amounts we see throughout the rest of the week. This is a slightly more complex analysis of the data patterns of the system, it helps the system operators to determine specific details about where the spike in usage is coming from and how they might be able to better direct the capacity changes to the system to accommodate this exact usage pattern.

In this case, the response might be two-fold. First, the engineers need to design a flexible capacity approach that allows them to increase capacity to meet the immediate needs of the users during the beginning of the work week. Second, the engineers need to gather more data to get a better sense of what exactly the users are doing during all this flutter of requests. What are these giant data sets that they are bringing down early in the morning, just before the work week begins? What are these little posts they are making shortly afterward? And what are the other users looking at when they come in Monday morning and immediately start querying the data posted to the system by this smaller group of users?

Analysis would target the user characterization: who are the different personas involved in these activities? This might be gathered from the user data the system already has in whatever authentication profile or settings that are being stored. Or this might be gathered by trying to categorize those users through the role they seem to play in their requests throughout the week. Suppose, for example, that the system has no user data other than a collection of logon aliases and some group memberships that are used to assign rights and access privileges. Even in that case, with just the smallest amount of data available to the system, it may be possible to analyze user request histories to discover that some users are making repeated requests throughout the entire 24 hour day at 5 minute intervals, whereas other users might be making requests in a more scattered fashion with periodic breaks throughout the day, no overnight requests, very few and mostly irregular requests over the weekend. It might be possible to draw the conclusion from these patterns that some of these users are service accounts or automatic bots associated with machine processes running on a schedule whereas the others are human beings showing usage patterns characteristic of people with lives outside of work. It may be possible to discover on this basis alone that the large extraction queries and then rapid posting of many small requests are being performed by service accounts and that the subsequent read operations are being performed by human users. This discovery may not impact the initial decision to change capacity for a few hours each Monday morning before going back to normal, but it does indicate that the flood of requests is likely associated with an automated process that many different organizations are running at the start of the week. This may give the system operators an insight into an opportunity for a larger scale feature to provide if only they could understand what the users were doing and how that action could be automatically performed by the system without all the traffic back and forth.

In addition to having the engineering team introduce the capacity modification mechanism suggested by the preliminary analysis, the data scientists might suggest additional dimensions and attributes to associate with the events emitted by the system. The purpose of these new metrics or dimensions on metrics would be to provide more information about the users making the requests and more information about the type or nature of the data they are retrieving. Adding greater visibility into the system's actual content through its metadata might turn out to be valuable for making changes to the system helping it run better in the long term. The data scientist perceives that it is not just a question of getting all the nuts and bolts into proper order so that the system is available and reliable for all of its users all of the time, rather the system's ability to provide what its users need depends on greater accuracy in knowing who those users are and what exactly they are doing when they use the system.

This phenomenon need not be constrained to the introduction of new features or functionality. In the course of analyzing the quality of service, its reliability and its availability, it might start to become clear that 'bugs' in the code are not absolute. By that I mean that a consistent and reproducible failure in an area of the source code meant to do a job will happen regularly under identifiable conditions. These types of problems are relatively easier to catch and protect against through standard unit testing that describes the use case and verifies it as part of the daily build process long before the code ever sees the light of day. The much more common bug scenario is where the state of the user data is what triggers a condition that is not correctly handled by the executing code. In that case, when studying the reliability of the system it is possible to see good quality of service numbers but still discern the recurrent presence of undesired outcomes in parts of the system. Isolating the location in the code where this happens and inspecting the code with a specific exception in mind might be enough to identify the root cause of the error. There may be, however, other circumstances under which this is not possible. The pattern analysis may show that there is no regularity to the requests that cause the problem and no regularity to the known attributes of the users and stores involved in the request. The hypothesis might likely arise that there is something about the data structure to blame for the problem. Or rather, some idiosyncrasy in the data that is unhandled by the code as it executes. Since these cases are quite common in large-scale applications with complicated code bases, proper identification of the issue and root cause analysis of the problem might require increased telemetry and metrics. The metrics may require additional dimensionality capturing aspects of the user data or its metadata in the events fired when the errors happen. Such metrics are the eyes and ears of the process for trying to relive the logic of the system when the errors occur.

The tendency or dynamic is that the increase in statistical analysis is made possible because of the data that has been collected and that the statistical analysis itself is likely to lead to discoveries and hypotheses requiring more data to further inform the analysis. Metrics are witnesses who report what they see without understanding anything beyond their narrowly defined partial perspective. As the complexity of the system grows, there is more data to emit and the more data that there is to emit, the more useful additional data can be to understand the system's operations. This is not a pattern of addiction, it is not a need for data that increases as soon as it has been fed a little. Rather, the iterative pattern is the nature of investigation and discovery, it is what makes data science a proper science. The data scientist grovels through the data with the intent of discovery, with the intent of questioning the witnesses to obtain insights about the trends and tendencies at work in the system's operation. These insights will not necessarily emerge fully formed and thoroughly understood, they will evoke questions, incite further investigation, spell out hypotheses that need to be tested or verified. This will lead to further action, changes to the system's behavior, new operations, new emission of events, and new metrics. These new or extended metrics will provide further elaboration of the theory, further insight into the context and tendencies of the users' behavior when querying the system. And some of these tests are performed not only through the gathering of new data, but through the gathering of new data in new contexts added to the system's internal behavior and what users are able to do with it. The system undergoes modification in alignment with the discoveries. And these modifications might either promote changes in user actions as they request data, or they might forcibly make changes to the way the requests are handled. Even in the case of a bug that is triggered by specific conditions present in the data, the iterative changes either emit aspects of the data being transmitted to better understand the root cause of the bug condition or a slight change is made to the code to handle the data correctly. Both changes can be made in tandem. The former, in such a case, may provide enough data to monitor for regressions against the fix that is made to the handler in the latter approach. In all such conditions, changes to the operational nature of the system are made in response to discoveries made from the telemetry results. The point of gathering data through telemetry is not only to make the system visible in its operations, but to change the way the system runs, to modify the way requests are handled and produce large sweeping changes to the way requests can be made and what kind of data the request might receive in response. It is not just that revenue drives this or that modification, but that revenue is built into the entire operation as it operates. The purpose of the system is to learn more about what is happening to increase its role in what is happening: the money will follow.

In the case where we needed more information to understand service spikes at the outset of each week, the change to the system would provide better insight into the content of the requests that comprise the increase in traffic. Emitting a more detailed metric is the minimum change that might be made in response to an insight based on data accumulated already. Before concluding that more data needs to be gathered, of course, the analyst would have to bend and twist the available data as much as possible. This might involve introducing new and different dimensions into the charts or digging into parallel operations that might tell something through loose correlation. The data scientist can look in all the nooks and crannies of the data to discover patterns that might elaborate on the meaning of other elements of the system to provide insight into what is taking place in elements at the focus of the investigation. Only when they are convinced that the data needed to better understand what is taking place is missing will the information gathering mechanisms in the telemetry be modified to be more intelligent. The realization that the user data, the content of the transport, the content of the requests and responses, has significant impact to the behavior of the system, is a great boundary condition that, once crossed, opened an enormous set of doors for the analysts. Once the user data becomes integrated with the telemetry collection, the insights into the how and the why of the system's operation became significantly more detailed enabling far greater analysis and understanding of the processes. It is not just machine functioning that is measured, but user behavior integrated with that machine functioning.

Suppose, for example, that data capture on the giant data structure pulled down by the service account was inspected and discovered to contain calendar data on all the people within the requestor's organization. The service account is starting the work week by pulling down all the calendar items for the preconfigured set of people such that multiple agents cover the entire enterprise. Once doing that, we may discover that the post operations that immediately follow turn out to be task assignments for organization members: things they must do during the week based on the calendar content. Then, a closer look at the large number of users making requests after these posts, shows that they are the individual users checking their task lists to discover what needs to be done during the week. Once this information is gathered and the processes that it describes are understood, it becomes possible for the system to take on this task as an internal feature. Suppose for example that customers of the system did not need to configure their own agents to carry out the multi-step operation of reading and compiling tasks from calendar items. Instead, the engineers working for the system's provider company can build functionality and features into the system based on this analysis. Doing this improves the value of the system as product for its users and it improves the performance capabilities of the system so that it can more efficiently carry out this work at the start of each week. Because the system owns the processes that prepare the tasks, the calendars no longer need to be downloaded, rather the processing of calendar items into tasks can take place closer to the data store. This saves on bandwidth, a critical system resource, and thus makes the system run more efficiently with less likelihood of outages and issues causing availability impact for other processes. The engineers can distribute the load better than the random distribution that happens when customers undertake this operation themselves. And, of course, the provider can charge extra for this service.

For the data in conveyance to be useful to the data collection process, to the metrics emitted by the system as it executes, it must be conceptually rationalized. If the data is a giant blob (binary large object) of information with no clearly modeled structure, then its usage in data scientific queries will be expensive and limited in value. Queries made against data in aggregate need dimensionality and measurement to be performant and capable of correlation with other such queries against other facets or aspects of the data. For this to be the case in the user data, dimensionality must be applied. In the case of the example, 'calendar items' is already a high-level attribution assignment. The data is no longer a blob, it is a calendar item. We may not know much about the calendar item but knowing that it is a calendar item already tells us something more than we knew before. The object has been typed and typing data is a crucial first step toward better understanding it. Without looking at the calendar item, we might be able to surmise a set of attributes or qualities associated with such data. Calendar items have dates, for example. That seems like a simple characterization, but it is an attribute that nonetheless deepens the rational structure of the blob that is being passed around in user requests and responses. Furthermore, calendar items are associated with specific calendars and calendars are associated with owners. Knowing this information and scanning for deeper structure to reveal the character of these correlated attributes would help our engineers associate date ranges with the weekly processing that the new features for the system have to provide. Knowing the calendar upon which the items are based and knowing the owner of the calendar is useful in the effort to bind together the various users who have tasks associated with these calendar items. The calendars may be owned by organizations and organizations may have members. The calendar data can be checked for correlation with the groups and roles that are part of the authentication configuration.

All data collection repeats this pattern in analysis: starting with data in some raw format, rationalizing it into a modeled or rationally understood structure, and using that structure to produce greater understanding of that data's production and use. The data scientist, starting from the perspective of a partially self-understood system made visible to itself through telemetry, has the capacity to expand its realm of action into all the operations and contexts taking place within the system's posited universe. All the data that flows through the veins of the system is susceptible to greater understanding through this rationalizing process, through the process that the analyst introduces to enable greater clarity in their view of the detailed nature of the processes and operations as well as the actions informed by humans to trigger those operations. The system is a tool for users, it is being used by the requestors to carry out actions of significance to them. Understanding the flow of data and all its meanings introduces a greater understanding of the users and what they are doing when they access the system. This means that the conclusions of the analysis are not just aimed at providing greater efficiency for the system but intend to improve the usefulness of the tool so that it can provide greater efficiency for its users. What the users are looking for, what they are doing, what they are trying to accomplish when accessing the system, changes and is better rationalized through the improved understanding that the emitted telemetry data provides. The surveillance mechanism provides a payoff for the users, helps them in the achievement of their ends; and the deeper the visibility, the greater the payoff. This is at least in part due to the after-life metric emissions append to the user's actions and then in part due to the organization of that after-life into higher order understanding by the analyst.

The story goes that Claude Shannon played a game with his wife in which he opened a book to a random page and put his finger on a random point focusing on a single letter. In the game his wife had to guess which letter it was and then keep guessing the next letters based on the accumulating string. Shannon thought the exercise revealed how some of the bits of information were redundant since, based on context, it was clear what the next letters might be. A sentence such as 'The lamp is on the t", for example, might make it easy for a player to guess the next four letters without hints. The presence of past data and the existence of context and rules make the subsequent data redundant. If we had to store the data, based on the rules of this game, it would be easier to make the system more efficient by reducing the amount of data to only that which was crucial for setting the proper context. This kind of efficiency is possible because the sentence is relevant to rational world structures. Lamps do not tend to rest on textiles or toast. They tend to be set upon tables. The world model that puts lamps on tables is therefore a critical efficiency mechanism to reduce the amount of information that needs to be stored in order to play back the sentence on another occasion. To the extent that the system understands the user's context and what the user is trying to get the system to do when they query the system, the system can be made more efficient in anticipating reasonable next steps. When the engineers automate the production of tasks as part of the normal Monday morning weekly preparation, the system will need to consider the fact that lots of users will query the system for that information once it has been produced. To anticipate this usage pattern is to get ahead of any problems it might create. Rather than make the users come to the system to retrieve the information for their tasks, once we are finished producing them, it makes sense to generate messages to deliver directly to the users so that they have their task list waiting for them when they arrive at work. Their patterns of action have been anticipated for their benefit (how convenient) while improving greater stability for the system (required resources more evenly distributed and predictable and –likely– cheaper to deliver).

Rationalizing the process made this efficiency possible. And not just the process, the objects and operations that are part of the process have provided greater depth of understanding. Having more meaning has made it possible to automate the process efficiently and remove the stress that the random behaviors of the individuals put on the system. The analysis uses the history of the system's execution to project changes into its future, to make modifications to it. Modeling the user data made it possible to see more clearly what actions are being performed and gave greater depth to the history of the system and project a more precisely targeted future onto it. The system's changes can be made with the material needs and concrete usage of requests and responses in mind. Those changes can more thoroughly address the larger context of the system's usage and exponentially improve its scale and scope. This becoming "self-conscious" of the system through long term analysis, this making visible of the system to itself through telemetric emissions and collection, feeds back meaningful data into its future operation. It is a form of self-consciousness to the extent that it provides the system a way to adjust and correct its actions to better realize its purposes and outcomes: the aims or objectives that it is designed to fulfill. The system benefits through greater optimization and capability by seeing clearly into itself and into what its users are doing with it, what kinds of requests they are making and what kinds of responses they are getting. Self-consciousness, per se, has an optimization and rationalization role to play in the progressive transformation of the system. The purpose of introspection was to comprehend the meaning of the feedback at the level of self-conscious structures, at the level of models and meaningful actions. As such, an introspective system that uses its statistically understood self-surveillance to grasp the meaning of its operations and procedures becomes better at meeting the needs of its users, better at anticipating those needs, and better at providing solutions to their problems.

Adding the feature to automatically produce the tasks and send the content of those tasks to the users will increase the benefits users get from the system. The customers who had built such a system for themselves will no longer need to maintain those systems and customers that did not have the skills or budget to produce that capability for themselves can now make use of the implicit function added to the system they are already using. The introspection of exact usage via the system's self-surveillance, long term analytic insight, and projection into new features and capabilities will increase the value proposition of the service. This will increase the value for existing customers as well as increase the likelihood that new customers will want to buy the service for those benefits. This description could function as marketing for the service, the terminology bears nothing insidious in it. Or so it might seem. The telemetric perspective that enabled self-surveillance and the introspective rationalization of the system is not trying to induce harm on the user population, it is trying to provide benefits. The customer response to the introduction of new features and new behaviors might suggest that there is consensus to this effect. The telemetric has improved the lives of the users by peeking into what they have been doing and using the information gathered to better respond to needs expressed in behavioral patterns reflected in the data they have been accessing and producing.

Mathematization or Absorbing Content into Self-Surveillance Behavior

No doubt the modest optimizations of the system made possible through the analysis and modeling of the user's data would have given the data scientists and the engineers all sorts of ideas. If the knowledge that the users were accessing calendar items and generating tasks from them provided small improvements to the functioning of the system, imagine what impact could be achieved from a broader scale set of analytics into the entirety of user behavior when accessing the system. The targeting of the calendaring items initiated because of a glut on system resources taking place at a specific time and place, but the telemetric event emissions making the system's operations present to long-term analysis need not be focused on only those areas where the system is starting to choke, the entire operational working set of requests and responses can be understood this way. And the fact that data is calendar items, tasks, email messages, photographs, presentations, documents or whatever the specific data elements of the requests are, can become an essential part of the process of rationalizing emissions. Now that the data scientist has learned the power of understanding user data in making the system more efficient for both its own purposes and the purposes of those who would access it, it stands to reason that all elements of the messaging relay could be vastly improved through greater visibility of user behavior and the objects they work with.

This phase, brought on by the validation of the data scientist's methodology, is the mathematization of the system. At this stage submitting the entirety of the system to rationalization will become a high priority for its owners. The intent becomes the eradication of all 'blobs' so that the carrier or transport is aware of what kind of data is being transported in all cases. Furthermore, the system must be modified to emit metrics and events that surface the behavioral and operational nature of the actions performed. The user dimension and the user data facets of the operations must be fully rationalized and introduced into the telemetry, into the self-surveillance of the system for the sake of making its operations transparent to itself. When a photograph is requested, we should know the context of this operation. Who or what is the subject of the photograph and what else is the user doing when they retrieve it, what larger context operation is the request connected to? If the system can understand this larger context or provide enough information for the long-term analytics to understand it, then the system will have a more thorough view into its operations and –more importantly– the future actions that are likely to follow, just as we saw in the game Claude Shannon played with his wife. Although here it has become apparent that it is not just storage that might be redundant, the implementation of the game shows us that system processing can be streamlined.

The data scientist need not have an in-depth understanding of human being and the types of enterprises they are engaging when they come to the system. Exhaustively modelling the user data and understanding the behaviors at work in requesting and acting on that data is fully within the data scientist's reach without any prior domain knowledge. That has been the odd quirk of data science as science thus far: the data scientist need not bring any knowledge to the table regarding psychology or business practices. Instead, they study the data and cultivate the domain knowledge from it. The application of modeling techniques to the data and the introduction of dimensionality so that a calendar item is thoroughly understood as a calendar item is as much an exercise in data structures as it is an exercise in calendar items per se. It may help to know ahead of time what calendars are for and how people tend to use them, but this can be gathered from the operations being performed and –to some extent– the actual usage of calendar items is a more realistic portrayal of their meaning than some set of vaguely catalogued conventions brought to the analysis before studying the data. It is enough to know that the data is characterizable by types and that any type is associated with a set of distinctive attributes or qualities that define that type. Furthermore, the type will be associated with various behaviors that are part and parcel of it. Calendar items may be associated with reminders where the association is transitive, and the characteristics of the reminder tell us about the calendar item: remind the user about the event 15 minutes before its start time in case there is something the user must do to get ready. The reminder can be turned on or off, dismissed or snoozed. These constitute attributes and behaviors specific to calendar items and do not require deep understanding of what the event on the calendar is. Suppose that the calendar event is a meeting that brings together several different users from the organization. There are a variety of attributes that can be immediately characterized: who is the organizer of the meeting, who will be setting the agenda, who is hosting the meeting, who has determined the list of invitees, and where will it be held? Are there any materials associated with the meeting? A document or a presentation or a set of web addresses that provide additional data for the users to review during the meeting. What notes were taken during the meeting, what action items or follow up issues will the various attendees have at the completion of the meeting?

Even if the original design of the calendar item required enough knowledge of the domain to properly model the type, collecting and analyzing data about these attributes or characteristics does not require it. The analysts might point out flaws in the design based on their study of actual patterns, showing that the real-world usage is rejecting certain assumptions put in place by the designers. Individuals with modelling backgrounds, experts at rationalizing processes and finding the attributes in the data, can introduce greater intelligence into the system by quality checking the conclusions drawn by the designers and iteratively providing insights and adjustments to their proposed designs. Eventually, of course, the designers and the data scientists will all be learning from these outcomes, gaining a greater understanding of the domain, and how it works in practice. Data driven design embodies a way in which the domain migrates from a tribal set of practices and heuristics –based on loose connections drawn from survey results or direct engagements with focus groups of customers and what they claim to need from the system and its functionality– into a more rigorously understood set of objectives based on actual patterns of use. Sophisticated query languages have been developed for the sake of permitting the domain models to richly describe the data cultivated from usage. These languages allow a fully declarative semantics to be applied to the data, enabling generic operations and procedures while increasing the expert knowledge that the system has over the data. In the modern application environment, the designers depend more on the data scientist than vice versa when it comes to learning the details of the domain and how they should be used to impact product evolution.

The origin story that I am telling is not meant to completely discredit domain knowledge or assert that the data scientist is better off without it. Domain knowledge is crucial, but the point is that data science challenged the origins of the state of that knowledge. It suggested that we had accumulated a set of presuppositions and prejudices about connections and associations in various virtualized universes that did not bear out in the actual usage of those universes. The data scientist showed that many of our so-called experts in some domains were basing their knowledge on general impressions and intuitions about what the domain should be like or how it ought to behave not on the actual practices of people working within that domain. That information had never yet been gathered from actual practice, so it was only after the data scientists applied their techniques to the work being done that a proper evidence-based picture of work in the domain could be established. It is Moneyball, but everywhere.

Establishing the domain or the mathematization of the data model can take place without negative impact to the system since its power resides in its not having any specific insight into the data moving through it. As the earlier parts of the example showed, the system's efficiency and self-surveillance was based on its ability to handle all the messages and relays making up its normal load. The urgent requirement to introduce greater rational configuration of the user data cannot be achieved at the expense of this carefully engineered all-purpose design. The introduction of a modelling protocol and corresponding semantics enables rationalization to take place without crossing the boundary into the system's runtime logic. The system's formal nature is preserved while the material of its relays undergoes formalization. A new layer of formality is introduced: a type system that spans the entire scope of all user messages transmitted in its daily operations. Various modelling languages and protocols evolved over the early years of distributed systems, all with this intent in mind: enabling declarative modeling to take place in a realm of operations independent of the system's relay operations. The system code and its binaries are good at handling messages, just as they always were, only the messages themselves gained greater rationalization through the introduction of a new language for describing and capturing those attributes, making the data content more meaningful, and spelling out clearly what can be done with it.

This formalization of the user data does not just improve the system's ability to understand what is being done with it, it also gives added power to the callers of the system, the callers that make use of the service. Any caller can now clearly see the declarative model that is at work in the service and with which the client code interacts. This improves the ability to build applications that interact with the service as well as providing greater capabilities for understanding system actions. In the early days of such systems, it might be common to find only one single client application sitting on top of the service, a web site or web portal that queries data from the service and exposes it to users. With the advent of mobile devices and the various protocols for use in application development, the number of possible client applications has increased exponentially. These applications add behavior and capabilities to the domains of human activity that are interacting with the service and the system behind its endpoints. For these client applications to provide more insightful and improved operations, the application must have visibility into the types and meanings of user data that the system knows how to relay and work with. The system is as much a taxonomy as anything else. The model that evolved out of a deeper understanding of user data can be exposed using standardized protocols and consumed by such client applications for use with more flexible and robust functional logic that micro targets aspects of the system rather than through some monolithic rendering of it in a single portal. The ecosystem of possible work that can be done increases with this greater understanding of the user data and this greater understanding of what the backend services can do with it. There is not one application for all the location data that can be scraped from a cell phone, there are many different applications that can leverage this data to provide useful micro functions like directions to good music in the area, or restaurants, or traffic reports, and so on.

The behavior of engineers building system components and client applications is transformed by this greater rationalization of the user data. From the engineer's point of view, the user data retains its formal characteristics, name value pairs associated with the name of the attribute and its value in the current context, so the engineer can reason over the data structures using these formal characteristics. The emitted events come alive and animate a component by capturing elements of these attributes and values in the metrics dispatched back to the central analytics store. The data is still manipulated with purely rational means that do not treat this or that data structure or invoked behavior as a special case: the type system allows the engineer to introspect formally over the types and attributes, values and behaviors of the instances being passed along in the messages, without having to sacrifice its formal characterization. This is powerful because it does not bloat the code into special handlers for each kind of data or each type of user data that comes along. It means that the system need not be modified just because the model is extended or made richer in some aspect. Instead, the formal qualities of the data can be handled in generic form, and those qualities and their values can be surfaced in metrics in such a way that the data scientist can gain access to the richness of the data without the system engineers having to pay the price of greater ad hoc interaction with user operations. This, in part, is what I mean to invoke when referring to the mathematization of the system's operation. The user data has become fully rationalized and made visible to the system operations without violating the principles of generic or formal execution requirements held for good engineering practices. And it is this pattern that enables the dimensionality of the user data to be included in the metrics dispatched by the system to the analytics pipeline. The form of data collection need not change, only the dimensions and events need modification to carry the additional wealth of information now available to the telemetric eye that emits events while watching the system carry out its operations. This too is an aspect of the mathematization of the system. Not only does the formality of the system expand to include all previously 'raw' data, but the raw data becomes part and parcel of what the system sees when it makes itself visible to itself in the analytics store.

Data science does not undergo any fundamental changes when this greater wealth of data is made available to it, it just becomes capable of more powerful insights concerning the data's meaning and the future impact it can have on the system, now expanded to encompass both component operations and behavioral operations performed by users in the form of requests and responses. The greater the rationality in the data sets, however, the more powerful data science appears. Many sciences seem to have come into their own historically once the content of that science has undergone greater mathematical delineation. As the objects become more manipulatable through formal means, the ability to discover patterns in those objects and their associations increases. Relationships and associations left to purely empirical means are less powerful than when we compute over the possibilities through formal comparisons and relative analysis. I am not suggesting that the extensive use of statistics or statistical tools in data science was made possible by this increased rationalization of user data flowing through the system, only that this increase improved the capacity of already existing techniques to produce results relevant to those user objects. As a science of probability where correlation between past and future events are generated based on derived formulae, the greater the range of data that can be fed into the data scientific processing engines, the greater the scope of the output will be. This mathematization of the system does not change data science in any crucial way, it just makes it more powerful in terms of the impact it can have on system operations.

At some point, however, this power in data science came upon a set of limits. The data scientists discovered that their analysis, their efforts at correlation, were limited by the fact that the model has been materially introduced into the concrete behavior of the system. The power of this modeling was clear, but the analysis revealed limits as well. The analytics, as they increase in power, will accept the fact that modeling the processes is critical for that power boost, but it will also discover that the nature of that model will be intrinsic to the insights that the science is capable of, a projection of what is already known, and which might constrain what can be known. This means that if there is something in the model that structured the user action or user data too soon, the omitted characteristics of that structure will be invisible to scientific insight during the analytics phase. The data science will suggest that the model, although often helpful in guiding system operations and understanding, is often biased toward a specific way of understanding user activity. And this bias is both characteristic of the success of the rationalization and of a possible inhibiting factor hiding other types of experimental uses. The data scientists, fueled by the increased power of their insights over the mathematized system and its data, will wonder what could be seen if the system could be skewed with alternative models, alternative ways of looking at the user data, and the behaviors executed against them. The data scientist will raise questions against biases that are introduced into the telemetry and they will lobby for raw access to the data itself so that they can apply variable models, variable ways of understanding the data, during their analysis of it.

This characteristic does not necessarily invalidate the value added by rationalizing the data in the system. There is still obvious value in making the users' operations apparent to the system so that it can adjust based on meaningful indicators of traffic flow and the types of operations being performed. It does suggest that new stores of data will be required by the data scientist, stores with raw data that has not been pre-aggregated or pre-modelled with dimensions and measurements that –albeit powerful– predetermine what queries can be executed and what insights can be obtained. This accounts for the rise in giant ecosystems aimed at storing unmodeled data gathered from the applications at runtime. And these are not just storage engines. By 'ecosystem', we mean that they are platforms with different layers and tools corresponding to those layers. The tools are meant to power querying in as flexible a fashion as is possible, where the query can declare the model as preparation for the query, structuring the raw data into a temporary formation enabling new types of questions to be asked and answered. In these ecosystems, data is parsed in extremely high volume and so enormous amounts of processing power must be leveraged to enable the data scientist to perform the types of analysis that correspond to the queries and rationalization they are interested in. New distribution patterns, like map-reduce, would come along to assist in the high-powered processing that must be performed. The data in the analytics pipeline would be segmented according to some one of the proposed dimensions, the segments distributed across different partitioning approaches allowing the data to be separated, operated upon in its separated form, and then brought back together once the results have been computed. These can be very long running operations that use 100s of thousands of servers for many hours in the analytics space to both apply an ad hoc modeling principle to the data and then perform computations on it according to the form of rationalization projected.

This new approach to offline analytics is not meant to replace the application level modelling of the data that instigated it, it is meant purely as an augmentation. And even among data scientists, we would see hierarchies within the profession as some more junior level individuals remain strictly concerned with the simpler views of the data that are part of the ordinary runtime of operations. Individuals with more experience or greater mathematical backgrounds will focus on the more complex methods of cultivating insights into the data and the realm of operations signaled by it. There is a place for both types of scientists to contribute to the overall project, and history may show that the more experimental approach has a greater rate of failure and –when successful– a higher rate of return. The experimental results, when successful, may be far more sweeping in their impact, they may impact many more of the moving parts in the ecosystem, but the pace of innovation here may be slower. There may be only 1 or 2 interesting insights per year in such a realm whereas the more constrained work of the other data scientists may churn out dozens of insights. While the major insights might launch big projects to adapt to them, the smaller outcomes can be immediately piped back into the system with little turnaround and with moderate but noticeable impact. It takes all kinds and we should expect to see data science evolve into just as diverse a field of investigation as any mature science bound to established domains.

(Data) Scientific Realism or Making Telemetry Available to System Execution

Statistical methodologies power the analytic insights that adapt future activity to past trends, which means that the analysis is all about probabilities. The effort has been made to find aggregated patterns in the data. And by patterns, at least part of what I mean is that behaviors are studied for the correlation between the various facets of the activities to project future correlation between those same facets. In the initial phases, this type of analytics is done manually by a human using tools to extract the data, pore through it, and then suggest modifications that would improve the value proposition of the system to its users. One way this might play out is that the system will become more optimally responsive to the user's needs. If the pattern is discovered that shortly after a specific kind of request is made, another request is made for a data set that is moderately expensive to retrieve, the system can be programmed to start fetching the larger data set as soon as the first request has been received given that the patterns of past activity are being used to optimize system responsiveness to user needs. Users that ask for information about the location of a meeting, for example, may be likely to view a map of the location. The system can preload the map in the background while the user is checking results of the initial request. If the probability of the map request is high enough, it might be worth the risk to prepare for that request before it happens. This behavior may not be a simple 1:1 mapping, however. It may be that the system, through a set of requests made by the user, can start to build a story about why the user has engaged the system for its services. And based on these determinations, the system may begin to preload data that is likely to be of interest to such a user. Maybe the user is checking minutes or annotations that have been stored for past meetings. The system may have previously determined that users who start to cycle through notes for a few past meetings are likely to cycle through notes for all meetings covering relevant topics based on a system of tags that have been attached to those meetings. Furthermore, the system may contain information indicating that these acts of researching notes tend to involve tangential investigations into the attendees and emails that might have been exchanged between those users and the researcher. The system preloads data matching these criteria in preparation for the user's queries. The preparation can even be presented to the user proactively suggesting where to go next, helping guide the user along in their investigation.

In the simplest of cases, these patterns are discovered through investigations by data scientists on the metrics store through long-term analytics. It may have once been the case that engineers would have built simple mechanisms for capturing these sorts of associations in a configuration-based or declarative way. Thus, the system would look for these real time patterns as they were unfolding. The disadvantage here was that the system was only able to anticipate the specific conditions and the likely correlated direction of those sessions if the specific condition was put into the configuration data. This is not optimal. Clearly, a better solution would be to pipe the insights and patterns directly back into the system in automated fashion. The system could directly and automatically consume the insights from the data analysis and use those insights to optimize its response to user requests. This process would automatically complete the feedback loop between telemetry and system. To do this, however, an entire set of protocols were required. The process for discovering the patterns would have to spit out the findings from the data analysis in a way that the system could immediately consume, meaning that the output needed to be in a format that can directly integrate with the software system, it had to be a computed mechanism itself. The domain that came into existence to provide these protocols for using code to discover the patterns with an output that can immediately be fed back into the system as input for consideration in handling new traffic was called 'Machine Learning'.

It was named after a form of learning because it had to undergo a training process where the patterns in the data were made explicit. By executing over past data, the mechanism could see both the past past and the past future in the operations. Whenever such and such took place, the following events seemed to follow. Thus, a pattern could be discovered between one set of attributes as precondition and a corresponding outcome state. These could then be tested against other past data. The test realm would hypothesize the pattern against past data and then check to see if the supposed outcome was discovered there as well. Algorithms that did well against test data were ready to be unleashed upon the system, informing its operations in real time so that it might better anticipate the actions of users at lower cost to the system. Anticipation of user interest was not just an optimization, it provided opportunity for new features and capabilities. The system suggests things to users based on these patterns. Once the system discovered that a user has started down a path that fits with a previously understood search pattern, the system could suggest the types of things such users were likely to want to do next. It is not just that the system is able to prepare the data in advance for the sake of optimizing system response time, the system was now able to offer these resources and investigative pathways to the user as possible next steps. The theory here is that this provides a huge new feature to the users by helping to guide them through the procedure that they are likely in the middle of, providing them with a better experience and increasing their productivity. The user did not need to search around in frustration to find the things they need, rather the system made it apparent what types of things they were likely to need given what they have been doing so far. These things were then made available through APIs listing those things and some helpful hints on how to effectively pursue them. The client applications that make use of the system could then easily surface these hints to the users and help guide them through their operations more smoothly.

Machine learning does not require this automated integration into the system, but as soon as data scientists invented this approach to pattern discovery and application, the goal of the engineering discipline became the integration of this pattern discovery method into the system's operations. Having an intelligent feedback loop that is immediately present to the system's operations was something like the perfect completion of an ideal. The system had been evolving to become visible to itself. And this self-visibility through surveillance made for a more finely tuned machine. The manual processing of the feedback loop –albeit useful– was inefficient in that it required such a long cycle of discovery and integration back into the system. The system had immediate needs and all its operational members wanted to address those needs with the feedback in real time. Introducing the telemetric point of view into the system's execution is an expensive operation, we have described a process that requires skilled human beings and dedicated machines running the analytics pipeline and handing the results off to engineers to implement code that puts the results back into the system. The payoff for this expense had to be more than just aesthetic, it needed to provide a real monetary payoff. Cutting down the turnaround time between insights and implementation made whatever value the process could create more immediately available to the system. With slightly more investment than the cost of manual turnaround, the participants would be able to create a streamlined pipeline that can repeatedly provide insights back into the system for its use and improvement in operations. It made the system more useful to its customers, made it easier to sell to new customers, and reaped the profits in all of it. It was the next new thing with the 10x payoff, it was what everyone had been hoping for.

Machine learning played a twofold role in this desired approach: 1) it automated and sped up the pattern discovery and 2) it offered a protocol for direct integration into the functioning system. It could be said to have both increased the power of the insights that the data was able to provide, and it made it significantly easier to streamline the results of those insights back into the system for functional application. There is no need to wax poetic about the infinite possibilities of artificial intelligence, machine learning can be contextualized in much more pedestrian fashion. It was nothing more than a codified robot arm that automated processes that human beings were doing less efficiently. The discovery of a pattern was simply the codification of a probability of future events based on an analysis of past events. That probability is the pattern itself. Once the probability has been identified, it can be used to anticipate what might follow once certain events have been observed. The events themselves are those very same metrically structured operations that we were discussing in the previous chapter. In the telemetry, understood in aggregate, analyzed statistically, human beings were discovering regular correlations of one event with another shortly after it arrived on the scene. Machine learning enabled that process to be undertaken directly within the machine's operation as a network homed project whose output is consumable by other machine processes. The discovered patterns are an important advancement in data science, no doubt, but they largely repeat historical advancements throughout computing. In our sample system, automating notifications for weekly scheduled meetings was originally described as a manual discovery that was codified and added to the process by developers. This is an example of Human Learning. When the feedback is performed by a new component that examines telemetry data for patterns expressed in an algorithm or function and feeds the results back by handing the algorithm off to the operational system to be used in ongoing operations, human learning becomes machine learning.

It has become clear that the telemetric approach amounts to a specific way in which the system makes itself known to itself and uses that revelation to anticipate what is about to happen. The gathering of metrics, the emitting of events, becomes a way of auditing and tracking the history of everything happening within the operations of the system and capturing that for analytic purposes. The analysis is not idle, it is not a purely historical investigation aimed at understanding the past for some esoteric purpose. Understanding the past of the system in the fullest possible way is a necessary step for the system's augmenting its ability to anticipate the future, prepare for it, and use that preparation to guide its users along their way. Users are not likely to have kept some detailed history of what they have done in the past, they do not recall their past actions scoped specifically to their system use. And even if they do recall it, they do not have visibility into the aggregate actions of other users with similar histories and similar past operations. The system is in a unique position to present that possible future to the user and anticipate the normalized intentions and objectives of the average user while providing indicators to help them find their way to those ends. Anticipation is a key value proposition. Just as a waiter who anticipates every need as the patrons are experiencing them provides a better dining experience, the system provides better user experiences by anticipating their needs and offering them help in meeting those needs and achieving their goals.

The system which maintains ongoing analysis of its past operations for the sake of optimizing and improving its future operations and their responsiveness to user conditions is the perfectly self-conscious system. This does not mean that such a system has qualia or being-in-the-world in some science fiction sense, rather the point is that self-surveillance for the sake of achieving perfect visibility into the past and ongoing operations of the system, when turned into an automated feedback loop that informs future operations anticipating upcoming actions, has harvested qualia to articulate a form of self-consciousness. The self-conscious being is the being that consumes past intentions for the sake of informing and anticipating future intentions. The feedback loop of experience that brings what we have seen into an effective relationship to what we are likely to see in the near or distant future is what constitutes a self-conscious being: a being sufficiently aware of its past states to be able to use them to guide and govern its future states. Human beings do not achieve this however hard they try or however close they may come. Human beings are individual, and they forget things, fail to notice patterns, and fail to apply what they have previously come to know to what they are about to do. We are full of imperfections and finite scope. The machine system can be significantly more thorough in these matters so long as storage and processing time are both plentiful, and the feedback process is properly linked back to the system's future behavior. I cannot stress this point enough; the definition of self-consciousness here does not include qualia or any of the human characteristics that we often associate with consciousness except as material content or captured ingredients for later use. I'm not trying to make an argument that is relevant to modern debates about the 'hard problem', instead I am focusing on a dynamic or pattern of activity and giving it a name: the rendering exhaustively visible of past operations and events so as to discover patterns and processes at work in those events for the sake of anticipating future events to respond more effectively to them. I describe this as the system's self-consciousness to reflect the extent that it is a higher order configuration feeding off human consciousness captured in representative metric emissions, analyzed, and fed back: it is the telemetric understanding.

The self-conscious system does not posit its goals through its self-conscious operations. The objectives of the system are implicit in its self-consciousness in so far as the data emitted proposes a set of interests to the system, shows what is happening to it, and is discovered by it as it functions. The data is fed into the circular mechanism of its self-consciousness and in that process, we might say that the system discovers its interests, what it is on about, what it is being asked to do, and what it is doing, based off the emissions of data it discovers in the course of its operations. Whether we call increased attunement to those interests' power, or understanding, or wealth, is arbitrary at this point. The accumulation of data based on the precondition of data accumulated is the positing of objectives for such a system. Self-consciousness in this form does not drive the intention of ends and objectives, it realizes or discovers them as the circulation of emissions back into the projective process of doing more in line with what has been done already. Its bond is with the aggregate. Only now, with a purpose. It may resemble a hive mind.

In the discussion of anticipation, we noted that the purpose of automated responses that anticipated user actions based on their previous performance would be based on the learned patterns that existed in the aggregate and accumulated data from all users. That is, the system studies the likely routes through data by the various users that have come to the system in the past and anticipates the pathway of the current user based on those patterns. As mentioned, this would likely generate a table of probabilities where the various things the user might do will fall into ranges of likelihood based on the pathway of their actions thus far. With each action that the user performs, the probability associated with their next action will change because the training group that their pathway will be associated with will change. Users who requested a picture form one group, but as soon as the user goes on to provide a caption for that picture, the user may become aligned with a different group. The likelihood of what they will do next changes with any change to what they have done. The premise here, of course, is that the only meaningful context for understanding the relationship between accumulated actions and the likelihood of the action to follow is the pattern of those actions. This is behaviorism pure and simple. All sorts of vital data about the users are not yet known. Of course, were such things to be known, like motives, or memories, blood pressure or national origin, these data points would be added to the pattern matching mechanisms and would account for slight adjustments to the groupings into which the user is placed as they traverse the digital clickstream. Everything human in this mechanism can be reduced to a data point in a behavioral chain of actions and operations. The more of this data that could be collected by the system as it is being used, the better the system would be at accommodating those idiosyncrasies in the user that affect their grouping and the concomitant prognoses accompanying their actions.

Although more and more data from the users would surely impact these anticipated behaviors, provide varying probabilities for them, and enable the system to do a better job of helping the user along in whatever it is they are trying to do, the organization of the operations is such that the principle is fundamentally correct with varying levels of accuracy. That is, the methodology of describing human behavior on the system in the aggregate and using that aggregated understanding of patterns and operations is not itself failing when what has been anticipated does not occur, rather the amount of data injected into the system must have been insufficient to properly group the user's actions. The process of reducing human action to statistical probabilities based on aggregate analysis clearly demonstrates its own validity even when it makes a mistake, even when it gets it wrong. The misstep is not because there is something flawed in the process, but because there was insufficient data available to make the prediction the system was trying to make. Since the anticipated behavior is understood in terms of probabilities, there is always the likelihood that the user will be headed in a different direction from that which was anticipated. Eventually, the process will get it right, or rather the likelihoods associated with different possible futures will increase and the chance of misstep will be reduced as the groupings become more precise. Thus, the performance of the mechanisms will appear to be improving over time. Since the process itself has this notion built into it, the mechanism trains on some of the data, tests against other of it, and then is unleashed upon the actual world, making its predictions in real time. The process itself is designed to be progressive. Training is something that can go on forever with constant improvements based on more and more collected understanding of the population using the system and whose data is collected for anticipating future actions.

The final principle of the telemetric is the declaration that statistical aggregation for the sake of pattern discovery yielding anticipation of user needs and interests is a well-conceived methodology for understanding the process at work when users come to the system. At this point, I'm not prepared to make any judgements about this methodology. If the system operators and engineers along with the data scientists are just selling a product with an interest in meeting customer demands, then this methodology may be revolutionary and one of the greatest inventions in human history. As a technical solution to a set of problems specific to service implementation and system maintenance, this would seem to be one of the most exciting possibilities to come along in the digital age. The invention of the smart phone, by this reckoning, was not earth shattering on its own but because it was a mobile telemetry device. Telemetry might be viewed as the culmination of a long held human interest in all knowledge systems. The physical scientist gathers data on the nature of the physical universe and uses that data to anticipate the future events and measurements that might be made. The same principle of relevance is adhered to in telemetric computing. Some prediction might prove inaccurate based on existing data and that does not invalidate the scientist's method, it just means that there was data relevant to the circumstances that we did not yet have. The context of the base data and the context of the prediction were insufficiently correlated. There were missing factors, aspects of the situation that were not understood, data that had not yet been gathered thus showing that the anticipated events were different than expected because the circumstances were not similar in the properly significant ways. Our failure to see that was because we lacked sufficient data to make the proper connections and establish the proper patterns. This methodology predates the digital age, but it has been perfected by the advent of large-scale computing systems. And the key to this perfection has been the ability of the processes being measured to emit those measurements directly from within the process itself. Remote measurement is, in a sense, a limitation of most scientific measurement and it accounts for many of these gaps and blind spots in the understanding of the circumstances. What is key about telemetry is that although the data is processed and munged, prepared for feedback, in a faraway processing pipeline, it is gathered from right up close and during the processing. The most effective form of telemetry is the form that emits metrics from within the midst of the process being measured.

The application of statistical aggregates to currently occurring actions would be far less likely to be successful if not for this characteristic of the digital telemetry mechanism. The aggregate data is applied to the case, but the aggregate is also an amalgam of measurements taken amid the operations similar in kind to the current one. As such, the association of the aggregate to the current action can be correlated through the attributes yielding the notion of similarity. The means of the correlation are those very dimensions used to emit past metrics. Because the aggregate data is aggregated in terms of the same sample properties and attributes associated with the current action, the correlation is easy to manage. The failing instances are due to one critical dimension or another that is needed to set the instances apart or bring them together. This allows for comparison of like constructed events and so increases the likelihood of accuracy in the anticipated action. The power of science improves with the power of measurement. Better measuring devices are those devices that can get closer and closer to the thing itself as it goes about its business in the way that it does when there is no external observer. The more the gauge can penetrate the inner workings of the thing being measured, the more robustly it can enable the definition and the clarity of the measurement. And this amounts to greater visibility into its attributes, greater associations with other like things as determined by measurement, and greater insight into what the real causal powers are and how they enable accurate anticipation of future outcomes.

Forget about how you feel about this, it does not matter. What matters is what you click next. Once we can accurately gauge your feelings, we can then form better associations between what you felt and what you clicked next. Collect similar events from millions of feelings and millions of clicks and the work to do for the next person that comes along will be that much easier. There is nothing truly new in you, nothing that could possibly surprise. If there were, that would simply be a set of data points that we need to figure out how to capture and then incorporate into the patterns of past behavior associated with future actions. After that, this thing that is unique to you will be properly correlated with the thousands of others who share in this very same 'unique' expression of some specific human behavior exhibited while interacting with digital entities in the digital world. The self-fulfilling prophecy of the data is to show that there is nothing in you that is unique. As soon as such is suspected, a new dimension or dimension value will be added to the emissions and accommodated in the analysis to ensure its capture in an aggregate and common pattern by telemetric techniques. Should the uniqueness persist despite the adjustments, the system will flag it and launch a set of automated responses aimed at preventing further activity until an investigation can be performed. The hive must absorb whatever transcends it. Anomalous behavior gets lots of attention from automated systems and many maneuvers can be applied, in one way or another, to remove the anomaly. This would seem like an appropriate segue to a more explicit discussion of privacy, since what is private is meaningful to the telemetric system as its limit. And this limit position of privacy may be comparable to the way in which anomaly detection identifies outliers to the aggregate computations of user behavior. In either case, there is something going on that does not fit and to which the system must adjust if it intends to recuperate the content surfacing as either private or anomalous.

Chapter III: Telemetry and the Privatization of Privacy

The telemetric gathers and then it anticipates. The gathering permits the large-scale data pipeline to provide a funnel for the generation of software that can execute procedures based on knowledge of how the past has previously come to fruition anticipating how the present will play out. What does this have to do with privacy? The real value and power of the telemetric came to be via the inclusion of user data asked for and provided. At first glance, this suggests that the surveillance mechanisms of the system at work needed to set their sights on the user's body of information flowing through it and then absorb that data into itself incorporating it through users' actions and the system's response. Whether or not this technical mechanism is relevant to issues of privacy would depend on whether the user emissions captured and analyzed were private and whether the process of capturing and analyzing them could be done without infringing upon their privacy. Raising concerns along these lines would surely send this discussion down familiar pathways.

Before venturing off in that direction, I would like to pause and consider the title of the chapter: the privatization of privacy. One immediate interpretation might suggest that privacy is for sale, that it is only available to people who are willing to pay for it. This is a common theme of scarce goods. It may have once been thought that the laws of the land in some nations offer protection against violations of privacy, but in a fast-changing high-tech world, privacy is becoming harder and harder for the public sector to safeguard. In response, it may be that the governing body must cut and run, must privatize the protection of privacy as they might privatize retirement savings, leaving persons to fend for themselves. You can have as much privacy as you can pay for. Bear this in mind, but it is not the only thing I mean to call out with the title of this chapter.

In addition, 'privatization of privacy' is the conceptualization of the notion of privacy, the definition of what it is in the average understanding of it. When some people discuss the widescale matter of whether a computer system respects or violates their privacy, they are privatizing privacy by thinking of it as property. The data that is coursing through the veins of the large-scale network distributed system is my data, the user may think. And I own my data and the system should not be doing things with and to it without my permission, without my knowledge, and without my consent. Enter the EULA, the end user licensing agreement. No one forces the user to log on to some computer system, no one makes the user access the internet and look for relevant web sites and resources via their favorite search engine. No one forces you to post pictures of your lunch or dinner with friends letting everyone know where you are at 2:19 on a Friday, these are services that are offered to the user and the user willingly takes advantage of those services. Or so the reasoning goes. And the provider of those services, when the user initializes their relationship, offers the user an agreement that they can make before proceeding to access the service. Since you want an account, you must agree to the terms of service. And in the terms of service, that is where the critical exchange of information takes place. You agree to let the system function in the way that it does, and the system agrees to let you access it and make use of it. The privacy condition has been met because the user has agreed to let the system access the data that is moving along, to analyze that data, and even to use that analysis to better anticipate the user's needs and desires.

Given this framing of the matter, it is possible that there are companies that provide services for sale and who strictly honor the terms of the agreement and only use the data in the agreed upon fashion, where that means using the collective data to monitor and maintain its proper functioning and to ensure that the system is always available and reliably handling the users' requests. It may even mean that the company is running all that user data through metric pipelines and analytic stores accessible to data scientists all for the sake of understanding how to better serve the customers' needs in the future, how to better anticipate what features they would like to see, and what actions they would hope to perform. And this can be done in a responsible manner. The corporation may mask the individual user data as it is captured, meaning they might remove anything that is personally identifiable. For example, the specific name or email or logon information of a user may no longer be associated with the data collected from that user. The system, when emitting the telemetry events, may have known that so and so was currently on the system, but in the data pipeline that relationship is removed. The person might become a simple demographic, an aggregate person: a 30-year-old male living in Florida rather than some specific person living at some specific street address. In many cases this aggregate representation is enough for anticipating the needs and wants of the user. The company may not need to know anything more specific to be able to better serve that user at the time they are navigating the system, trying to get what they want through an on-line action.

Of course, providing a service costs money and any company that provides a service is not doing it for the sake of humanitarian or charitable causes. The service and data infrastructure to provide a large-scale system may be vast and require capital investment. The users would be expected to pay for this service and in return they would be treated to the terms of the service agreement that constitutes a contract on the reciprocal scope of usage between parties. The user pays for the service, the service provides the goods. They honor their agreements and property rights on both sides are respected. The point to gather here is that the customers are the people who pay the service for the goods they receive. In so far as the customers and the users are the same people, this clear delineation of the privatization of privacy will hold. The provider will honor the customers' property and give them what they are paying for so long as they are in good standing and their accounts are up to date and paid in full. In the event, however, that there are participants who do not pay for the service they receive, they would likely receive different treatment and, along with it, a very different licensing agreement. Such users might still think in the way that the actual customers think and assume they are getting the same kind of treatment, but that would be confusion on their part. Buyer beware. This is an oft-remarked point. If you are not paying for the service but others are, it is entirely possible that you are not the customer but the product. We are getting ahead of ourselves.

The root of any confusion or misunderstanding between the users and the system itself originates in the privatization of privacy, the conceptualization of privacy as property. That data that is in the system, that is being moved around, is my data and I trust that all that legalese in the terms of service is meant to assure me of that fact. The user thinks of the data as though it were their car or their house. The whole social fabric is aimed at protecting their possession of that property, their ownership of it. They can trust that the whole of modern life is looking out for their property ownership in the form of their photos and documents, of all the artifacts that they post or present on line, and that the social forces of security will come to their aid in the event of a breach just as they would if someone broke into their house or their car. My things are secure because my society has pledged to keep those things secure. The issue of note is not just that the data in the system is equated with familiar property items like houses and cars, but that the data itself is understood as things: photos and documents. The user thinks that the data the system has access to, the data of primary significance to the system, is all the artifacts and digital things that the user has put into the system or that the user retrieves.

And the description of the metric emission process from the first chapters may reinforce this notion up to a point. The items in the body of the responses and requests coming from and to users are the artifacts of user creation and consumption. A meeting request is an artifact, digitally created, but it represents the product of an action, a scheduled event with a set of invitations associated with it. The invitations have been sent to persons with names and accounts. The body of the meeting request might include an agenda that has been written by the meeting organizer, there might be a document, or a presentation associated with the agenda and available to invitees for inspection prior to the meeting. These do seem like objects and do seem to be human created objects at that, embodiments of human effort and action.

What could be simpler and more straightforward, our digital world is familiar to us, there is nothing questionable or strange going on here. We make things, we produce objects and we either produce them with the services we are using or produce them outside the system and then upload them for storage or to share with others. The document that we upload to include with the meeting agenda is something we may have fashioned on our own computer or tablet and then we upload it just as though we were taking money to the bank or putting our summer lawn furniture into the shed for safe keeping during the winter. Nothing could be more ordinary. This is all stuff and it is our stuff and we have placed that stuff within the service for care, in exchange for the monthly payment we make. And the service promises to protect and honor the terms of our agreement by keeping these objects safe and using them to improve our experience with the system. And this last need not be insidious, it may be for the sake of improving the security of the system. The system analyzes ordinary usage of those artifacts to better anticipate actions involving them. This permits the system to understand normal usage patterns and so might allow it to flag abnormal patterns and protect against them. The telemetric is a service to the user in this common understanding of the relationship, it is there to provide the terms of the service to the user and it is there to protect the users' interests.

But that notion of 'normal usage patterns' already signals the sense in which this privatized notion of privacy, of privacy as property protection, becomes muddled. Recall that the goal of data collection and analysis in the telemetric survey was to anticipate the future to provide better services to users. The analytic insights and measurements were not merely aimed at peeking at user data for the sake of some voyeuristic interest, the system wanted to understand what the users were doing to optimize their ability to do it. The concern here is the users' behavior, what they are doing, what they are looking for, what they are interested in, and where they might be headed. Although the access to user data during the servicing of requests and responses does give the system access to the artifacts associated with the user, the primary interest is not the artifacts as such, but what those artifacts reveal about the users interacting with them, about their patterns of access, about their actions and interests, and about their objectives. A meeting invitation with an agenda that includes a discussion of a conference that is about to take place in a foreign country may lead to some of the invitees purchasing airline tickets to get there, it may lead to hotel reservations and rental cars, it may lead to reservations at restaurants. These are opportunities for understanding the future actions of the users, these are opportunities for the system to anticipate the users' needs and offer them directives on how to meet those needs. Maybe this amounts to simple advertisements that can be presented to the user indicating special fare rates on the airlines or advertising hotels near the conference center or restaurants within walking distance of the hotel. And users might be happy to have these advertisements available to them once they realize they need to plan for the conference. A good system might know where they are going before the users do. Maybe the users did not read the agenda for the meeting, maybe they did not connect all the dots to see that they would be attending that conference. History might have clearly shown any learning algorithm that was paying attention to the data what the likely outcome of the meeting was going to be. And the algorithm, unlike the attendees, has nothing else to do. The machine, therefore, might have learned of this first and the machines would be able to trigger responses that would make sense of this information. And so here come the ads, and they would be welcome. Business travelers want convenience, they are not interested in shopping the best deal to get to the conference, the plane tickets may need to be refundable because the attendee must be able to cancel at the last minute and it is better to pay a higher fare to have a ticket that can be reused without penalty later. Maybe there are cheaper hotels further from the conference, but they will not house other attendees and networking with other attendees is one of the primary reasons people go to such conferences. The targeted advertisements that are injected into the system would inform the user of their options, meeting with these various well-understood criteria, and the users may gladly act on them, just as the system expected them to.

The service's interest was not the artifacts per se, but the actions and intentions of the users. Part of the reason I have been careful to tell the story of the service through its telemetry is to make this clear. There are solid technical reasons to want to understand the user behavior of customers accessing and uploading resources via an online service, and those solid technical reasons provide tangible benefits to those users.

Although the privatization of privacy concern is being undermined by this emphasis, there is not yet anything fundamentally hostile between the service's interests and intent and the user's interests and intent. The users may think in terms of their property, their artifacts and the protection of them, whereas the system may think in terms of the users' actions and their relationship to those objects and what the objects indicate about their future actions. The service cares about the users' actions and behavior while the user cares about their objects and artifacts. These concerns need not be at odds with each other. The offset of these two different concerns may be the origin of a perfect agreement where each party is able to gain from the exchange precisely because they are interested in acquiring different things. If the service provider wanted access to the artifacts as such then the two parties would be less likely to find agreement. The users would want to protect exactly that which the service wants to gain access to and understand. Instead we have a symbiotic relationship between the service and its users, they get what they want, and the system gets what it wants. It seems to be mutually beneficial to all involved parties.

In some of the preceding, we have been spinning a yarn that has described the telemetric system as self-conscious in that it uses data collection to complete a feedback loop allowing it to anticipate the upcoming requirements placed on the system by its users. The yarn is not meant to be specifically Hegelian and I have not done anything to show the dialectical relationship between emission, gathering, and feedback. Consider this a purely pragmatic appropriation of the Hegelian journey in his *Phenomenology of Spirit* starting with an initial stage that spells out how the system becomes aware and then how that awareness becomes an awareness of itself to further that initial position once it came upon a set of limiting problems. Again, I do not mean to suggest awareness as the direct presence of qualia of its own, although it is most decidedly a consumer of qualia in the aggregate. And because it actively aggregates in analysis, it is a producer of a collective configuration of such mental states. The service at work in telemetric activity is not aware of a world, per se, but it embodies such awareness and is engaging in operations that have been programmed as well as absorbed into it, it has a behavioral set of procedures and those procedures mimic the pattern of consciousness completing itself in self-consciousness as it gathers together metric emissions into an intelligent store of analytics.

And now I would suggest that the next stage in this dynamic is for that system to fully realize itself in Reason. Again, just a pragmatic repurposing of the Hegelian picture here. Through this privatizing of privacy, the system and user form a larger machine, a rational machine whose goal is the rationalization of the users' behavior. The user who gets immediate access to airline fares based on the meeting agenda is a user who is entering a more rationalized universe of possible actions. They are fulfilling the action orientation of a feedback loop. This user need not go search erratically for what they need but will have the most optimal offerings presented to them immediately after the meeting. This offering that meets with the interests of the system and the interests of the user is a more rational system than any service that stays out of these matters and lets the users fend for themselves when arranging flights, hotels, and restaurant reservations. A world in which our actions are so matched with the computer systems that provide us with services, is a more rational world, a more efficient world, a more optimized world in terms of our interests and in terms of fulfilling our needs and desires. Call it a telemetric enlightenment that surpasses the rationalization of bureaucratic organizations.

It is a huge injustice to the Hegelian system to suggest even the most remote resemblance between what I have shown as the Reason of the system and what Hegel understands as Reason in its various forms. There is no simple correlation in the system's design with an observational reason that moves through an attempt to realize cognitive rationality before arriving at the kingdom of ends. It is, however, fair to say that I do sense in the operations of the service on behalf of its user population a set of utilities for improving the conceptual order that governs objects, persons, and networks of persons. And to that end, I do see a loose architectonic association that can be made to Hegel's project. The rationalization process that we have inherited from Weber via Habermas can be understood as the implementation of a richer and more thorough conceptual model for understanding the events, things, and actions that make up the world. It is possible to see this richer conceptual model not only in the attempt to know these processes, the critical attempt to understand historically what these processes mean and amount to, but in the things themselves. The digital object, the digital artifact is richer than the thing that is extended in space. Or rather, it can be ascertained as richer because it yields more thoroughly to the telemetric viewpoint. Because we can emit metrics directly from the processes at work in creating and manipulating these objects, the objects themselves are not only more easily accessible to our understanding, they are more conceptually robust. The process of measuring is done for precisely that reason. Augmenting the things of the world, the events, and the actions with deeper conceptual dimensions, with greater meaning, and with more clearly ascertainable attributes, the telemetric makes the world richer by spelling it out more clearly and in greater detail. The rationalization of the world through dimensional modeling, through conceptual modeling, is a process that happens through measuring and analysis and the purpose of this measuring and analysis is to make the world deeper. And the 'Reason' that I mean to invoke as the next stage in the system's odyssey is exactly this stage of development, the stage in which the objects, actions, and events become more rationalized and more thoroughly available to conceptual insight and introspection. We see more, once we have gathered and analyzed, because the process of gathering and analyzing not only sees more but produces more to be seen. When the users are introduced into the mechanism and their needs are symbiotically intertwined with the operations of that distributed computer system, the users themselves become more rational, their actions more rational, because they are more rationalized: they become more, there is more in them to be seen.

As far as this goes, you might think I could have called this chapter 'the rationalization of privacy'. To the extent that this rationalization occupies the user and thus the human subject of agency within the telemetric system, however, it is a personal state and so –in a sense– privatized. I am not therefore, exclusively interested in rationalization from a social point of view, but equally from the personal impact that it has. Such connotations of rationalizing privacy need not be insidious or terrifying. It may be as simple as riding a bicycle. The picture rendered thus far need not pose an imminent threat, the relationship proposed is symbiotic: this is an idyllic possibility showcasing the perfect cyborg creation, human beings augmented to live better fuller more optimized lives because of the role that the machine plays in completing the rationalization of ends. This would be the highest end and aim of Enlightenment rationality, exactly what the ideals of the 18th century had planned for us, the realization of our highest hopes where true liberty has been bestowed upon the individual through telemetric augmentation.

'Privatization' is suggestive of many such nuances offering up something of a family resemblance. The objects and artifacts of the rationalization process are our property. The interests and intent of the service provider are their private concern. The actions perfected by the feedback loop that realizes the Reason of the system at work are the proprietary intellectual property of the corporations that offer those services. Reason, as we know it in this inquiry, is not the rationalization of privacy but its privatization, privacy's becoming private, a private concern for each of the parties involved, where rationalization merely describes the scope of that privatization as part of a larger machinic dynamic. And through this combination of subject and substance, privacy appears as something that can be alienated from us. If it is something we own and are paying for, then it is possible for someone to buy it from us or to expropriate it with a trick or some dirty dealing that imposed terms we did not understand or through some agreement we did not realize we had made. This is not merely the alienation of private property, it is the alienation of privately appropriated Reason in the form of telemetric rationalization. The more isolated and alone that we are in this alienation, the more intricately intertwined we have become through the operations of the telemetric systems that host our lives.

Taking a commonly held view of privacy cast as property and suggesting the connotations that have been historically aligned with property might serve to reduce our understanding of privacy rather than increase it. It may be that the use of the telemetric patterns to illustrate the place of privacy in such a world obfuscates what privacy might mean and I have ventured down a path that plays fast and loose with the likelihood that we all have some common notions here and that we can use those common notions to flesh out what so far has only been suggested. And this perversion, this operation that proceeds without a detailed understanding of 'privacy', is also what I mean by privatizing the private. Each of us may have a vague sense that we have collected over the years of what privacy might mean and that personal notion of privacy is tacitly leveraged and made to seem as though we are not talking nonsense when we talk about it. Each of us, on our own, contributes connotations and associations that give meaning to the term. A further privatization of privacy. It is possible that, at this point, we are not even discussing privacy at all, but rather property as whatever belongs to someone, whether that someone be a human being or an organization or a software system stretched out over space and time. If this chapter can bring the reader to a position where you are no longer clear about what we mean when we talk about privacy, then it will have done its job. The privatization of privacy suggests that privacy itself is a muddled notion and when we engage in activism and legislation to protect it, or actions that violate it, we may not be clear on what we are doing.

Social Media

If the usage pattern we have provided so far is the idealized version of the system, then social media might be considered its anti-pattern. In the social media systems in popular use as of this writing, users and customers are not one and the same set of people. The users are those who inject data into the system and provide the access patterns on the data, the customers are those who purchase access to that data for the sake of producing and targeting the population with additional content. What made this anti-pattern possible, however, is the data that makes up the social media system. Social media usually involves networks of accounts that are associated with each other through a link that might be known as 'follower' or 'friend' (sic, the synonymy is astonishing here). This is a special association that each account makes to indicate that the two parties involved have an interest in each other's content. Content from each account is made visible to the other account and the other account can express reactions to that content. A follower may 'like' the content, may comment on the content, or they may choose to share it. Visibility for all content is constrained but in a typical scenario an account makes content visible to all its followers. If a follower expresses any kind of reaction to that content, the follower's followers may now have access to the content and its commentary or reactions. Network node expansion of content visibility allows accounts to effectively socialize via the process of content publication and reaction. We publish on such a network by producing content or by reacting to someone else's published content. Accounts may work to grow their network of followers to include friends of friends and so on. Social media are therefore sometimes thought of as social networks where the account is associated with a set of identifying attributes. Like the "real life" social world, anomalous account structures are introduced through bigger and bigger groupings with a determinate subject matter enabling member accounts to associate based on a category of interest. The accounts in a group need not have explicit follower status relative to each other, they can be a more loosely based grouping that does not require explicit node to node relationships. Instead they increase their network size through membership and access to the content posted to the group by other members. We do not follow each other in this case but come to publish each other's activities by both following The Black Cat Appreciation Group or some other such topically driven association. The result is the same: when you comment on how cute the kitty is, you expose your reaction to the other members and when they like that you think the kitty is cute, the exposure increases beyond the group.

Typically, the content producers on such social media applications are not paid for their efforts and they are not charged for the service provided to them. Their interest and usage must be driven by their desire to partake of that service. It is therefore crucial that the owners of these media ensure that the feature set and capabilities of the platform continue to have and gain accounts that are active in the production of content. To achieve this, the service providers use their telemetry to ensure that users are engaged and that their usage is optimal. By and large, the allure of the platform has been historically established. Human beings are social animals and they form social bonds of different kinds throughout the course of their lives. Clearly, we do better in groups and so exhibit the tendency to establish clans, tribes, towns, cities, nations, and civilizations. Social media did not need to create this characteristic, it merely tapped into it via a distributed system spread out across the globe. It provides highly mobile contemporary humans the ability to keep in touch with people they have moved away from and people from different phases of their lives. It also permits them to branch out and associate with new people who are associates of their existing associates. It permits them to connect through categories of interest and subject matter. People who like cats or physics or balloon racing can virtually form associations. The content of their virtual association will likely allow for real groupings. People who get to know each other through the social media following and publishing mechanisms may arrange to pursue their interests in the "real world". All sorts of possibilities are at their disposal. And it is the social aspect of these systems and the allure of the social that draws in the users in the first place. Social media can be a great place to extend your area of influence as well as reestablish contact with lost associates. It can be a virtual world for social interaction or it can be a useful instigator for real life connections.

This is enough to get users to sign up and onboard to the system. Keeping the users interested and helping them to grow their networks is part of the application's anticipation of needs and desires. The engineers building the system want to promote the social experiences that draw the users in; and the interactive nature of publishing multi-media content, along with the system of reactions, is how that works. There is great entertainment value in the content published by the people that you follow and there is the capacity to provide instant feedback to that content. Hierarchies can easily be established through these mechanisms. Users may quietly or loudly compete for followers, posting content that is more likely to draw attention. They extend their reach by meeting new people and inserting their reactions into each other's networks. This gives them the opportunity to connect with members who are already part of those other people's networks. The positive feedback mechanisms of commentary, sharing, and liking enables the content to get instant feedback as to just how successful the publisher has been in gaining this attention. If a lot of people like it or comment on it, then not only does this validate the interest followers have but it extends the content's visibility to those people's networks of followers. This competition over followers and their reactions to the content provided to them can lead to increased usage of the application. The more positive feedback that the user receives for their activity, the more likely they are to continue to participate in the publishing and consuming processes that make up the platform. As networks grow, more content from more corners of the world becomes available to the user, and there is just more to do in the application. The more people you follow, the more people that follow you, the more there is for you to attend to when you log onto the system. This creates a very powerful incentive for users to keep returning and to stay longer when they do.

The users producing the content do not pay to use these services. In some cases, the services have multiple billions of distinct monthly active users. A distributed system with data storage and telemetry pipeline to support billions of monthly active users is extremely expensive to own and operate. The number of servers required to perform the various roles involved would be in the 100s of thousands, the routing requirements would introduce large scale networking system requirements, and the storage would require additional servers to persist user data over time. The telemetry system's data processing pipeline might be equal in size or even larger than the system providing the application. This all entails large scale energy use and engineering payroll expenses. A company that could maintain such an infrastructure would have to be able to pay out billions of dollars in expenses each year. Although they do not have to pay the producers of their content, they do not have any revenue directly from their publishers either, so where does the money to sustain all this infrastructure come from?

Using broad strokes, it is fair to say that proof of concept for some social media was done with limited expenses and small populations of users. As proof of concept, they may only have had access to the personal investments of the founding members of the corporation. In this proof phase, they were able to build lists of publishers demonstrating value to investors. The population or audience had been assembling for business opportunity. Even without an explicit description of what those business opportunities were, it was clear early on that having many thousands of people accessing your system could be potentially valuable. The venture capitalists saw potential in this gathering of population and so invested in the system and in the efforts to figure out how to turn the user population of publishers into revenue streams. One clear historical model made the business model for these proof of concept systems easy to describe: television. In television, the population of viewers was a captive audience and so ideal for advertising. Many businesses were eager to put ads for their products in front of viewers of otherwise free content on TV. The same principle holds in the case of the social media applications. The owning corporations could make good on their assembly of publisher populations by getting customers who wanted access to those populations. They would begin to sell access to their publishers and gain revenue that way.

To the extent that the individual accounts publishing content provided personal background data, the accounts became gold mines of demographic information. Even without too many of those attributes, it would still be possible to build demographic descriptions of the users based on their various behaviors. It is not necessary for a user to state that they are interested in hot air balloons if they are members of groups that accept content on hot air balloons. This interest can be inferred from the various activities of the users. Even in platforms without groupings based on interest, there may still be data to be extracted from associations. If a popular TV actor has an account on social media, the followers of that actor must have an interest in them and in the TV show or any of the other projects the actor might be a part of. This association based on behavior is enough for declaring a new form of demographic grouping. The potential for user segmentation and conceptual modeling in the social media space was significantly higher than it had been in the television platform because the users were more active, their behavior formed part of the system, and their reactions were being captured as part of the ordinary functioning of the system. It would be possible, based on this behavior and the content provided, to form demographic groupings of the users that broke down into finely detailed categories. The key demographics in TV might have been something like 'females aged 18 to 35'. In the social media application where users are publishers of content that includes pictures of themselves, descriptions of places they have eaten, activities they have been performing with their friends, music that they like, and books that they read, significantly more fine-grained descriptions can be formulated. Thousands of attributes can be modeled against a user based on their various likes and publications. And the customers of these social media can target their access based on these attributes.

A restaurant located in Dallas Texas and offering Tex-Mex fare would be ill advised to advertise broadly to the whole world. They would be far better served in the expense of getting their message to a desired audience if they could narrow down the population to those who are most likely to be persuaded to go to their restaurant. This might include people who live in and around Dallas and who have shown a disposition for restaurant patronage or they might more loosely focus on any people who are likely to be in and around Dallas in the near term and who are likely to visit a Tex-Mex restaurant. These attributes can be cultivated from a variety of other behaviors that may have been discovered in the larger data segments produced by the application. That is, the data science for the system may suggest that certain users, because of various publications or reactions they have made, show that they are likely to be in Dallas and likely to enjoy Tex-Mex food at a restaurant. This may be due to an explicit expression of interest in such things or it may be due to an inference based on common characteristics and past behavior of other users with those same characteristics. To trap all this data properly, the restaurant might need to have a presence in the content delivery areas of the system so that users can express reactions to the restaurant based on actual visits. This establishes a set of users with computable characteristics as having visited the restaurant in the past and enables an inference to be drawn from them to others that are like them but have not yet 'liked' the restaurant. Other characteristics might come into play such as what other restaurants they have liked, for example. If those other restaurants create a profile that matches the profile of a current user being evaluated for inclusion in the target population, then the data science might suggest that the user should be included in that target group and shown the advertising content for the restaurant. The association may be weighted, and a threshold set for triggering inclusion in the group once the volume of connections has passed it.

This turned out to be a revolutionary business approach. Because the networks were social, they were able to cultivate behavioral information on users' preferences and interests that were deeply personal and indicative of some of the core motivators driving the person to act. The data collection would be as thorough as the person's participation in the social nature of the network and so the demographic target construction would be as broad as the publication tendencies of the users. Some users might be more two-dimensional than others based on their usage and this two-dimensionality might place limits on how broad an understanding could be achieved of this person's demographic situation; but the data within the scope would still be valuable. Some users might show only a news-centric interest in the platform whereas other users might only be involved for the sake of pursing some hobby they have. The data collected from such behaviors would surely be less suggestive than that available from users with a more widely encompassing pattern of usage, but it would be valid nonetheless in so far as it goes. The hobby itself, or the news stories of interest, would still supply information about the user and what sorts of targeted content they might be likely to react to. Some of the nuances of the analytics might not be available for such users. It would be more challenging, for example, to make inferences about what other hobbies such a user might have because there would be less range in the statistical comparisons to make between this user and others, but it would still be possible to target that user with those items that are clearly deemed of interest to people with those specified hobbies. The corporations providing the platform hope that users are not so disciplined in their usage. Even someone who mostly publishes news stories, may have a soft spot for cats and be unable to refrain from showing that interest in some of their on-line behavior. The minute such a person does so, they go from being someone who is interested in the news feed from the New York Times and Washington Post to someone who is interested in those news feeds who also likes cats. And this might be enough for the feedback analytics to suggest that such a person might also be susceptible to other kinds of content that people with those interests frequently share. The targeting might be less accurate in the cases where there are fewer points of contact, but the feedback may increase in value as connections increase. The publisher is likely to have established followers and is likely to occasionally react to publications by those followers. The nature and frequency of such reactions will provide additional attributes for the user and thus permit better targeting of content.

In addition to gathering behavioral statistics on the publisher population, the system would also be gathering behavioral telemetry from the customers who are paying to use the system. What they are paying for, what they produced, and what response they got from their targeting are the likely outcomes of these analytics. The variation on the idealized form in the social media system is masterful in so far as the customers and the publishers are both producing content. At some point there may have been some distinguishing markers to separate the one form from the other, or the one point of origin from the other, but in other points in the history of these systems, the content from the customers was published in formats that resembled the content of the users publishing their content without the ability to target it. What the customers are paying for is therefore twofold: 1) the ability to target their content and 2) the ability to publish content as though they were ordinary users of the system who were not specifically targeting that content. And because the absence or presence of targeting was not critical to the underlying infrastructure that carried along the content coming from these different sources, it was possible to capture telemetry in much the same way for the paid content as for the free. The data could be collected using the same dimensions and attributes that the other activity was using in the metrics emitted for analytics. This meant that the paying customers could easily get feedback as to how their targeting was working, how successful it was in not only reaching their desired audience, but how successful it was at engaging their attention. This feedback on their investment would help them improve targeting of their future content, they could plan based on the specific feedback they received. They might notice that users with specific attributes are more likely to react to the content and they might notice that some targeted areas are unlikely to get any discernible reaction at all. Because it lacked this interactive feedback, the historical paradigm was quickly usurped: TV advertisements had nothing like this, focus group and audience had become one and the same.

This kind of analytics makes it possible to offer additional features to the paying customers and one of the more popular features that has come along is the notion of experimentation or A/B testing. In experiments, the customers might declare a control group based on some attributes and they may declare a test group based on an alternative set of attributes. They can then hit the different groups with content to check to see the impact it has on the two different groups side by side. To the extent that this feature is lightweight and easily manipulatable, it may be possible to rapidly iterate on different experiments covering a variety of different attributes. Such variation and experimentation are an important feature because it allows customers to discover important correlations between the various targeted attributes in terms of both contents offered and responses received. If the two groups have most attributes in common, for example, with only one or two key properties separating them, the responses by the different target groups might tell the customer a great deal about the population and how it responds to different content. This testing can be driven by either variation in content across the group or variations in group definition across the content. In both types of experiments, the customers can learn how to craft their content and craft their targeting capabilities to get the right message to the right groups with a high probability of getting the right response.

The types of reaction that are possible are therefore not simply features that the publishing audience might cry out for, the paying customers would like to see finer granularity in the types of reactions as well. We have seen in some social media systems that the variety of reactions to content has increased. And the public sense of the impetus for these 'improvements' was that the greater variety of response types allows users to better express themselves when reacting to their followers or followed users' publications. It seemed obvious that this was a new feature that seemed to add value for the publishing users, but it is far more likely that these new reaction types were added for the sake of the paying customers and the feedback they would be able to receive from the content publications. If a customer can distinguish between strong reactions and weak ones, between humorous reactions and angry ones, sad and joyous, etc. then the customer makes a better determination of the impact of the content on the targeted audience. They can also make more informed comparisons between the two groups that have been configured for a given experiment. It is possible, for example, that two different groups or two different content publications to similar groupings might both cultivate equally numerous responses and an equal amount of attention from the various users targeted by the publications. It may be, however, that the one group responds more strongly to the content than the other. One group might simply like the content whereas the other group loves it or is angry about it or is shocked by it. Alternatively, we might see that the same content causes sadness in one group of people and loving enthusiasm from another group. This allows the paying customers to gather significantly more useful outcome data from their experiments. It may teach them how to energize a set of people or it might teach them how to harness the energy of another group and redirect it into a related area, an area that is best served by some product they are interested in selling or some set of ideas they are interested in promoting.

We will come back to this later in the chapter. For now, the point is merely that the split system services both network targeted data from users and demographically targeted data from customers through the same patterns of collection and feedback. The telemetric system can be used for both purposes since the different populations are themselves clearly delineated via account attribution. Their content moves through the same veins, crosses the same nodes, and infiltrates the same parts of the system's delivery mechanisms. The varying persona that make use of the system are themselves classes of users that differ only based on demographic attribution that the system has captured in its internal models. This means that targeted publication and network publication can be variant system behaviors with one underlying set of design principles. It means that the system can expand the capabilities and functionalities of the delivery mechanism for the sake of shifting and changing business models and usage patterns that correspond to the interests or desires of the company that built and operates the system. And although we are investigating this phenomenon under the auspices of social media, there is nothing specific to social media in the system itself. The example from the first two chapters was more of an enterprise productivity tool in which collaboration between different users was the focus: users who sent each other email or shared documents or scheduled meetings. Even a system such as that might undergo a transformation based on the 'social media' pattern of splitting out user types and the kinds of feedback that are relevant to the various categories of users.

Let's say in the earlier example that the employees of a large corporation are using the system to collaborate in doing their jobs, in sending email to other employees, sharing documents and spreadsheets, scheduling meetings, planning agenda, going to conferences, and so on. The managers of these employees may have an interest in evaluating their performance and so may wish to plug in additional analytics that measure the activity of the employees as they use the system. This way, the managers know which of their employees is more productive, which more effective, which has more points of contact in different parts of the organization, and which of them are not really doing anything at all when they say they are 'working from home'. It is not critical that my assertion here be supported by historical evidence to prove that tactics employed in social media caused collaboration-based enterprise systems to begin surveillance programs on their employees, maybe the ideas were in the air at around the same time and so the various systems began working toward these patterns of surveillance coincidentally. What is important to the argument is that the variation in the user attributes and the ability of the system to track users differently depending on the various roles they might play (paid vs nonpaid or manager vs nonmanager) opened up the possibility of the anti-pattern that permitted the system to accommodate wildly different interests and personas in the rationalizing activities of the system at work. Artifacts and objects are not the only rationalized outcome of such systems, the users at work in the systems are further rationalized too. They are understood better, their relationship with their content, with the objects of their inquiry or manipulation are understood better. The users are just as much a part of the system as the messages that they are sending and receiving. And the segmentation of users into different personas enables new scenarios that can be promoted and fleshed out in the operations of the system.

At least part of what I mean then by 'privatizing privacy' is made clearer here. The system segments its user population for a variety of reasons, separates these users into different personas, identifies them along the boundaries and attributes of their different qualities of usage as well as their different professed qualities outside the system. Age and sex might be examples of the latter, customer and network publisher might be examples of the former. If there is a common abstract base class of logged on users that make requests and receive responses that can be surveilled by the system via metrics that are ingested into the analytics pipeline, then all the different personas and scenarios can be used to construct different points of view on the system and promote different sets of interests there. The data specific terminology here might suggest 'normalization facilitating efficient polymorphism'. At bottom, they are all users, but they implement different specific behaviors on the common interfaces that they execute as they work within the system. Such polymorphism encapsulates varied interests that may correlate to distinct revenue streams or business segments. And the segments are all intermingling within the same system, pushing and pulling on the same knobs, but with different outcomes and results leading to different inputs feeding back into different subsequent actions. The system is privatizing in the sense that it is building subdivided views of the population, subdivided arenas of actions and access, reporting and understanding. There is a sense in which we are brought together by the system in a common network of users with the capability to relay messages back and forth, but the telemetric reason that characterizes the system also segments those users. It separates them into subgroups in terms of both objects and artifacts of concern as well as in terms of their actions and behaviors, the feedback the users are receiving, and the modifications to future input that they are encouraged to produce when adjusting to the reactions of their past activity. The anti-pattern that we set out to describe under the guise of 'social media' is therefore just the system's implicit rationalizing effect for the sake of breaking out the user population into conceptually clear subsets and subgroups. This rationalizing effect privatizes in that it segments user classes, but it is also a privatization, as I said, in terms of the forcing function that leads users to evaluate the impact of the system in terms of the property rights and material interests that are either protected or violated by the system's operations. We will turn to this aspect next.

Securing Private Resources

The 'social media' anti-pattern yields a growing awareness of a conflict in interest thus making it clear that there are indeed interests at work in all usage of the system. Interests, per se, become something to be concerned about, something to worry about, something above and beyond the simple usage of the system itself. The users become aware of a set of 'issues' at stake. They may have thought their interests were purely their own: whether the requests they were making got appropriate responses, whether their needs and desires were met reliably, and whether the system was available whenever they needed to access it. These were the only concerns the users had, purely technical in nature. When it surfaces that the users' interests are at odds with each other, that some users are making use of the data produced by other users for the sake of projects at cross purposes to them, the customer grouping and the user grouping show themselves as different personas. To get what they want to get from the system, the two groupings need to be concerned about the other group and what they are doing. The users become worried about the customers because they fear that their interests may be more crucial to the service provider. They sense that the system is exploiting the users for the sake of the customers, they worry that they are only there for the sake of making desirable resources available to the customers in exchange for the price they pay. This sense grows the more powerful and the wealthier the service provider corporations become. The symbiotic relationship is at risk because the system facilitates users with conflicting interests.

The social media corporations underwent a period of massive growth in which they became extremely wealthy and powerful. Yet, the user population must have detected that this wealth was difficult to explain. The users are not paying for the service so where does the money come from, what exactly is the source of such enormous profits? The users learned that the companies were selling 'advertising' and this simple notion is treated as the familiar norm that it had become through their experiences as TV viewers, radio listeners or magazine readers. Even when the peddlers are selling political orientation or discord, it still tends to be normalized as advertising and the power of targeting based on interactive data harvesting is reduced to an almost trivial detail. The users realized that the service providers are selling advertising on the social media sites and applications to enable third parties to sell them stuff. This is all too familiar and not likely to raise alarms. Advertising, so they might think, is not necessarily insidious. It is a minor presence, just a reminder of the various products and services that exist out there in the world and that the users themselves might be interested in. The advertising might even be helpful, if it turns out that the users happen to be in the market for the kinds of goods and services they offer.

Slowly, over time, however, the users started to notice that the advertisements in social media were different somehow than the advertisements they were used to in radio and tv ads of the past. The users started to notice that the advertisements they were seeing through their access to social media had been accurately guessing the things they were looking for or shopping for in recent days or weeks. The targeting feature of the social media is better developed than anything they had ever seen before and so it is possible that the greater accuracy of the targeted advertising was making an impression on the user population at large. This may account for a background sense that there was something different about this new media and its type of advertising, that there was a war raging for the users' attention.

It is also possible that the user population was vaguely aware of the fact that these growing corporations functioning as service providers in the new media were somehow different in kind than the advertising infrastructure they were used to from print and other media. In television, we had the networks who were offering programming, we had the companies that were offering products, and we had the advertising agents who were preparing the ad campaigns. The common understanding of the social media service provider was altogether different. They were 'tech' companies who were responsible for building new technologies with the latest and greatest innovations. And yet, their revenue streams made them look like magazines, radio stations, or television networks. Because these service providers were offering multiple services with multiple applications that accessed comparable social networks of users, they did not look like the classical example of the television network. They may have seemed more pervasive and ever present in the lives of the users so this role as advertising platform may have yielded a disconnect between, on the one hand, the users' perception of the service provider and the role they had in their lives, and on the other hand, the reality of their business model and its various methods of obtaining revenue.

These factors may have been vague notions, not entirely well formed and clear in the general user population. Alongside these perceptions, there were the various news stories that came up now and again describing government access to surveillance metadata from phone systems as well as government and other third-party backend access to some social media data to assist in investigations into terrorist activities or to understand electoral trends and tendencies. These were new notions that started to find their way into the popular press and ordinary ways of talking about things. People in the user population would have heard about 'metadata', 'backend access', 'third-party use', and 'cracking encryption codes' in their everyday access to news stories. The general sense of mass surveillance was becoming more and more common. People knew that there were eyes on them. And it may be that they were not much worried about the operations of law enforcement because, by and large, they may have thought that law-abiding citizens did not have anything to worry about from such surveillance. After all, they were law-abiding, and the law was only out to get the bad guys. The relative merits of this kind of thinking are not the key issue here, rather what is interesting is that there was a lot of conversation in the air to make users aware that people were keeping an eye on them and that something weird was going on with these new platforms that were becoming an increasingly popular part of everyone's daily lives. The user population was becoming aware that they were the targets of a variety of different surveilling interests which meant that they started to form into an independent group with a set of interests of their own.

The powers that be, in this case, might be a big corporation providing a service, it might be another big corporation providing internet access over the phone or wi-fi network, it might be their government searching through 'metadata', or it might be big corporations targeting them with advertised information while they use the various services that were more and more a part of their daily lives. All these different powers had a bigger role in their lives and played a bigger role in distinguishing the user group as such, as a group with a distinguished set of interests. The service provider and their customer base were just the beginning of a much larger set of concerns. The entry of the government and the network providers made the system of connections more complex. Even a simple search of what was available on the world-wide internet, put the user's metric emissions, free of charge, into the hands of one of the largest and most powerful corporations in the world.

Alongside these powers taking on bigger and bigger roles in users' lives, there were all the stories about data breaches and data leaks. Big retail chains were being hacked, credit verification companies, and all kinds of smaller online organizations. Data breaches were everywhere. And the data that was leaking were personal asset information about the user population, credit card numbers, email addresses, passwords, and even social security numbers. Identity theft became a common plague on the popular consciousness and people were growing in their awareness of their vulnerability to online crime because of the complex ecosystem of data collection and usage evolving in their profiles and lives. Users were not only aware of the powerful presence of giant organizations, but they were becoming aware that this presence made them vulnerable to crime and criminal activity in an unprecedented way. Everything, they started to realize, was online. Their property ownership records, their rental history, their medical history, even their job history. All the critical services that made up their lives were present in the internet-based data stores and systems. And although it is not technically correct to associate all of these system as one and the same, with the same infrastructure, or the same mechanisms at work for making associations or taking advantage of personal connections, the average user may not be technically savvy enough to make the kinds of distinctions required to separate the various systems and to accurately measure the risks of one type of software system over and against another. In the popular mind, the information systems were all lumped together, and the data content was amalgamized into a common image or picture of the purpose of these systems and the risk that they posed to the users' lives.

It may, therefore, have been this amalgamized or generalized sense of the threats and risks of internet security of personal data that yielded the extremely 'privatized' interpretation of the issue that the users sensed they were suffering from as they navigated the hazards in the new online ecosystem. Although the digital universe may have crawled into the users' lives through more frequent social media use, it was a fallacy that the data systems that abound in this amalgamated view were one and the same with those social media systems. Just because a utility company adds online services that allows its customers to pay their bills and check for outages over the internet, using some application on their phone, or even through an automated menu system over the telephone line, does not mean they are using the telemetric pattern of metric emission for the sake of feedback aimed at anticipating future activity. The telemetric is not a priori present in all things digital. A data breach at the department store that did not correctly secure credit card information at rest need not have any relationship at all to government access of phone metadata for the sake of creating detailed understanding of networks of communication that might reveal a terrorist plot. Nor do either of these necessarily associate with the kind of data publication that goes on in a social media site and the metric emissions that include such content in creating a user profile for sale to advertisers with an interest in targeting their content to the most likely set of buyers. These events may all happen on the internet and they all happen to involve new technologies often clustered together, but that does not make these areas of life and operation the same thing. Just as going to a department store to shop for clothes and using the telephone to call your friends and colleagues are not usually activities associated with each other when done in analog, there is no reason to immediately associate the practices of these two realms when they occur digitally. People had bought into the computer age and they made associations driven by the economics of this infrastructure even when the domains of life and action were radically disjointed. Although the use of a common identity service to secure access to various software systems might associate the data (login with Facebook or with Google), this nuance suggests the importance of details in discriminating between the various systems at work and the relative risks and threats that they have on personal status when used.

Consider an example: a system in which the published data is a variety of different test results from cancer patients and the set of medicines that have been used to treat these conditions. The purpose of the data system is to provide expert assistance to diagnosing and prescribing oncologists. Based on collective evidence and analysis, the physician may be able to more accurately prescribe the treatment with the highest chance of success in curing their patient. In this case, the various parties may be physicians and patients who directly pay the service provider for the expert system's computing cycles. No clear division of interests may exist in such a system. Suppose, however, that the actual payments are coming from private insurance companies who start to do a little analysis of their own and discover that the treatment of cancer is extremely expensive, so the insurance company starts to determine that profits can be improved by cutting down treatment in some cases. Say the service provider starts opening its platform up to an insurance company mechanism that will inject alternative content into the system based on additional metrics and analytics. The insurance company may bring in data about rate adjustments and lifetime payouts to patients. They may discover that in some cases, the correct treatment for a patient is not cost effective and that *nearly* correct treatments are far more profitable. The results for the patients collaborating with their physicians to arrive at the best form of treatment for the disease may now be skewed by the insurance company's injected characteristics. This system has introduced a new set of interests into the analytic ecosystem and this new set of interests will impact the outcome for all parties. If the public at large were to find out about this, they might become incensed and stop patronizing insurance companies that subscribe to this service (supposing the absence of any legal constraints so the only means available for public outcry is market-based tactics and strategies). In addition, it may be the case that such public outcries not only target insurance companies that inject their analytics and metrics into such systems, the public could determine that all systems with the same higher order function are essentially flawed in the same way. Such snap judgements may be made without any investigation into the details of how the various user profiles and roles work in those other systems' operations. Rather than focusing their objections on the role that insurance companies play in such medical analysis, the public might mistakenly broadcast their objections to the entirety of the technology itself: the use of expert systems to assist in the diagnosis and treatment of various forms of cancer.

This conflation of the amalgamate based on the participation of the different walks of life in the underlying communication technology, might make it increasingly easy to concentrate the interests of the users into a more clearly delineated area of concern with a clearly delineated form of concern: the data producing population and the artifacts that they are producing. The perception, in the cancer database example, might be that the user data published to the database is being exploited by the insurance companies in ways that are hostile to the interests of the patients and physicians. In that case, the outcry may involve the call for regulations to keep the patients' publications private or hidden from those users that the patient blocks. The publication is conceptualized as the placement of an artifact into a system for various parties at work promoting idealized interests in the domain of life targeted by the system constituting the line of business declared by the service provider. These are my medical records, these are my photographs, my treatment plans, my data and only those parties who I wish to see that data may be allowed to see it. The separation of interests in the systems' operation made this manner of conceptualizing the privacy of the various users into a natural outcome. The publisher is the provider of the data and the publisher thinks the data is their data and so they should be able to control who has access to it. This is true if the data consists in pictures of the person's body (like CT Scan data or MRI imagery) and it is true if it is a set of medical diagnoses, just as much as it is true if the data is a bank account number or a social security number. Data is a thing and it is owned, it is the property of some person who is somehow described by it or bound to it. What could be more truly mine than my body? And if a set of high-resolution pictures are of my body, my organs or even conceptual maps of my sequenced DNA, then how can these things, these artifacts produced by a technical process, belong to anyone other than me?

You may determine that your physicians can see the output of your tests, but suppose you do not wish to have your insurance company see that output? Clearly you have an interest in preventing such access since they are running analytics to determine if the cost of the proposed treatment is worthy given the likely outcome and the likely financial payout (the future premiums you pay). They can claim an interest based on the amount of money they stand to lose under the conditions. The public might claim that the insurance companies are being opportunistic, that such companies have good grounds to use statistical data to determine the amounts for premiums, but that they abuse those reasons when they use a different set of principles to drive their profitability through influencing diagnoses and treatments. These latter should remain outside the domain of expertise of an insurance company and left to the physicians and patients. That is, the insurance provider may base your payment on some generalized rules of what the projected outcomes in your health history will be, but when those actual outcomes occur, they cannot demand a recalculation of the odds to determine whether their costs can be reduced. This would amount to two radically different ways of thinking about the relationships and about the data involved in the computations. In the one form, we are concerned with access to the artifacts, whereas in the other we consider the methods of computing or performing judgements based on the meaning of those artifacts in various of the domains in which they are relevant. We can admit that an insurance company has an interest in the artifacts without granting authority to influence the actions on them via an opaque set of algorithms and analytics that they hold as their intellectual property.

Consider the social media site that is selling advertisers access to the publishers' personal data as it is stored in the social media analytics store. Somehow the data for sale belongs to the publishers of that data and the sale of it violates a privacy right that extends to the property belonging to a person or user. The users agreed to some terms of service that have accounted for these scenarios and so the users have effectively given permission to the site to share their data with the customers who are paying for that access. When the focus is on the artifacts and the interests that the various roles have in those artifacts, the agreement is the central determining factor in how things are adjudicated. If a user has entered into an agreement that permits the service provider to make secondary usage of the published materials, then clearly a transfer of property right has been made and the service provider is well within their rights to resell access to those materials. Likewise, the insurance company that is paying for the procedures and treatments agreed upon by the physicians and their patients has entered into an agreement with the patient when they started their exchange of services: the patient's purchase of the insurance to cover the procedures and treatments that their physicians prescribe. These agreements become the fundamental basis for determining whether interests are properly served and properly realized within the confines of the system.

A privatized realm of interests and artifacts is the obvious result of these sorts of service level agreements that bind together the various user roles that take part in the day to day operations of the system. And the privatized realm of interests and artifacts is supported by various systems and services. All sorts of variable characteristics are squashed together to create this amalgamate of computing ecosystems. Whether we are considering the social media case or the government metadata case, the retail store transactional database or the medical records store; in all cases, we apply these same set of conceptual schemes to understand the relationships and try to think through the merits of the interested parties and the case they might make to further their interests through the actions and operations they have contracted to perform. These are civil relationships governed by civil agreements. If a person chooses to upload their personal details and artifacts into a system, the expectations are set by the agreements between the user and the service provider. If that service provider clearly articulates the terms and conditions of the operations that will be performed on that data, then there is no legal recourse to users for changing the nature of the agreement. They can, however, turn to the dynamics of the market place to boycott services with bad agreements or unfair practices. Failing that, or failing enough mass momentum behind that, there is little more that can be done.

The ordinary notion is that published content is the property of a user. Securing published content becomes a matter for property rights and the agreements that cover their exchange. The telemetric logic does not rise or fall on this and so it is often expedient for service providers to advocate this ordinary notion and work to secure user data. Since the telemetric hinges on being able to apply aggregate analysis to the specific on-line behaviors of individual users, the property rights of those users can be easily admitted so long as the provider continues to have the right to flow data into their telemetry system for the sake of analysis and feedback. Property concerns are a ruse in so far as the primary focus of the telemetric lens is human behavior.

Modifying Human Behavior

If I have a picture stored on my phone or tablet and someone hacks into it and copies the picture to their own storage, it certainly makes sense to think of this as theft. Something that I own has been breached and some other thing in my possession has been taken from me without my consent. This makes sense. In another case, however, I may give the Target corporation my credit card number, expiration date, security code, and my address to purchase a set of bath towels. Doing so may involve agreeing to a set of terms and conditions. Target may then put this personal information into a storage system somewhere and it may be breached by a hacker. In that case, the hacker has stolen from Target, but Target may be in violation of the terms of the agreement I made with them. I did not give them permission to share this purchase data with other people. Target may claim that they did not intend to share the data with the hackers, that they were victims of a crime and I was merely the victim of a transitive crime. Not a crime perpetrated by Target, but by the hackers who stole the data from them. I may decide to argue that Target has been negligent in the way they handled my data and this negligence made it possible for thieves to steal it. These are two different ways to conceptualize privacy issues as property concerns. Target, for example, could bite the bullet on the issue and claim that my data was not stolen at all, that the only theft perpetrated by the hackers was the theft of Target's property from its storage systems. You gave us the data, they might say. Once you did, it was our data. And no doubt the terms of the agreement you made with Target might include language to this effect. Often terms of service indicate a transfer of ownership of some kind when a user hands over their data to the company providing the service. Google, for example, has gone so far as to claim that the results of any search constitute its free speech. The underlying point here is that the data that Google has crawled the web to obtain from websites all over the world is something they think belongs to them and that their rendering of that data to searching users is their own business. I am not yet ready to look at the legislative details of these sorts of debates, but at this point I want to make it clear that thinking of privacy concerns as essentially property rights that are either transferred or preserved in the various agreements made between users and service providers leads to a specific way of thinking about the problems that arise. The courts would be able to designate who owns what, what terms of service signal which transfers of ownership, and so on. The federal government of the United States, when it intends to gather data from large scale storage systems, considers the search warrant that it requires to be something that it needs to present to the storage system owner and not the person whose data they are sifting through. Property rights become the singular concern of privacy discussions and whether a person has a right to privacy becomes a matter of whether the agreements they have entered, or the constitutional rights they have been afforded, cover the case in question insofar as it is a case of contested infringement upon their property.

Although an interesting line of thinking when evaluating the power of the telemetric, this is not the point that the telemetric itself is interested in making. The systems we have been grappling with thus far have not been about property. The systems I have been describing have been services provided by companies to various kinds of users, some for pay and some as simple publishers. And these systems are managed via anticipatory feedback loops to optimize functionality for the sake of enabling user procedures and operations. That has been the language we have been using all along. It is crucial to recognize that such systems are primary enablers for specific user behaviors and for actions that users are trying to perform for the sake of a set of projects they might have. Business users or individual consumers are trying to do something with these systems: contact friends, collaborate with colleagues, establish themselves as organizational leaders or provide insights into their community. They are doing things and they are interacting with others who may be assisting in doing those things. It may be the goal of these actors to get others to do things they would not otherwise have done without the interactions taking place in the system they are using.

When I organize a meeting and invite several of my colleagues to it, in a trivial sense I am attempting to modify their actions. I am trying to get them to do something they had not previously thought to do, come to a meeting to discuss a set of issues. It may be that their job description or role in the organization is such that going to these sorts of meetings is exactly what they are supposed to be doing, but this meeting that I am inviting them to, is not something they had previously concerned themselves with. Only once I make the invitation, do they accept and arrange their schedule so that they are available to come to this meeting at the appointed time in the appointed place and with the various other meeting attendees. This is impact. The single user is trying to have an impact with their actions, they are trying to organize the actions of others who are likely to let themselves be organized in these ways. Whether these be community projects, entertainment events, or business opportunities, the agents are collaborating on various projects and ends, having mutual influence over each other and even inspiring each other to become more involved in certain kinds of activities. Generally, these sorts of events and actions are transparent. The meeting organizer or the event planner is not trying to trick the other users into coming to their meeting or event. The agent is likely known to the others and is likely making their agenda crystal clear to the other agents while trying to collaborate with them.

As zones of interest, however, secrecy is likely to find a place in the system. Because users are carrying out different roles and have different personas, it is entirely possible for some of those agents, representing some of those roles and personas, to be more opaque in their methods. Maybe the invitees do not want to come to these meetings typically or maybe people are hopelessly overbooked and unable to make time for events that are not as high a priority as other items on their calendar. There is no reason to suspect that on-line collaboration will be free of the types of conditions that ordinarily impact upon human collaboration. And the wide spread presence of network connectivity in human life might exacerbate the traditional problems in such collaboration. Meetings may be scheduled all over the world and may involve remote conferencing, meaning that on any given day I may be invited to significantly more meetings than I could possibly attend. The greater availability of information and networks of contacts makes it possible for me to be exposed to more social events than my ancestors were ever exposed to. And this puts huge demands on my time and my attention. It is a common topic in popular literature to discuss the battle for users' attention that goes on in the various services and information systems that make up our daily technical ecosystem. It is a commonplace understanding as well that agents will use all sorts of tricks and surreptitious mechanisms to get users' attention and try to bring them into whatever events and operations they are trying to organize.

These opaque practices still qualify as efforts to modify actions, to get generally agreeable agents to focus themselves in one direction rather than another. A person who has a job description requiring they be involved in certain kinds of projects is likely to attend meetings involving those projects. If there are too many meetings to choose from, it may be necessary for the organizer to spin things a bit to make it seem more essential that the invitee attend this meeting as opposed to some other. The agents are generally ready for these kinds of actions, but the choice of one generally appropriate event over another is being influenced based on opaque means. That is the scenario I am trying to distinguish: where the dark ops of the agent are not deeply insidious, but gentle pointers to try and move the person a little bit in one direction rather than another. This is not 'behavior modification' in the sense that this section is opening for discussion, rather this is something I'm trying to separate from behavior modification by calling it 'influence'. And some forms of influence are open and transparent and other forms are trickier than that. It is possible to imagine the spectrum as going quite far in both directions, with extreme candor on one side, and fraudulent or violent coercion on the other. It is possible, I suppose, to kidnap a coworker and make them attend a meeting that fits with their job description. It would be an extreme reaction given the circumstances, but I am using the case to illustrate that the means being used may be coercive, but the ends ultimately fit within a set of possible ends that the coerced agents already accept as their own. Coercive influence does not force the coerced agent to do something out of character or beyond their current role, it forces them to prioritize something that may not otherwise have been a priority.

When the social media customers produce advertising content that attempts to gain the attention of the users of that system, the service provider and the customers are likely to claim that they are trying to influence the users in the way I am describing. Not that they would admit to coercion, only that they would admit that the purpose of their advertisements is to influence the actions of the users in a way that benefits the customers' ends. The user may have been determined already to be a conference attendee and as such would likely need transportation to the host city and a hotel room near the conference center. There are many flights on many different airlines that the user could purchase, there are many different hotels near the conference center. What the advertisers are trying to do is insert themselves into the attention and decision procedure of the conference goers so that they choose the advertiser's product over some competing product. Again, nothing nefarious going on here, just an effort to quietly impact the decision making of the user to benefit both parties. A win-win situation, if you will, the users get what they want because they get easy and convenient information about services they need to acquire, and the advertisers get what they want because this is exactly the kind of sale that they need to remain successful. Even if the business lies about their service, even if they make their advertisement look like a recommendation from a friend who has attended conferences in the host city before, these lies are still merely influence, the user was already predisposed to go to the conference and is being offered a service to get them there. Coercive influence is still just influence. And all influence is perfectly consistent with respect for property rights and the privacy of users seen in that light.

Under conditions of capitalism where property rights are sacred, we must nevertheless learn to be savvy consumers and question the influences that are put upon us via advertising. "Let the buyer beware" has become common sense and so we must grow up learning that people are out there trying to commandeer our interests and influence our decisions for the sake of their own interests and ends.

Influence takes a dark turn, however, when the influence is pushed onto the customers in their relationship to the users. That is, suppose that the system is examined from the perspective of the customers working to influence the users through advertising. When the telemetric pushes the customers down various pathways and moves them toward specific decisions and conclusions about campaigns to run, tactics to use in getting their word out, or decisions to make regarding which groupings and domains the targeted user population are placed and what means are used to get their attention, then the system is influencing the influencers in a way that is unknown to the user population at large. In a service providing diagnostic assistance funded by insurance companies and promoting their interests, it may be that the physicians' decisions about treatment may be impacted by results and mechanisms implemented for the sake of the insurance companies. They may protect their financial interests by changing the options available to physicians for the sake of their own profits. The insurance company's interests may impact the mechanisms available to the physicians acting on behalf of their patients enabling one set of treatment and eliminating another. And this decision may be driven by cost and the likelihood of recuperation of the benefit based on future opportunity. In that case, the actual behavior of the physician will be impacted by the system. When conflicting interests are at work within the various personas making use or contributing to the system, those interests may be codified into different system elements and those encodings may impact the outputs available to other users operating in different roles within the application's logical space. Opportunities for treatment plans or pharmaceuticals may be eliminated due to the realization of a set of interests that are at odds with the interests of the other participants. The patients are the publishers of data and they are the subjects of the diagnoses and treatment options. They may not be the authors of the rules used to feedback appropriate responses to a given set of conditions. The system may be designed to serve the interests of its paying customers and this may amount to a restriction on options available to other groups of users when the paying customers think they should not have such options.

In the supposed case, it is possible that the rules developed to control the outputs of the physician diagnostic utilities are not known to the customers nor to the users of the system. Since the service provider is a private corporation, the rules and algorithms it uses for operating its service are considered its private property, its intellectual property. In that case, the corporation does not need to make these rules public. Should customers and users decide to use the service and enter into an agreement with the service provider, that is a free decision that they have made in accordance with standard market principles of shopping the best service at the best price. In this case, the insurance companies themselves may not know the exact rules of the service, they may have merely provided feedback and insights via focus groups, leaving it to the service provider to determine from that information what the rules and practices should look like. In that case, they are willing to follow through on the agreement given some vague assurances and then, based on their regular evaluation of their profit and loss sheets, they will determine whether they need to shop for a new system or not. If the system meets their fiscal needs, they will stick with it. And this kind of negotiating and marketing goes on all the time in the tech industry so likely the service provider, if it is a successful company, has learned the rules of the game and will be extra efficient at translating its paying customers' expectations and interests into rules for use in the system's smooth operation.

Behavior modification, in this case, is the tacit goal of the system serving the interests of its paying customers, it is their behavior that is modified, it is their goals that are changed. The patient still aims at getting well, but the doctor's practice has been transformed and healing has become harvesting. It is hard to see how this would translate into more open systems. In the case of a diagnostics and treatment service, it may be the case that the domain is closed. That is, the physician uses the service and so there is no other place to go for that same set of goods. The patient may choose another medical professional, but that provider just subscribes to a different system and if the system is run by comparable rules, it may be that the results turn out the same. Within the context of the patient's relationship with the physician, there is no immediately obvious alternative. It may be that a patient does some independent research or makes use of doctors that do not adhere to any system of this sort. In those cases, the patients may discover alternative treatments that are known to be highly effective, but it is likely that they will also discover that these outside options are not paid for by their insurance company. If they have the means to pursue these options on their own, of course, then they are free to do so. The purpose of ignoring options deemed unfeasible by the customers of the system is to make it appear that options discovered elsewhere, outside the system, are exotic and this will justify their omission from any company sponsored set of options.

Targeting 'advertisers' in a social media system may function in the same way. Take the example of the business traveler who has been allocated time and resources to go to a conference. Based on the meeting agenda and some pointers indicating the likelihood of this, advertisers start spamming the user with information about hotels and flights.

Suppose, however, that many aspects of the conference are being broadcast using virtual meeting spaces accessible over the internet. It may be that the value that the user's company wants to get from the conference can be met by a limited session attendance and does not require that the employee travel to the conference site. In this case, the advertisers –when they do not highlight this option and suggest to the system engineers that visibility of this option would be detrimental to their interests– indicate the need to downplay or dismiss this opportunity. They might even go so far as to put together some content that actively targets likely consumers of this option with negative testimony and dismissive advertisements that try to persuade the user that this option is undesirable. In this case, the advertisers, still straddled by the zone of influence, are at least attempting to move into the area of behavior modification and are trying to prevent users from acting in a specific way that would promote their own interests, but which are detrimental to the interests of the paying customers.

Obviously these two examples are very different. In the case of the conference traveler, it is possible for outside influences to be leveraged so that the user can investigate alternatives that go against the interests of the service's paying customers. Since the travel industry is not paying for the actions chosen by the users, they are in a much weaker position than the insurance companies regarding patient treatment plans. If the vested interests of the travel case were enacted by the employers of the users looking for methods to get to the conference, then the cases would be more alike. The variety in examples is meant to show that influence, coercive influence, and behavior modification live on a continuum created by tension of multiple interests present in the telemetric system where some of those interests impact the choices made at the expense of other interests. Depending on the specific dynamics of those interests, the systems may operate as a full-on behavior modification mechanism aimed at clarifying options for one segment of users as dictated by the interests of another set. On the one end of the progression this may amount to gentle influence that is mutually advantageous for all participants, on the other end will be the complete closure of reasonable options to drastically curtail the action potential of the impacted users.

But does this really qualify as behavior modification or are we just talking about increasingly coercive forms of influence in which the interests of one group of users become ever more invasive upon the actions and interests of other groups of users? The influence exerted may go so far as to be predatory, singling users out for the sake of exploiting weaknesses that make it more likely that they will bend to the influence and go along with the advertised or published ends. Nonetheless, this influence is not behavior modification since that requires deeper changes to the actor, the user must undergo a more fundamental change. A person who is afflicted with a medical condition and who goes along with a snake oil salesman because they think the bogus tonic will cure them of their condition, is not undergoing a behavioral change. Although they may be influenced in such a way as to pursue an end that detracts from their ability to get what they want by either depleting them of financial resources or time necessary to pursue those ends, the promoted ends are still in keeping with their own objectives. This kind of fraudulent coercive and predatory behavior is nefarious, but it does not amount to low-level behavior modification.

Behavior modification requires that the agent adapt its ends to promote the ends of other interested parties exerting influence over them, it signals an actual change in goals. And the goals that undergo change may be direct or indirect. A direct goal is something like 'being cured of a medical condition.' If someone were to convince the person to stop pursuing a cure because they convince the actor that the condition is a desirable thing, then the person's fundamental behavior will change, since pursuit of that goal will cease. The person will stop trying to find a cure and stop doing anything to resolve the old condition. The person might start looking for new ways to exacerbate the condition, to make it 'worse', because they have been convinced that worse is better, is an improvement to the condition.

A milder form of behavior modification against direct goals is when a person has a vague sense of a need but does not have any clear sense of how to meet that need. Say a person desires to get out of their impoverished state. They do not have any specific goals for doing this, they just know they need something to change their plight and help them obtain upward mobility. Should a targeting advertiser determine this about the person, they could lavish the person with advertisements and content that shows them specific actions they can do to meet this need. This may have an enormous impact on the user as it may change many things about how they conduct themselves and what sorts of actions they engage in. Suppose, for example, that this desperate person looking for a way out of their poverty never thought to attend a university to acquire credentials of some kind, and the targeted advertisements convince the person that some online university is the perfect solution to their problems. The person may never have had any educational ends per se, may never have even thought that school was a way out of their circumstances, but the advertisement convinces them that this is the way to get what they have always wanted. In this case, an indirect goal such as 'improving one's financial state' is supplemented with a specific direct goal that did not previously exist. This kind of behavioral modification is common and not necessarily contrary to the person's interests, we do it to our children all the time. When the goal sculpting interests are predatory, which means they are only there to promote the interests of the targeting publishers and not the consumers of that content, then we would distinguish them from another set of goal sculpting interests that might be in the interests of the targeted users. When physicians change their goals from healing the sick to building a thriving practice and when that is done in the context of insurance companies providing the instruments and infrastructure to make this a natural transition, the physicians are undergoing behavior modification. We will come back to this later; the primary point I wanted to make now is that goal modification is what constitutes behavioral modification and behavioral modification is not necessarily contrary to the interests of the agent whose goals are being modified. The questions that will occupy us in the second part of this essay will aim at trying to understand how to go deeper in sorting out and evaluating 'interests' in such a telemetric context.

Suppose that the impoverished person, rather than being convinced to attend an online university, is convinced to engage in criminal activity. The person may have had strong civic resistance to such activity, may have believed in the rule of law and the importance of following the laws even when violating them would be to their own immediate advantage. It may be that such a person has religious beliefs and affiliations that have led them to believe their own and their family's long-term well-being depends on following certain laws of morality that prohibit certain kinds of activity. In this case, the interest group that is targeting this person with the intent of changing their behavior is trying to get them to do something that will promote a direct goal at the expense of an indirect goal, violating their principles for long lasting well-being for the sake of a more immediate need of social improvement or survival. People might be misled into thinking this is the only way out, that they must compromise their ends to make it in the world today. They change some of their most fundamental goals and behaviors aimed at achieving these goals because they think there is no alternative to doing so. In the end, behavior modification may be either a change to direct goals or indirect goals. Changes of this kind are not new to computing systems or social networks or any of the technology changes that are characteristic of the recent era. Human beings have been susceptible to these sorts of influences as long ago as the 4th Century BC in Greece where Socrates corrupted the youth of Athens by teaching them that the goals laid down by their fathers were not nearly as meaningful and important as the goals laid down by a love of wisdom.

One way in which both direct and indirect goal modification can take place is through the phenomenon of 'gaming the system'. As users become savvy to the mechanisms at work in any service that they are making use of, they may start to modify their actions to get some other good that the system has to offer. In a social media context, this may amount to something as simple as 'liking' things they do not 'really' like for the sake of building up their personal profile and preferences in such a way that they will become members of different distributed groups or populations. They may wish to appear to be a different sort of person to gain access to things that the system has to offer to those sorts of people. Suppose, for example, that someone who does not like cats even when they are at their most playful, starts to 'like' lots of published content about cats. The person might start to publish cat content themselves just so that they can get a large following. People on social media, and the internet generally, seem to flock to cat related content and so anyone who gets into the cat racket is likely to increase their visibility. The increase of 'followers' or audience may have financial rewards via some advertising click agreement and so the person satisfies their own goal of making money via a goal they did not previously have: cat enthusiast. Now this person is scanning the world wide web for all sorts of cat content that they can share or publish. Maybe they get a couple of cats of their own so that they can stage videos and photos aimed at capturing that characteristic cat-ness that so many people are drawn to. In this case, the person finds themselves to have been altered into a cat enthusiast despite having had no prior interest in cats. I may be making a kind of Pascalian assumption here, claiming that the absorption in all feline matters will eventually lead the person to develop genuine feelings for cats, but I do not think the case depends on it. The artificially induced cat lover will have spent a large portion of their life immersed in the society of cat lovers and whether there was genuine feeling behind this is not critical to the point. If their actions were impacted, and their behavior modified, they qualify. The telemetric system does not care if you really like what you 'like', it only knows that your behavior can be aggregated along the attribution boundaries formed out of your purely behavioral liking.

The public debates about social media and information systems often focus on the opaqueness and conflicts of interests in the influence and attention game, but with this focus on 'gaming the system', I am trying to point out that some of the worst effects of these systems can occur when they are not completely antagonistic to one's interests nor are they completely hidden from view. With a small set of insights into the algorithms at work, the more sophomoric the better, and the indirect relationship to other vaguer goals that the user might have, like social mobility and wealth, the more powerful the sway of behavior modification might be. When a person does not fully understand the algorithms and analytics that are at work but does have some *partial* understanding of them, that they are there, that they are being measured, that they are being surveilled evaluated and clustered, then they may be at their most vulnerable. Because in this case, unlike the case where the user is oblivious to these sorts of mechanisms, the user will adjust, will make changes to their actions, to accommodate or impact the ongoing surveillance thus impacting the way in which they are rendered within the system. Gaming the system in this way depends on partial understanding because it may be that the interests at work in presenting content and options or opportunities may have accounted for these types of adjustments. The mechanism may have been developed to anticipate the kinds of gaming behavior that will ensue from one type of telemetric feedback and will institute changes to the mechanism to direct and exploit these adjustments. It is entirely possible that the modifications taking place in gaming the system (or managing the metric) are themselves in the interests of the publishers that the user interacts with. Full understanding of the metrics and mechanisms would suggest that the person is a statistical aggregate and their attempts at gaming the system are just ordinary patterns detected by the system. A person who understood this aspect of the system might be unwilling to participate in any kind of commerce within it because they would see the extreme disadvantage of their individual actions against such mechanisms aiming at distributions of the population in general.

It would not be too much of a stretch to think of this form of behavioral modification due to gaming the system as the 'panopticon effect'. In Bentham's prison design, the point was to modify the behavior of the prisoners, to reduce the need for guards and surveillance, because the individual prisoners take on that role themselves as a compensatory reaction to constant monitoring. And this belief was partially true. The design of the prison was such that they certainly could always have been seen: the observation tower was dead center and in plain view of the prisoners and the lights were always on. They could not see into the tower, so they did not know whether anyone was observing them or not. The trick was that they had partial knowledge of the surveillance condition and this partial state led them to fill in the blanks in whatever way they could. They were influenced to fill in those blanks in the way most advantageous to the prison authorities, so they were led to believe that they were being watched all the time. This was the necessary condition for them acting as though they were being watched all the time and becoming their own guardians controlling their own behavior without being told to do so. This was exactly the behavioral change the designer wanted to affect through the internalization of interests by those who otherwise have decidedly different interests under the circumstances. And again, it has nothing to do with who owns the artifacts in their personally identifiable form.

There is a gaming the system element to the panopticon effect but the analogy with the information systems in our examples is more explicit than that. The system functions like a panopticon and functions more and more like one the more rationalized the user practices become. As has already been suggested, the subjects of the service, the users who are the target of its operations, are acting with only partial understanding of the mechanisms at work. The system's reason is not therefore implicit in any one actor but spans the entirety of its operations. The customers or other interested parties also obtain only a partial understanding of the system. This is true by definition, since it is their stated goal to exploit the mechanisms at work to gain a greater understanding of the operations taking place. In addition, the service provider may be hiding aspects of the functionality from them. All individual agency is taking place in a condition of partial understanding. All who participate will have learned throughout the course of their usage the basic conditions of visibility that are part and parcel of the operation. Some therefore may identify as guards within the system and some may identify as prisoners. Others may identify as subjects in the sense that they will think that the entire operation is there solely for the sake of their own convenience. The social media, for example, is not meant to harvest their data, but it is to be used by them in achieving their social goals and carrying out social action in a well-established public space. There will be variable identifications by the various interested parties, all of which will only be a partial picture of what is going on. Regardless, the telemetric will have become part and parcel of each of these points of view. Something remote is at work. No user within the system, the corporate account that uses targeting advertisers and then evaluates the revenue impact of their efforts on sales or just collecting the ad revenue generated with click bait, the socializing user who is sending semi-public short messages to their followers or posting pictures that their followers can react to, all partially aware of something far away that is taking place. The virtual character of the space is made apparent through the medium and the distance between themselves and the content they are pushing around is apparent in the operations they carry out. And they know it is being metered as well. Metering and measuring in some vague sense have entered everyone's characterization of the collective action. Whether it is the internet provider company that charges them based on throughput or the service provider who charges them based on the amount of processing time they use to target their potential customers, they sense the measurements. The constant talk in the media of the surveillance and data collection gives them some dim or vague understanding that the measuring goes deeper. We may not know all the details, but we sense it in all our actions.

The phenomenon of a new technology impacting the behavior of individuals making use of it, and even individuals who do not directly use it but live in a society in which it is widely used, is not new or exclusive to computing systems. Train travel, the telegraph, the printing press, many technical devices that have entered human civilization in large-scale usage have impacted human behavior and may have changed who we are in profound ways. The existence of a change such as this does not immediately foretell its significance, nor does it immediately imply a value judgement of a specific kind. The impact may be to vastly improve human beings. The printing press may have been one of the single most important inventions for making democratic forms of government possible, not to mention more participatory and liberating forms of religion. The printing press was a communication mechanism, as was the telegraph and the trains. These inventions made it easier for human beings to communicate in more complex ways across greater distances. The existence of the internet, the existence of services on the internet to enable communities of users to form and express their interests in juxtaposition to each other, as buyers and as sellers, none of these conditions automatically suggest either dystopia or utopia. They do, however, provide the basis for changes to human behavior and so it has always been fair to ask whether they have facilitated changes for the better or for the worse. And it would be well within any historical analyst's rights to assert and try to prove that these advances have been to the advantage of some and contrary to the interests of others. The meaning is not inherent in the advance, it must be teased out through efforts at understanding the changes themselves and the various nuances to them across the population and within all the various impacted segments.

In this case, the panopticon effect that is at work in the behavioral change as the injection of the telemetric into the fundamental attitudes of human beings involved in the newly constructed systems, reveals itself as the new way of being rational. This telemetric panopticon need not be new or original to the existence of internet-based communications systems, but the system with its deep dependency on telemetry, with its immediate dependency on variable interests at work within the systems providing services, and its feedback loop that adjusts and corrects services to cultivate desired responses from the user communities, marks a significant leap forward in this effect. Sciences, since the Copernican revolution, have striven to be the remote measuring devices that establish the view from nowhere to which all human life must adhere. Data science is a notable leap forward in that project with its attempt to subject human action to the telemetric delineation of action aware of itself as constantly surveilled by partially understood mechanisms and modifying itself into the orientations and manners of the system at hand. This self-adjustment does not hinge on an obsessive compulsion to be online all the time, to see what's going on at every moment, although there may be some users who have succumbed to that. It may be reflected in slight adjustments in people who do not spend that much time engaged in this new public sphere. They may sense that their ideas are only valid if they reflect correctly the community viewpoint that they have come to know in some service they access on occasion, they may become unable to apply rigorous means to determine their own point of view independently of the feedback they get instantly from various networked media. Additionally, it may be the case that their experiences of the world, of eating dinner with friends, of hiking through lush country side, of enjoying a celebration with family, can only be meaningful if captured by a photograph that is then circulated amount 'friends' or followers and liked. These snippets, so my hypothesis goes, are indicative of the panopticon effect brought on by this telemetric pattern that, even if not new, has rapidly increased in speed. The rationalization process that may have previously taken years to insert itself into institutions and actions, can now bring about major changes in days and weeks.

A deeper look at this aspect of the rationalization process of the telemetric, this so-called panopticon effect which is so eminently more important than any property at stake, will have to wait until the second section of this essay when we focus more explicitly on phenomenology. It is important to remind you here that the background concern of my hypothesis is privacy. And the position that I had hoped we would achieve by the end of this chapter was that we were no longer entirely sure that we knew what privacy was. The insertion of remote measurement into the actions of system users has, I hope, at least suggested that what is most our own can be inhabited by something that observes it and computes its operations from afar. In so far as this is possible with only a partial understanding of its means of effecting us as well as its ends and how they work upon us, those characteristics of our action that may be the most secure, the most private, the most closely held as truly our own, may be undergoing modification for the sake of an agenda that is very much a public matter, something that comes from somewhere else, something that is not our very own at all. If a child had the level of sophisticated self-consciousness and critical awareness that you and I have, how would they experience the parenting techniques of those acting on them for 'their own good'? If what is most my own is being toyed with from out there, how did I ever come to think of it as my own? The innuendo is far reaching and far-fetched, so let us see.

Chapter IV: Publicly Determined Privacy

The achievements of reason mystify. It has never been entirely clear whether the philosophers describing the rationalization process meant it as a 'good' thing or not. This is not an accusation against Weber, Adorno/Horkheimer, or Habermas, and I do not mean to suggest that they are ambivalent, although at times they seem to have been. Rather the descriptions of the rationalization process have always spelled change such that when a process increases in rationality, includes greater distinctions and nuances, becomes more taxonomically organized and apprehended in its associations of disparate components and elements, it seems as though something has been lost in such progress. And being lost might mean no longer visible to eyes now trained to see only what has been organized in more rationally complex fashion. Was this the desired end? Or was it a side effect? The communities thriving within a remote service may be more rationalized than the pre-existing forms of community and they make it hard to see something at work in those communities since it is no longer at work in the newly modified forms. Even our immediate neighborhood might have undergone changes interwoven with the rationalization of communities on-line. We can no longer see what used to be so very much present to us. And this invisibility is the residual of the rationalization process. Or so it seems. Can analysis get at what has been lost? We can measure the social practices and compare them to past measurements and we can study the results of cleverly designed surveys and data analytics. Will increasing the amount of remote measurement bring back or reveal what has been lost? Will storing more data to describe the rationalization of the comparative values between what it was like then and what it is like now show any loss from the rationalization process? How can increasingly organized deliberation give proper visibility to those things which organized deliberation has eclipsed?

The shadow that haunts this investigation is the Hegelian holy trinity recast as a fourfold dynamic. The emission of data functions as the outreach of consciousness, the collection of that data into an analytics store that processes the results in search of projected patterns is the self-consciousness of the system at work, the feedback loop that automates the analytics back into the system adjusting for the sake of more efficient and optimal operations is its reason. The holy trinity has been completed and should now move into the space of spirit where the goal is for the kingdom of ends, the community writ large, to achieve self-understanding by traversing a similar movement passing through religion and into absolute knowledge: subject become substance must now yield a substance become subject. This is a perversion of that holy trinity, no doubt. Hegel always lost me in the advent of spirit, it felt anti-climactic. I could not see anything there. And that is his point and it is because I was a bad or fearful reader, unable to go beyond my own subjectivity and get caught up in the higher orders of substance, because I could not see the world from the perspective of the civilization or the species, that I had gotten lost. From my point of view, the work petered out, appeared to go nowhere, and my attraction to most of the philosophy inspired by Hegel was largely driven by the sense that many of those thinkers had the same reaction and had likewise gotten lost. From Hegel's point of view, this was easy to anticipate and many of these reactions were strategically placed into early sections of the Phenomenology: this one realized master slave, this one realized unhappy consciousness, etc. Yes, the point being that they could not traverse the realm of spirit, they could not follow the narrative on its course past the realm of reason. And why is that? Well again, only a hypothesis, but something was being lost as we traversed the path of rationalization, something that I have spun inextricably into the workings of reason. And that something lost is nagging at us, nagging at them, nagging at me.

The fourth phase then may be the phase of the residual and the impact it may continue to have on the system at work once it seems to have been factored out of consideration. It tries to articulate that which is silenced by the rationalization process, but which continues to resound in the inner workings of the system, unravelling it in places here and there, revealing its incompleteness if you will. I am going to boldly name that which remains in the case at hand: privacy. And this is a ridiculous decision since it is an immensely overloaded term that has become privatized and nearly meaningless in what has transpired here so far. Even if you have stayed with me at every step and have come to be perplexed by the word based on the discussion in the previous chapter, it is outrageous to invoke such a term to name this architectonically induced dynamic that resists analysis: the so-called residual of reason. And it is outrageous because it is such a common-place term that can be found everywhere in our contemporary discourse. Laws are in place to protect it, there is a news story about it nearly every day, there is so much written and said about privacy that it must be one of the most visible and heavily rationalized notions of all. I accuse reason of mystification, but it is I who mystify.

This is the point, the obsession with privacy is an indication of a deep problem. Suppose that we are obsessed with it because we do not know what it means, because we think we have missed something in it, something about it has and continues to elude our efforts, our endless efforts, at speaking its name, at clarifying its nuances, and at apprehending the important role it must have in our lives. We obsess over it because we worry about it and our nagging worry is the emergence of the deep fear that we do not understand it in the least. We sense that it is the most important thing in the world and we equally sense that it is nothing at all. We fear that it will be taken away, that it will be exploited, and that our very lives will be ruined by its violation, by nefarious interests doing violence to us through its invasion. We fear that it does not exist because it is being violated, that the violation has caused it to disappear, and we fear that we have lost it. And deeper down, we fear that we have lost it because we never truly had it, because it never had any real status in the world. How, for example, would you protect hydrogen? Does hydrogen need protection and is there any way to do it if it needed it? If the answer is that such questioning is absurd it may be because things that have concrete natures and which exist without a doubt are not in need of protection. Yet privacy needs to be protected. And if it needs to be protected, maybe it is not very stable at all and does not have any grand concrete and indubitable status in the world? If we do not protect it, if we do not look out for it, then it will not exist, it will disappear and be gone forever.

We talk about privacy because it is under attack and we worry that it will be lost in the attack, that it is something that can be lost. And the fact that it can be lost might mean we do not know what it is and are unable to identify it. Clear enough and maybe not so ridiculous after all. Suppose that all this talk and all this concern has ultimately been for the sake of rationalizing privacy. We legislate it and we analyze it, we discuss it in detail during congressional hearings, we go to all sorts of lengths to demonstrate in favor of it. We introduce more thorough analysis of its meaning and the ill effects of its loss. And what if this rationalization of privacy ends up removing ourselves from that which made it so valuable in the first place? What if I am right in my fears about the rationalization process being the orchestration of a loss, an operation that hides, that makes something once widely known into a secret, makes it something so private that it is no longer known at all? This would be horrifying because it would mean that there is no turning back once the rationalization process has begun. Once we sense that privacy is under attack, we begin to study it for the sake of understanding it, what it is and what parts of it need to be protected, as well as what the best way to protect it might be. We engage in a public determination of privacy. And if this 'logic' is correct, nothing could be more of an assault on privacy than this process, because the process of rationalizing something may be the process of making it public and privacy is conceptually determinate only through its juxtaposition with the 'public' sphere. Rationalization may have a two-fold negative relation to privacy. Rationalization makes 'private' the residual of any phenomenon that undergoes rationalization, first off. And then secondly, privacy itself is hidden from view by the rationalization process. Publicly determined privacy, therefore, threatens to make privacy so private that it can no longer resound, no longer play a role in any one's life, and no longer be the critical origin of experience and understanding that we sensed it might be.

This suggests that publicly determined privacy both hides privacy and, in a subtle sense, prevents it from ever being completely hidden, completely removed from view. The fourth movement of this anti-dialectic, the fourth phase, is the phase of the hidden effect. The suppressed or transcended phenomenon that continues to exert itself even as it has been pushed out of the clearly organized realm of the rationalized notion. The very thing that nags at us in the process of rationalization, that shows itself through its loss, through its privation, has been protected by rationalization. It may have been entombed in a crypt that hides it from plain view, but doing so has preserved it, allowing it to have a constant and intense presence in the workings of the very system that has hidden it away? Or this mumbo jumbo might be an overly optimistic spin on a great tragedy and, as with the true dialectic, the way of the fourfold is a constant progression of tragic experience, experience that never quite gets at what it needs, never quite realizes itself? It is too soon for these musings to be anything more than suspicions and hunches, conjectures or hypotheses. I want these possibilities to come along with us as we start to question the publicly determined privacy that has been growing in focus with steady pace for over a century and, with the advent of distributed computing systems and instant worldwide node to node communication, seems to have come to a boil and become a matter for widespread concern. Even people you would not expect to say such things are asserting that privacy is a fundamental human right, that it is the most important thing there is for human beings.

What then are we saying that privacy is? This is not a question meant to get at the distinction between a public sphere of action and a private sphere, a distinction that has been historically useful for separating the domain of government oversight or of properly political action from the spheres of action where such notions are no longer appropriate. Domains of privacy were often used as mechanisms for granting some individuals unmitigated control over other individuals in the family or workplace, for example. And by 'unmitigated' I mean free of government oversight or action. The law held no place in these private spheres and so the traditional patron of such domains could control them as they see fit. In a sense, these domains were constrained by public intervention into those spaces for the sake of the persons operating there and controlled by said patrons. The governing power might gradually impose itself upon the so-called private sphere for the sake of protecting those individuals who were suffering abuse at the hands of the 'masters' of those domains, either the head of household or the head of a business. The advent of domestic legislation and labor rights legislation, regardless of how slowly or unfairly they may have developed, still marked an imposition of government power into the so-called private sphere for the sake of individuals there who were suffering violations of their person by the protected authority of that private domain. The public took an interest in the private domain for the sake of safeguarding the interrelations of persons there. This is not meant to be an oversimplified assertion that public authority invaded the privacy rights of some for the sake of the human rights of others. Rather, the invasion of the private domain for the sake of protecting the rights of persons acting in that domain was established early on as something radically different than a matter of privacy. The private domain, as such, was not secured through the private authority's right to privacy, it was the private authority's right to property and the control of it that instilled the security. Contemporary liberal legal approaches might suggest that this right is no longer to be protected when it permits assaults on the rights of persons who have entered relationships with that authority in the context of those property rights. The right to be master over one's private property might suffer publicly imposed limits to the authority of the property owner over other persons who assisted in the development of that property, in economic matters, or who cohabitated on the property, in domestic matters.

In 1890, Warren and Brandeis wrote an article called "The Right to Privacy". Their primary concern appears to have been the untangling of privacy rights from property rights. The rights to property may have been useful for establishing constraints on the private sphere, beginning the process of setting limits in private that were drawn by the presence of protected persons acting in those domains. And the notion of property would be strategically advantageous to this end because it allowed the courts to crack their way into the heavily guarded private sphere without appearing to be violators of personal rights, but protectors of it. This approach should not be confused with an equation of privacy rights and the operations of persons within the private sphere. Furthermore, the association of the two notions should not lead courts to think purely of privacy rights in terms of property violations. The authors were worried that the application of property law to privacy law would require that the violation of those rights be clearly shown to have a material value. That is, to claim that one's privacy was violated, the use of a property right to justify privacy would require that the violated person show that the violation caused some material damage, some financial impact from the violation. If a person was not hurt financially by the publication of a privileged communication or an unauthorized likeness of their person in the form of a candid photograph, then the established property law would not afford them any protection from these acts.

> This development of the law was inevitable. The intense intellectual and emotional life, and the heightening of sensations which came with the advance of civilization, made it clear to men that only a part of the pain, pleasure, and profit of life lay in physical things. Thoughts, emotions, and sensations demanded legal recognition, and the beautiful capacity for growth which characterizes the common law enabled the judges to afford the requisite protection, without the interposition of the legislature. (Warren and Brandeis, 1890)

They therefore meant to start from what was referenced as the right "to be let alone". This was merely a guiding principle as it immediately runs aground. In the case where I write a personal letter to a friend and the friend turns around and publishes it, it becomes immediately obvious that "the right to be left alone" is inadequate for understanding what implied rights there are here. After all, I have published my thoughts and sentiments in the personal letter, does not this publication constitute a waiver against the right to be left alone? In such a case I may have freely revealed myself to my correspondent, but the authors intend to suggest that this was an emission made in confidence and so I have not automatically surrendered the right of publication to my friend when sending the letter. To conceive of the transfer of rights here in terms of an implied contract is to continue to operate under the rights of property rather than some independently understood right to privacy. And no doubt they wish to steer clear of this, but the going is hard, and the lion share of their article is spent grappling with these problems. It becomes increasingly difficult to safeguard privacy as the right to be left alone when the law seems to want to protect privacy in matters of communication from the people to whom those communications are made. Generally, the right to privacy comes to an end when the individual whose rights are in question chooses to publish the details under consideration. If the publication is directed to specific individuals who are the recipients of those publications, an immediate constraint is placed. The individual is meant to be able to both publish and keep private an emission that is made to a specific audience. The courts struggle with a way of conceptualizing this emission that continues to bear the limits of conveyance. Some utterances, it would seem, would have to carry the implicit connotation of restricted re-telling or forwarding of the messages conveyed to intended parties or persons. How to tell the difference?

Arguably, the confusion in the courts, and the limits that they tried to impose through the notions of an extended interpretation of property as something more than just owned goods, but including one's letters, emotions, and feelings in so far as they may be embodied in written documents, shows that the court is going to great lengths to artificially conjure a right to privacy out of a right to one's property and the control of it via implicit or explicit contract. The gaps that remain if this approach is used expose us to all sorts of undesirable effects: for example, the legal requirement that unauthorized exposure might mean one has a share in any profits received from the publication and not that someone could be punished for having published the material in the first place. It seems then that the existing law that can be applied to matters of privacy is inadequate for so doing, there are plenty of gray areas and strange goings on that cannot be addressed through existing decisions that focused on contracts and property. The authors might be making the claim that the right to privacy is its own area of law, not reducible to existing precedent, and that it must be carved out by judicial action to bring clarity and conceptual integrity to it. On this reading, the article does not so much as define privacy as it hints at the need to define it. And presumably what they mean is that the courts must begin to flesh out the notion of the right to be left alone and what it means in terms of specific case law.

> We must therefore conclude that the rights, so protected, whatever their exact nature, are not rights arising from contract or from special trust, but are rights as against the world; and, as above stated, the principle which has been applied to protect these rights is in reality not the principle of private property, unless that word be used in an extended and unusual sense. The principle which protects personal writings and any other productions of the intellect or of the emotions, is the right to privacy, and the law has no new principle to formulate when it extends this protection to the personal appearance, sayings, acts, and to personal relations, domestic or otherwise. (Warren and Brandeis, 1890)

The right to be left alone is something to be newly developed and they suggest some limitations on it that might help with that effort. Ultimately, these limits are not intended to begin the process of setting precedent, but they are meant to be existing principles that should guide such efforts. 1) violations are permitted when the public has an interest in the material that would be revealed. 2) No violation exists if the material is introduced into a public concern directly. 3) Oral gossip is not a violation. 4) Self publication ends the right. 5) Truth does not justify the invasion. 6) Absence of malice does not justify the violation. These limits, seemingly of variable type and orientation, seem to exacerbate the problems rather than begin the process of clarifying them. Earlier, the comment had been made concerning what the courts had an interest in protecting: "The principle which protects personal writings and all other personal productions, not against theft and physical appropriation, but against publication in any form, is in reality not the principle of private property, but that of an inviolate personality." (Warren and Brandeis, 1890) Maybe this is not meant to be a serious remark as they are working through some of the limitations in the current legal status of privacy, but it seems that – in the piecemeal fashion that they are using to flesh out the notion of a right to privacy– the notion of 'inviolate personality' would be a notable contribution. Why does privacy matter? Why consider it? Why take the time to write about it, to legislate on it, to execute judgement upon it? The right to be left alone is the only persistent standard and the notion of an 'inviolate personality' would be a meaningful development of that right. What's more, the focus on non-monetary damages to the person through the exposure of their innermost feelings is tantamount to an interest in protecting the personhood of the person from painful attack on their personality using their own emissions embodying the nature and material expression of that personality. Except for items 5 and 6, which are meant to curtail certain defenses that might be made when charged with violating the privacy of another, the first four items in the list seem to immediately indicate the larger significance of this protection of personality.

1) When there is a strong public interest, the person's right to privacy can be ceded. Apparently, the significance of the person's right to be left alone or right of inviolable personality may be superseded by the interest of the public in evaluating public acts like running for office or actions performed against other persons. And the flipside might be that the master of the private sphere might be deemed to have support of a public interest when lording over the subjects within his place of business or domestic abode.
2) Communication that is pertinent to libel and slander should not be protected. You do not get to maintain privacy so that you can go around causing damages in private to other people based on false information.
3) Oral publication of true but private material about a person is not

protected. You cannot prevent people from gossiping about you so long as the form of that gossip is transient. The phrasing might also suggest that insightful but abusive speech is not to be covered by a so-called right to privacy.

4) If the persons themselves publish the material or give consent to the publication of the material, then there can be no violation of their privacy in further communication of those materials.

There may be a prima facie reasonableness in each of these limiting cases, each might make sense on their own. Make no mistake, they are each potential violations of personality. Brandeis and Warren are not suggesting that these would not be violations of privacy per se, they only lay claim to the fact that they are permissible limits to be set on it. This is the important facet here, they are not interested in establishing an ontological fact about persons and their personalities, their lives in all its nuanced meanings. Rather they are trying to establish a legal right to claims that may be made against one's peers and against one's compatriots. The law should protect the inviolate personality, but only up to a point. Personality as the personal lives and all the emissions that articulate those lives, is worthy of protection, but not absolutely. There are limits to it. When the public has a strong interest in violating the individual's private life, the public may do so. The second limit could be an elaboration of the first. The public interest in protecting against libel and slander may extend to private communications that are instances of these legally defined and actionable circumstances. The suggested limits to place on the right to privacy boil down to the public interest to violate privacy, the indication that informal transient violations are okay, and that it is okay when the person violated says it is okay. The first of these cases may generalize social prejudices and inequalities into the protections of private persons as extensions of their social roles. The second of these cases suggests that there is a kind of inevitability in the violation of another's privacy, that there are human foibles that we will never be able to factor out of our experience of living among our peers. They will talk about us, they will gossip, they will have water cooler conversations about their colleagues and associates. This violation of our persons cannot be factored out, it is with us in so far as we live with others. And then, finally, in the third case we must be taken seriously as subjects of our own private states. If we knowingly and willfully permit the transmission of private materials to public audiences, then we have given consent to the violation of our person, and we must be allowed that right. The right to violate your own privacy is a given, an obvious recognition of your autonomous nature.

The Right to Privacy is not the final declaration, really it is just the beginning of a judicially based discussion on the meaning of a right to privacy and how it can play out in civic and social life. What is most interesting about this, for our thesis at least, is that it asserts a right and attempts to place limits upon it that are largely driven by conventions, and therefore leaves it largely unclear what privacy is. There is hand waving in the vicinity of the right to be left alone, but very little of the discussion tries to explore what that might mean. Instead, most of the focus is on the publication of transmissions or emissions that would damage the person, where damage is established as emotional rather than strictly financial or material. In general, the article gives the sense that a set of conventions are being established and those conventions are in lieu of being able to clearly articulate what is the human facet or feature under protection. No doubt, this article opens a powerful domain for future exploration, but it does so in the same moment that it both presents a set of conventions that should be considered when exploring that domain and hiding any ontological claims that might be made about its subject matter. Some might be more confused about the meaning of privacy after reading about the right to it than they were before doing so. What is it that I have a right to exactly when the authors talk about protecting my right to privacy? What does the right to be left alone mean when talking about whether my friend gets to publish on their internet blog my personal correspondence addressed to them? The authors have boldly begun the process of talking explicitly and at length about the meaning of privacy and the right to it, but in so doing, in attempting to explicitly initiate a rationalization process that works to clarify the notion of privacy and the appropriate legal conventions to ensure its protection, they have made the concept itself more obscure.

Following the publication of their article, the notion itself, as they hoped, started to find its way into more and more court decisions, while becoming a frequent topic of philosophical investigation. In what follows, I want to start looking more closely at some of these decisions. We will be doing so with these established notions in mind: 1) how do the decisions advance the notion of a right to privacy, a right to be left alone. 2) how do the decisions advance the notion of privacy itself as a quality or state of being. And 3) how do the decisions advance the rationalization of the rights and material qualities to both expose them to legal precedence and to hide whatever it was that made those characteristics worth protecting. We must bear in mind, based on our earlier discussion, that it is entirely possible that the increased focus and discussion on the right to privacy has made it increasingly difficult for the subjects of these laws to have anything like privacy. The thorough establishment of the right might make it more difficult to exercise those qualities protected by it.

In the philosophical investigations that ensued, debates raged around whether there really was anything to the notion. Many argued that the set of conventions were themselves the establishment of the thing as a juristic construct, others thought there was a cluster of notions brought together into the legal determinations but each of those could be reduced to other notions like property rights or the right to liberty. Although not universally accepted, many philosophers were adamant in suggesting that there was nothing there to protect, only something to establish via protection. If privacy were reducible to liberty, for example, then there is no meaning to the private above and beyond the general protections that all of society or all of philosophy might wish to make on behalf of the basic freedoms of the individual actor. In this way, nothing at all is hidden by the establishment of such privacy rights since there was nothing there to begin with. Rather, legal formulations might be better or worse attempts at deriving a set of decisions about privacy in the process of establishing its existence. Other philosophers took issue with this form of reductionism, claiming that there is no phenomenon that one could clearly point to as the one and only materialization of the privacy attribute, but that it was possible to talk about a non-reducible set of concerns and interests that amounted to a recognition of something inalienable and in need of protection in the human person. In their view, what we would expect to get from an analysis of the precedents is an understanding of the various distinctive properties and qualities of privacy and the various ways to ensure the protection of it. Regardless of whether privacy as a notion is advanced in coherentist or reductionist terms, the philosopher's position would be that one way or the other, the discourse on privacy gets at it, captures it: either by reducing it to those more fundamental qualities and the rights protecting them or by showing the various aspects of the qualities that come out in the multitude of coherent ascriptions to and about it. In no case would the philosophers on either side of this dilemma tolerate a residual: a hidden component or aspect that is pushed out of sight and eclipsed in the very act of revealing or rationalizing the matter at hand. The philosophers, like the judges, are telemetric through and through in their approach to privacy. They think we can talk about, think about, and judge about privacy as part of the larger project of making it clear what we mean when we refer to it, the right to it, and the protection of that right.

If there is an actual thing called privacy, then even if we lose it, there is a chance we will someday find it again. If there is no such thing and if its existence is a fortified convention, then its loss will require that we create it again. In so far as creation is not all at once, but results from an ongoing historical set of judgements and determinations, then recreating it will amount to something different, some variant form of it that could only come to be once the lost form has been reinvigorated and used as inspiration for the re-creation. In that case, what is it exactly that will inspire?

In many ways, ontogeny does recapitulate phylogeny. The person and the personality they embody, to the extent that they can be conceived as a system like a distributed internet service, can be understood as an instance of the pattern we have been presenting in the first three chapters. The person is a mass of multiple emissions of signals where these signals might resemble metrics describing or expressing the operations performed. Anyone who receives such a signal can reflect upon it, analyze it, and consider what patterns are present in those actions and what personality traits they exhibit. People who live and act are rendered visible to those others with whom they live. Even in the most natural of states, a person is under scrutiny by those who are close by, by family members, by friends, by colleagues in the work place, by cashiers at the grocery store, and by the sommelier at the restaurant. And all recipients of this emitted signal can provide feedback to its origin. All who function as audience to a person's actions may react to those actions, and this constitutes a feedback loop to those emissions and reflections. The person will respond to what people do in response to their emissions, in response to their actions. If it is part and parcel of a human life to be engaged in such interpersonal relations, what are we trying to talk about when we start –amid all that community and society– an investigation into privacy? And do our efforts dig deeper holes that hide considerably more than they reveal? Does the rationalization process, the telemetric, get at the essence of something or does it hide it, cover it up?

Reasonable Expectation of Privacy

In Olmstead vs. The United States of 1928, the Supreme Court of the United States (sic, the US permits its supreme court jurisdiction over cases brought against the United States) permitted the recordings from tapped telephone conversations to be used as evidence against a defendant. Now Justice Brandeis dissented on the grounds that the mechanism used for surveillance was insignificant and the root of the issue was an unwarranted search. His position was forward-thinking in its comprehension of the technological possibilities of surveillance and the impact such mechanisms might have on individual rights. Furthermore, Brandeis was clear that such surveillance, regardless of what technical means were used to achieve its ends, constituted a search of one's person. The dissenting opinion, although not directly establishing a precedent, appears to have set a standard for the association of privacy with unwarranted search and seizure. And not just that, but with such an emphasis, repeated from the earlier article written with Warren, Brandeis was attempting to situate this protection in something broader than a concern for one's property and the right to it: "the sanctities of a man's home and the privacies of life" in the much-referenced Boyd vs. The United States spells out a concern for privacy independent of one's personal property rights over their own home. And Brandeis underlines this with a sweeping emphasis on an aspect or element of human life that goes well beyond property and into the very rudiments of a person's mental state:

> The protection guaranteed by the Amendments is much broader in scope. The makers of our Constitution undertook to secure conditions favorable to the pursuit of happiness. They recognized the significance of man's spiritual nature, of his feelings, and of his intellect. They knew that only a part of the pain, pleasure and satisfactions of life are to be found in material things. They sought to protect Americans in their beliefs, their thoughts, their emotions and their sensations. They conferred, as against the Government, the right to be let alone – the most comprehensive of rights, and the right most valued by civilized men. To protect that right, every unjustifiable intrusion by the Government upon the privacy of the individual, whatever the means employed, must be deemed a violation of the Fourth Amendment. (Brandeis dissent opinion in Olmstead vs. The United States)

I do not think Brandeis is being fully transparent here, but he seems to have hit upon something that touched the imagination of later justices. There is something about human life that evokes concern, that merits protection. And Brandeis wants to locate these protections in the Fourth

Amendment's protection against unreasonable search and seizure. To be scrutinized in one's person, in one's affects and sensations, in one's own most personal endeavors is, he posits, an imposition on the security of one's person. Such, he thinks, is best conceptualized as worthy of protection against any unreasonable search and seizure. The Fourth Amendment, however, is a double-edged sword in that the assertion of a protection against *unreasonable* search and seizure indicates that there is a *reasonable* search and seizure. If one's privacy hinges on the Fourth Amendment, one's privacy hinges on conditions that make it reasonable or unreasonable to invade that privacy. The amendment sets up the position specified in the earlier article with Warren and suggests that it is reasonable at times to violate an individual's privacy, to encroach upon their personhood and violate the security of that person no matter how quintessential that privacy is to that person's liberty. In dissenting to Olmstead, he seems to be asserting a set of human concerns meriting protection, but the questions that are raised outnumber the questions answered.

In Griswold vs. Connecticut of 1965, there is further cryptic mention made of protection for "those liberties that are 'so rooted in the traditions and conscience of our people as to be ranked as fundamental.'" Privacy was a shadow present in the Bill of Rights. In our freedom to associate, in our freedom from Military quartering in our homes, from unreasonable search and seizure, from being forced to enter testimony against ourselves, etc. privacy is there in the background. This shadowy notion, unclear and not explicitly called out, hovers behind all the other rights spelled out in detail. There is some traditional sense of something basic to human life that achieves its highest realization in liberty and lies behind the rights that constitute it. It is as if the very personhood of a person, the very subject of a right, is at stake. The private person, the human being alone with their conscience, is the bearer of all and any rights that a constitution may proclaim worthy of protection. To the extent that this is the case, the right to privacy and the protection of that right are essential to all other rights, indeed they are fundamental and structurally supportive of them.

> The foregoing cases suggest that specific guarantees in the Bill of Rights have penumbras, formed by emanations from those guarantees that help give them life and substance. ... Various guarantees create zones of privacy. The right of association contained in the penumbra of the First Amendment is one, as we have seen. The Third Amendment, in its prohibition against the quartering of soldiers "in any house" in time of peace without the consent of the owner, is another facet of that privacy. The Fourth Amendment explicitly affirms the 'right of the people to be secure in their persons, houses, papers, and effects, against unreasonable

> searches and seizures.' The Fifth Amendment, in its Self-Incrimination Clause, enables the citizen to create a zone of privacy which government may not force him to surrender to his detriment. The Ninth Amendment provides: 'The enumeration in the Constitution, of certain rights, shall not be construed to deny or disparage others retained by the people.' (Douglas opinion in Griswold vs. Connecticut)

I am not trying to take either a reductionist or a coherentist stand here. That is, I am not trying to draw out the decisions in such a way as to show that the so-called right to privacy is really a phantom epiphenomenon that is easily reducible to other rights. Nor am I trying to show that there is some coherent notion that is being multiply described by a variety of different statements and approaches to construct a unified notion of privacy upon which to establish its protection. Rather, this seems to be the very origin of that debate and I am more interested in the sense at work in these various claims that there is something that is under it all, something that we care about but which it is troublesome to name. The justices are not able to put their fingers clearly on the thing that they are working with, but they sense that there is some traditional and fundamental concern, something that the framers of the Constitution cared about and something that we today still care about. They are trying to gesture in the direction of something that is quintessentially human. Something that may have been well-described by these remarks from Planned Parenthood vs. Casey:

> At the heart of liberty is the right to define one's own concept of existence, of meaning, of the universe, and of the mystery of human life. Beliefs about these matters could not define the attributes of personhood were they formed under compulsion of the State. (Kennedy opinion in Planned Parenthood vs. Casey)

Not liberty itself, but something that lies at the heart of it. Something that underlies all concerns for liberty. The rights of autonomy rest upon this: the personhood of the person in its most essential and natural state is the subject upon which all further conceptions must be laid. For simplicity sake, I would call this the 'personhood of persons' and submit that it is shadowy and obscure, not at all clearly delineated. Yet, it was this shadowy personhood that the precedents set in Griswold and the dissenting opinion in Olmstead were aimed at articulating. The Supreme Court, following upon the Warren and Brandeis brief, was making headway toward trying to clarify that essential right that was the substance, the underlying substrate, for privacy rights. And they may have been working to bring to light the very personhood of persons that are the subjects of all rights as provided by the Constitution and its Bill of Rights.

Anyone schooled in legal hermeneutics would recognize the pattern here. There is an implicit or hidden notion and it is the intent of the interpretation of the law to try and tease out what is hidden there. The teasing out is the bringing to light, the making clear, and ultimately rationalizing the understanding of the previously obscure notion. The efforts in the various decisions are aiming at drawing out that human quality that is the desired bearer of the right to privacy. The effort to draw it out is the effort to make clear what human beings are.

Almost forty years after Olmstead, in the 1967 case of Katz vs. The United States, the notion of 'reasonable expectation of privacy' was asserted as a fundamental criterion for defending the right to privacy. And it seems entirely possible that the justices, with that decision, were making it clearer that the personhood of the person, in its most intrinsic qualities, is responsible for providing guidelines for the phenomenon in question and the rights that rest upon it. Katz went into a phone booth to place some illegal bets. Unbeknownst to him, the FBI was eavesdropping on the conversation and used the evidence gathered during that session to convict Katz of illegal gambling. His challenge amounted to the claim that the eavesdropping was a warrantless search of his person and so a violation of his right to be protected from unreasonable search and seizure. Justice Harlan writes in a concurring opinion: "My understanding of the rule that has emerged from prior decisions is that there is a twofold requirement, first that a person have exhibited an actual (subjective) expectation of privacy and, second, that the expectation be one that society is prepared to recognize as 'reasonable.'" An interesting aspect of this landmark construction and development of the notion of an expectation of privacy is that it situates privacy itself directly in the center of the person's state of mind, what the person expects. Granted, the notion of the expectation is immediately constrained by the notion of a socially reasonable expectation. The justices do not want to reward ignorance of the nature of things and of the common presumptions of life. They do not mean to suggest that some crazy person who expects privacy everywhere and anywhere is correct to hold such an expectation. Rather, the notion of a reasonable expectation must be immediately introduced to ensure that the state of mind of the individual person is a state acquired through deliberate and informed response to the realistic conditions of the world. And given that, it seemed perfectly reasonable for a person enclosing himself in a phone booth to expect that any communication taking place therein would be protected from prying ears, be safeguarded against an unwarranted search or surveillance.

In Byrd vs. The United States of 2018, the Court ruled that a person driving a rental car had an expectation of privacy regarding that rental car even if the person is not explicitly present on the rental agreement. Byrd was not a declared driver with the rental company. And the agreement indicated that only drivers specifically on the agreement could drive the car. The government's case was that if the person driving the car was not explicitly covered by the rental agreement, they could not be protected from a warrantless search of that car. The court ruled that the expectation of privacy itself was not relevant to the fine print on the rental agreement, rather a person driving an automobile has an expectation that this car provides a boundary that enables the person to exclude others from that space. The car is a zone of privacy and the driver has an expectation of privacy regarding whatever of their belongings they have placed in the car. In the decision, expectation of privacy is the primary driving force behind the decision and the way in which the decision is characterized. Clearly, the standard of expectation has become deeply entrenched in the Court's understanding of privacy even if they seem prone to attach those expectations to spatial parameters.

The movement in the direction of the expectation of privacy, confirmed in Byrd, further supports this notion of the personhood of the person underlying whatever it is that the right to privacy is meant to protect. The reasonable person is being touted as the judge and jury of that determination. In so far as you are reasonable, and your reason reflects the standards of reason adhered to by your society, by your civilization, you may proceed with the confidence of the Constitution in expecting that the security of your person is protected from any unreasonable search and seizure. In this moment, the justices are asserting the executive power of the reasonable person, and the notion of privacy as the terrain of action is held up as a standard and guiding light.

In the very breath that gave such an outstanding responsibility and power to the person living their private life with the greatest confidence of the nation's protections, the decision may have taken away this same right. Should the person lose that expectation, should the reasonable person suffer deterioration in what they come to expect from their society, with that expectation so goes the right itself and –so it would seem– the very thing itself. By resting the right on top of the expectation of the condition, by interpreting that personhood as an expectation, the right itself becomes malleable to whatever extent the expectation itself is malleable. Reasonable expectation is meant to describe the state of mind of an acting person, the conditions used to bolster it are the person's intentions as they step into the phone box, as they close the doors and dial the phone. Should the mental state be eroded by modern forms of communication, the modern capabilities of the surveillance state, and the existence of special courts that grant eavesdropping warrants behind closed doors, then the expectation will decrease. And if the expectation decreases, if people expect to be surveilled, to have their government, employer, or their social media service provider, spy on them, then not only is the right ceded, but the thing itself is reduced. If privacy is nothing other than the reasonable expectation of privacy, then the thing that is protected by that right is diminished by the *social* determination to protect it. Therefore, privacy would appear to be an invention established only by convention. The protection of it feeds the expectation of it and to the extent that it is less protected, there is less of it. Only the legal status guarantees the existence of the thing itself, privacy itself, it has no status beyond its being protected by law.

Note the impact this has on the Fourth Amendment. The search and the seizure that is permitted is reasonable search and reasonable seizure. Whether or not a search, read surveillance, is reasonable may be said to depend on the expectation of privacy that the population may have under the given conditions. A reasonable search does not violate the right to privacy and the right to privacy is a socially condoned reasonable expectation of privacy. The transformation of the right to privacy into a personal state of mind throughout the society at large means that the reasonableness of a search is broadened in scope because the expectation to avoid it has been reduced. The police or the government or some corporation can watch everything you do if you do not expect anything different. People of different ages and generations and cultural backgrounds may therefore bring different expectations to the table and so provide for altering norms over time within the society as the population ages or as one demographic takes over a greater proportion of the country. The right itself is no longer built into an inalienable foundation, but rests on an historically moving target that permits the modification of that right over time. And the deterioration of the right has a recursive relationship to itself that could cause a fast rolling cycle to occur, leading to the complete dissolution of the right. At least, this is what the language of the Katz decision would suggest. The decision both attaches itself to the historical emphasis on the personhood of the person and distances itself from the ontological spin previously assigned to that notion: an inalienable respect for 'personality' became respect for a state of mind.

I am not advocating judicial realism here. I do not mean to give the be all and end all of privacy to the courts. The dynamic that I see at work is the establishment of the right to privacy on the firm foundation of the personhood of the person as that shadowy figure that underlies much of our constitutionally protected rights and which has been gradually transformed through the emphasis on expectation. The personhood of the person, therefore, is not the solid ground that 18th century Americans thought it was. Instead, this seating of the privacy condition on the intangible humanity embedded in our aspirations and hopes, in our pursuit of happiness and our liberty, also enables a moving target, something that might evolve over time. True liberty only exists when it has been sanctioned with social rationality, but privacy is constrained in socially recognized boundaries. The existence of wire-tapping made famous in the Olmstead case was unknown before that. And the changes to the modern communication ecosystem have gone exponentially far beyond that. The existence of devices that fit in the hand and which can be carried around and used from anywhere and the existence of devices that can fool those devices and get them to connect to a pirate cell tower that intercepts their calls, both coupled with a government that has proclaimed itself to have a powerful interest in thwarting terrorism through the collection of metadata, may diminish privacy itself in unprecedented fashion. Thus, we reduce that most quintessentially human characteristic that we were trying to protect, in the very act of protecting it. If the hidden something, I know not what, that lies underneath the powerful urge to protect privacy is the personhood of the person, then the diminishment of the right to privacy, which amounts to the reduction of privacy itself, turns out to be an attack on personhood. They did not have to articulate this essence in terms of expectation but having done so they put the essence at risk.

The concurring opinion from justice Harlan does not just point to a random notion of expectation, his phrasing is more precise and oriented toward a twofold representation of the matter. First, the person must exhibit an actual expectation. And by this I take it to mean that the person goes beyond a simple utterance and acts in such a way as to indicate that they think they are in private, or that their privacy is to be respected. The fact that Katz entered the phone booth and closed the door indicates that he was truly expecting that the content of his conversation would be exclusive to himself and the person on the other end of the line. In addition to this actual demonstration of an expectation of privacy, Harlan says that the society must be prepared to recognize this expectation. The former of these points corresponds to what I have been suggesting is the traditional gesture in the direction of the personhood of the person, the acting individual shows an actual expectation that their actions are taking place in private without vulnerability to surveillance. The latter of these points indicates that the society at large must be willing to recognize this action. The phone booth has a door on it and that suggests that the general understanding is that phone booths are zones of privacy, zones where the society has recognized the expectation of the individual person and affirmed the legitimacy of it. The disappearance of phone booths, the advent of cell phones in which most every person has had the experience of hearing both sides of a conversation taking place in line at the grocery store, may suggest that society is no longer prepared to recognize the privacy of the conditions under which a phone call is made. Since everyone is using cellular mobile phones and since the most popular models of mobile phones are often audible to people other than those currently using them, it may be reasonable to conclude that no user of a mobile phone has a legitimate expectation of privacy while using it to make phone calls, although texting and other uses would not necessarily be covered by this same reasoning. If they have no expectation of privacy in line at the grocery store, then if someone employed by a law enforcement agency is listening in on their conversation without a warrant, this cannot be considered an unreasonable surveillance of their person.

The move to expectation as a human mental state is the coming to fruition of the sense that Justice Brandeis had in wanting to associate the real and tangible right to privacy with some underlying human quality that a free society would want to protect and foster. It is supposed to bind the notion to the blessings of liberty and the greatness of freedom by elevating the human agent into a crucial and defining position for this right. Additionally, Harlan's clarification on the second aspect of this point is to embed the individual person in the center of their social relations. In the moment that we assert the liberty of the individual human agent acting on their expectation, we also refine that attribute via the society's willingness to recognize the reasonableness of it. This inextricably binds the person together with their social setting, what they may expect is what their society recognizes that they can reasonably expect. The mental state of the human being is always mediated by the collective recognition that justifies that mental state. People are not isolated and alone on a desert island when projecting these states of mind into their actual setting and projecting their sense of space and their propriety over that space upon the world. Instead, they are acting on socially recognized principles. The person has this right to privacy in so far as they expect to have it and they can expect to have it insofar as their society recognizes their expectation.

When we say that privacy is a publicly determined privacy, there are two senses involved. First, there is the notion that the judiciary branch is creating the right by asserting its protection of it. Secondly, the expectation itself is insufficient on its own but must be confirmed by its social recognition. The underlying unknown something that drove the justices to try and carve out a feature of human being that they could build precedent around and establish as a bona fide first class citizen in the realm of rights guaranteed by the Constitution, did not just attempt to articulate the personhood of the person, what is most characteristically essential to the person and thus worth protecting and defending in a democracy devoted to liberty and justice for all, it also attempted to situate that person relative to their society. Their intent was to make that person into a member of the society and dependent on it for recognition of the security of their person, for recognition of their expectation of privacy and hence their right to it. The hypothesis that I am working up to here is that the something hidden that is referenced by the right to privacy, the residual that is being pushed into the background by the telemetric, is not just the expectation of the security of one's person, but the embedding of the person's personhood into the society that is willing to recognize them. The justices, in their rationalizing decisions, may be acting on behalf of the telemetric. Beneath the intense efforts to judicially rationalize the notion of the private person, of the person with a right to privacy, of the person with an expectation of privacy, is the situatedness of that person amid other people. And not just randomly associated people, not just a loose association of other people, but a set of people bound together by a civil constitution, a doctrine that declares the structure of the government within which they live and the rights they can lay claim to within the society they have formed. The justices need not have been trying to colonize privacy, the telemetric is perfectly content to operate behind the backs of its agents. This is a dark picture, let's now see if there is anything hopeful to be found in it.

Subjective Expectation of Privacy

Growing up, our privacy is constantly invaded by our parents. They tell us what to do, they come into our rooms when we are not there. They might go through our drawers, look under our beds and in our closets. It is their house, they know everything about us. And then they send us off to school where the teachers tell us what to learn, what to read, what to think. All this is for the sake of proper child rearing, the proper raising of a child, teaching that child the basics for how to get along in the world, learning the multiplication tables or how many angles are in a triangle. Where is the child's right to determine their own existence, to define their world as they see fit, and to pursue happiness in whatever way they might imagine? This is not advocacy for liberal childhood education. There might be parents who do that, who let children have all sorts of private moments, secrets that they can keep to themselves, and so on. Even those parents do not let their children go hungry or let them sit in the filth of a dirty diaper. Parents invade the space and person of their children, it is natural and normal to do so. Where did this expectation of privacy come from? How is it that children come to have this expectation? Do they uniformly achieve it? Is it possible that every parent in the world somehow knows just the right way to cultivate socially recognized expectations of privacy? If there is a reasonable expectation, it might be argued that all the parents in the world construe this very reasonableness through a common inference. And having come to this common conclusion, the parents gladly begin the process of establishing and cultivating these expectations in their children throughout the course of their childhood. If not, then it is surely mysterious where this commonly held and socially sanctioned expectation might come from.

The Court does not address the privacy rights of children relative to their parents. There are cases of emancipated minors who argued that they needed to be distanced from their parents to prevent some harm, but these are not traditional privacy cases. We do know of cases where children's privacy in school is under discussion, but the cases there tend to be oriented toward spatial matters like searches of their lockers and their person while they are in school or as they are entering the school. No one has challenged the use of standard exams in the evaluation of children's progress at a publicly funded school as a violation of their privacy. Home schooling is a right that parents may exercise for the sake of their children, but that is not in defense of the child's right to privacy. If the school wants the children to study math or geography, there are no precedents suggesting that it is a violation of little Jane's expectation of privacy that she be forced to learn the fifty states and their capitols. And yet what could be more essential to the personhood of these children than the ideas and thoughts that run through their head on any given weekday? The state does not consider this to be a violation of privacy and we do not read about challenges to the expectation of privacy condition founded on the desire to protect children. Regarding what they will learn in school, children do not appear to have any expectation of privacy and so no one can be charged with infringing upon that right. Where do these expectations come from and how do they achieve anything like the uniformity the Court claims they should have if they are to be sanctioned as socially recognized, and therefore reasonable, expectations of privacy?

The absence of these concerns suggests that we are talking about the personhood of the person when we construct tests for the expectation of privacy in a socially sanctioned scenario. The cultivation of children to mature personhood is an introduction to a set of norms deemed crucial to socially acceptable and appropriate behaviors. The extra-curricular activities of children are the areas of their lives where they will develop many kinds of expectations about how people will treat them and how they are expected to treat people in return. Learning norm-oriented behaviors is essentially learning such reciprocal expectations constituting a concrete civil life. Say please, say thank you. People will expect this of you and you can expect them to be more helpful if you follow these norms of action. Strangers should not be talking to you on the street, offering you rides in their cars, other children should not be bullying or hitting you. These norms of behavior instill in children expectations of proper action, they teach them what they should expect from other people. Long before children could ever learn to have any such thing as an expectation of privacy, they must have first learned to have expectations, they must have learned to expect. And the fact that expectations are part of a complex process of cultivation and learning, of value and norm acquisition coached by parents and teachers, indicates that the proper parameters and limits, the boundaries, of personhood must be acquired. In the best of all possible worlds, children learn to respect one another and to expect this respect for themselves.

We are teaching our children that the world is a somewhat predictable place and that it is so because we make it predictable through our norms and values. By teaching them that they should be reliable and consistent with others, we teach them to expect that others will be predictable and consistent with them. We are teaching our children about expectation spun telemetrically. We are instructing them to be predictable and to rely on regularities. And we do this to be able to count on how others are going to behave and what sorts of actions they are likely to perform in various circumstances. While violating their children's privacy, reading their diaries, looking in their drawers, reading the slips of paper that are in their pockets, we are also teaching our children to have expectations, including expectations of privacy. It is entirely possible that children ultimately learn the greatest lesson about privacy when they discover that their parents have been systematically violating it throughout the course of their lives. The kid wants to use the bathroom by themselves and suddenly is upset that you have been helping them all this time. The realization that their parent has been violating their privacy is their realization that they have something called privacy to be violated. They are developing the expectation of privacy in the moment that they realize that it is not being respected. They see that their parents are telling them to behave a certain way toward their little friends, but how does that seem fair when their own parents do not respect these same norms regarding their own actions? The open rebellion of childhood begins and with that rebellion the establishment of boundaries. And with those boundaries, a sense of privacy and the expectation of it.

The association of expectation with learning the norms of proper behavior would show that expectation is the human intentional quality that was lying dormant until jurisprudence began work on the excavation of the true and fundamental personhood of the person underlying any and all claims to a right to privacy: the private person is pleonasm, person just means private person, someone with an expectation of certain norms of behavior that will include respect for one's own being as a thinking, feeling, and willing person. And who is the subject of this expectation? The expectation sets up a relationship of the person with themselves, it is an expectation about themselves and how they relate to others, to the rest of society, and to other private persons. The subject of the expectation is the one that is expecting but not in isolation, rather the subject of the expectation is oneself in so far as one is among others. As an expectation of privacy, this being among others is a differentiation of oneself from others, an expectation to be set apart from them, to live among people in such a way that the boundary between oneself and the others is respected by those others. In establishing the right to privacy as the right to be left to oneself in those cases when one expects to be left to oneself, or as the right to control the exposure of one's thoughts and feelings to only those others that one wishes to expose them to, the judiciary has established a zone of privacy that establishes the private person as a legal body, they have established the subject of all rights in its innermost constitution and not as it relates to owned property, a fire arm, the freedom to assemble among others, or to speak their mind publicly. Rather this established subject is the agent of a defining existence, a self with the expectations of having its own feelings and thoughts, its own life, its own sense of meaningfulness and belonging. And ultimately, a producer of metric emissions.

In the right to privacy, the person, the private person, becomes the subject of the law. The right to choose a method of birth control, to sit in one's car without fear that some authority will yank you out of it and rifle through its contents, being able to talk on the telephone without the fear that someone might be eavesdropping, or sitting in your house without concern that a nosy neighbor or passerby is picking up the signal of your most intimate behaviors when you thought you were alone. These decisions are aimed at establishing the private person. Not the person who goes to rallies and leads a demonstration against an unfair practice, not a person who has entered a contractual relationship, but the person on their own, in their own thoughts, and in their own mind. And not just that. There is recognition here that we establish close and intimate relationships with others by sharing these details about ourselves. We tell our closest friends and family things that we would never wish to share with the general public, never plan to tell our colleagues at work or the other shoppers at the mall. These intimate relationships, whether we establish close ties and connections, are the further establishment of the private person. The subject of the right to privacy is the person in their most intimate settings, their most personal thoughts and feelings. The law has been striving to establish the privacy of such personhood since 1890, so my story goes, and the establishment process begins to make clear what it is that Brandeis and Warren were concerned about when they referred to that something else which is at stake, something other than property or liberty, when we talk about the right to privacy.

In Descartes' *Meditations*, the cogito, the thinking thing, is established at the seat of knowledge. It is the bedrock foundation for any and all rigorous knowing that a person could hope to achieve. The fundamental nature of this cogito, this underlying basis, this subject, is the self-relation that is realized every time the subject asserts that relationship. It is a relationship that is not dependent on any other, not even the deity, for the self-establishment of its own existence with absolute clarity and distinctness. The subject of knowledge is the purest thing that there is. And there is something of that to be found in this judicial establishment of the most intimate personhood of the person inhabiting the notion of privacy. The judicial picture already represents a more complex being than Descartes thought possible. Descartes was suspicious of memory, for example, because the memories were less certain than the immediate now, the immediate assertion of one's own existence at a point in time as this individuated thinking thing. Memory was fuzzy, and we might misremember, or an evil deceiver might inject us with false memories of events that never took place. The legal precedents in privacy rights do not seem concerned with memory but with expectation, something that Descartes does not think we can rely on without rigorous investigation. The expectation is a mental state that the subject has here and now, but it pertains to something that is upcoming, something that has not happened yet. When we spoke before about predictability it seemed as if this was a way of characterizing expectation such that the things we expect, although they had not happened yet, were anticipated in such a way that the manner and nature of how they were about to happen had already been determined. The subject of privacy expectations is a subject that has an orientation toward what is about to take place and that orientation presumes certain conditions will hold and that these presumptions are socially recognized as reasonable. Although the expected event has not happened yet, the society deems that it is right and proper to expect it. The subject as private person is bedrock to the set of rights, but so much more than that through this insertion of expectation into the subject's capabilities.

The private person is not the only basis of their world as the Cartesian subject, rather there are a whole set of rigidly affirmed and protected conditions that the private person can rely on. The establishment of expectation is aimed at securing the private person's future-oriented point of view toward the events and transactions in their world. We are not only with ourselves, in the constitutionally protected democracy, rather we are with others who recognize our personhood, are willing to establish the proper expectations of our personhood in the establishment of the subject of the law as a private person. Human life, like all animal life, is filled with uncertainty, with the potential for rapid change and immanent hardship. Emergencies befall people all the time: health, legal, and financial emergencies, events and situations that require immediate attention. People must act and react. What sorts of things can they count on in such times, what sorts of predictions can they make, and what sorts of plans can they devise based on their awareness of the basic conditions of life that they are subject to within the society where they make their lives, where they live as private persons? The point here is not to try and establish any sense of a social safety net or set of higher order infrastructure that would help people with these emergencies. I'm not suggesting, at this point, that there is anything like an implicit right to health care embedded in the notion of a right to privacy. Rather, I am concerned with the juristic foundations of human existence and what is most basic and essential in it. A person trying to get help in a medical emergency must be able to count on some very basic principles when planning their actions. They must be able to rely on a social fabric in which private persons are treated in specific ways by the people around them. They must be able to rely on a set of expectations that govern the basic principles of interaction as they act on their plan to acquire whatever resources the current emergency requires. People do not conceptualize the right to be left alone in the literal sense that all other people should ignore them all the time, that would be radically contrary to anyone's interests when they tried to carry out some action or resolve some set of problems currently impacting their wellbeing. What the right to be left alone may suggest is that people should not try to impede others unnecessarily. We have the right to expect that people will respect our current projects and not go out of their way to get in our way.

Producing a domain of privacy that constitutes the personhood of the person is a required basis for all further consideration on this line of thinking. And such privacy not only establishes human beings with the capacity to respond to emergencies or just go about their daily business without fear of brutal intrusions, it is also the basis for criminal activity. Establishing the expectations of privacy as something reliable that can be counted on by any private person carrying out their personal lives, gives each of us room to breathe and makes it possible to plan to hurt one another. In secret, anyone may carry out some self-serving act at the cost of others, sometimes even through the violation of their rights or through some fraudulent manipulation of their ends for one's own benefit. And the judiciary has accounted for this through the standards of reason. In those cases where it is reasonable to look more closely at a person, where it is reasonable to get in their way to prevent them from doing whatever it is that they are doing, where it is reasonable to expose their private affairs even to the extent that those affairs will humiliate them, then it can be done using the established due process. We determine with our precedents on privacy not only the reasonable conditions under which a person can have those expectations, but also those reasonable conditions when privacy can be violated for the sake of some other good, some social interest that may include the protection of the rights of others or anything else that has been deemed a compelling interest of the state.

Human beings living as free beings have hopes and aspirations, they are future oriented. That is, the rights of such human beings are often meant to protect their dreams and aspirations, their pursuit of happiness. As such, their relationship to the future, their future orientation, is of central concern in the establishment of a society that safeguards their liberty and a just relationship between members of that civil society. By invoking expectation in the heart of a crucial right, whether it is a cluster of rights or a single thing, the Court is establishing this future oriented emphasis in its recognition of what counts as liberty and justice. We are free to act on our expectations, we are protected from infringement by others, and justified in having those expectations. The focus on expectation is a delicate thing since a change in mental states can result in a change to the right they support. People can be taught to expect less and to the extent that they do expect less, they are caught in a vicious cycle. Failure to *cultivate* the expectation, equally as much as the failure to protect it, would amount to an infringement upon the right.

The formation of the subject of the private person is the rationalization of the bearer of rights that is the subject of protection of those rights. The justices tasked with interpreting the Constitution have turned to the person beyond the realm of the political that was traditionally held as the proper target area of constitutional concerns. There was something missing from the letter of the law and that something that was missing was addressed by developing and evolving considerations of the right to privacy. Brandeis and Warren pointed to a residual that they sensed beyond the boundaries of the legal precedents. There was something vitally important to the founding fathers that was not yet properly conceptualized in the laws. And that something was to be the subject of the right to privacy. Privacy talk, therefore, is the publicly established set of discourse aiming at conceptualizing and clarifying the private person in such a way as to establish that person with a publicly recognized role in the society. From the point of view of the justices, this maneuver was crucial for protecting that personhood and for giving it a proper place within the law. Analyzing it, measuring it, conceptualizing it, talking about it and legislating it, incorporating concerns for it into rational feedback loops that drive legislative and juridical decision-making poses a great danger to such a private person. It brings the person, albeit in the abstract, into the light of day, exposes the private person in such a way as to make it something to be taken account of, something to be aware of, and to be on the lookout for. And this is a risky set of insights as it potentially exposes to harm that very thing that the society was seeking to protect. By knowing more and more about the private person, by raising awareness, we make the personal characteristics of each of us something worth getting to know, something that we now realize it is possible to get to know. In this judicial focus, we have made it possible to protect the right to privacy, but we have also made it clear in what ways that right can be violated by any interested party who would apply a magnifying glass to the actions and behaviors of other people.

We have found something hopeful in this judicial history and the something hopeful is the future oriented personhood of the person. In finding it, we realize that the discovery has made it possible for others to find it as well. And if there is no protection from those others, then even the search for the subject of all rights has been a part of some whole with an ulterior motive. The pattern at work here is telemetric. And this telemetric pattern is the same as that which we studied while laying out the structure of an information system that provides services to users and which may have been built and offered with the best of intentions. I do not suspect that the Supreme Court of the United States had a long-term plan to construct a set of concerns ultimately aiming at exposing private personhood to analysis and aggregate understanding so that it could become the subject of influence and manipulation by large-scale corporate endeavors offering up terms of service agreements in the new digital universe. We do not care about their intentions. This chapter turned its attention to privacy because the description of the system at large was making privacy into a concern. Privacy is a boundary condition of that system. We started out to discover the basis of all our thinking about privacy and intended to use it to place constraints on these computing systems and the way that they function. Up to now, at least, that does not seem to have been the case. Instead, what I fear that we have discovered is that the discussion of privacy carried out in the courts and in the public sphere has been for the sake of attributing various known and knowable qualities to that private person so as to establish them as the object of investigation and study that the telemetrically designed system is then able to target and absorb for the sake of furthering the ends of various interested parties inhabiting that system and realizing their goals in it.

The concern here –not demonstrated, just glimpsed– is that the establishment of privacy as such is for the sake of extending the reach of the service provider, of the advertisers and business opportunists. The more these interests know about the private person, the more they know what they are and what their limits and boundaries are, the more they can adjust their strategies and operations to influence those private persons more effectively. The fear or concern is that all our talk and worry about privacy is just a way to further expose the "private" to the realm of social relations that would market to it, prey on it, derive value from it, and exploit it in one fashion or another to achieve some end. Some have been worried about the diminishing expectation of privacy that could make our right to privacy evaporate before our very eyes. That is a problem that needs to be addressed, but we should also bear in mind that it is possible that consuming privacy as a profitable industry does not require that it be destroyed in the process. It is possible that the right to privacy may persist and even become stronger and better understood through the mechanisms that strive to better define it for the sake of protecting it. Indeed, this phenomenon may serve to produce clearer delineation and boundaries making it easier to get at the private person in various ways, easier to understand the motivations of such a person, easier to target their desires and sentiments, and easier to mold them into exactly the sort of persons that it serves someone else's interests that they become. By exposing personhood as a future oriented foundation for rights, the law exposes our hopes and dreams to colonization.

The Protection of Privacy

Protecting privacy means protecting the expectation of privacy. Since diminished expectations of privacy constitute diminished privacy, its protection cannot only be the protection of the good –as though placing guards around it will secure it– it must be the protection of the expectation itself. Since the expectation of privacy is the standard for the determination of privacy rights, anybody interested in protecting the right would have an interest in safeguarding the standard for determining the right. The purpose of the Constitution and the judicial action that interprets it would be to protect the subject's expectation of privacy in so far as it is recognized as reasonable. That should indicate that the Court would not permit any reduction of the expectation of privacy and any case in which such a possibility exists would have to be addressed properly. Empirically, however, it seems that the Court may be failing in this regard. People, immersed in the various tools of mass surveillance, have lower expectations of privacy. They seem comfortable with posting and sharing images and facts about themselves that did not used to be so normal and natural. Additionally, we have become quick to crowdsource the video capture of events taking place around us for immediate publication, and so seem less respectful of the privacy of others. If this is the case, then the expectation of privacy has been diminished. And if it has, then the high Court has failed to protect the expectation of privacy in the subject of the constitutional rights they have sworn to defend. If the personhood of the person lies in the private person as the basis of all actions subject to rights, if the Constitution charges the state with protecting those rights, and if the fundamental embodied basis of those rights has not been protected, then the entire Constitution might be in crisis. This must be a heavy burden for the Court to bear.

Another way of broadly conceptualizing this dilemma is to try and understand whether a publicly determined privacy can ever truly safeguard it or whether that very dynamic already corrupts it. I suggested the connection between the two by claiming the precedents worked as though trying to cultivate a hidden something lying beneath the other rights. The hidden something turned out to be the private person who is the subject of all rights. This is a publicly constituted determination of the personhood of the person. The question then is what effect the publicizing aspect of this cultivation process may have on diminishing any expectation of privacy. Is it possible that the rationalization process of the courts over the years has achieved a meaningful definition of the right to privacy at the expense of privacy itself? The long-fought battle to declare the right to privacy as the expectation of it is the emission of the private person as the subject of rational action so that analysis that completes a feedback loop can be constructed such that the expectations of privacy are decreased, such that the agent itself has less expectation of privacy.

Cameras are everywhere. We see this represented in movies and television shows and read about it in countless news stories. They pick up all the goings on in the parking garages or in the streets of cities. People install them around their homes to catch sight of any unwanted visitors that wander up to their door at odd hours of the day and night. If people are aware that there may be cameras all over the place, capturing all kinds of activity, are they ever reasonably justified in thinking that their behavior is not being recorded? There may be zones where the expectation of freedom from video recording is still present, but the number of those zones is vastly reduced from what it was in 1974. And that is just video. The home itself, albeit safe from video recording by an external agent, is pervaded by wireless communication that we use to connect to the internet and visit all the popular services that we cannot live without. It has become common knowledge that our Internet Service Provider is monitoring our click stream, keeping an eye on what websites we are visiting, what content we are viewing and downloading, whether we are violating copyright laws, and so on. It seems obvious that anyone would have a lowered expectation of privacy under these circumstances. There may be zones of privacy, but those zones are shrinking and the context under which the expectation of privacy is reasonable is becoming more and more narrow, extending only with the reach of our arms. The case law has not kept up and since the establishment of the expectation-based standard for privacy there has been an all-out attack on that standard. Why would not the expectation of privacy be exactly what would demand protection given its establishment as a standard? A civilization that loses protection for the private person as the subject of constitutionally guaranteed rights is no longer a civilization made up of persons with rights.

But how could a court protect the integrity of the private person? How could a court safeguard the expectation of privacy as the innermost sanctity of the person engaging in public affairs? Congress is not making any laws to constrain, Congress does not require video cameras on every corner and clickstream analysis on all internet surfing. All that is available to the Court is to draw out the private person into public and turn its character and qualities over and over until it has been properly analyzed for the sake of a clear delineation of all that it is entitled to under protection by and from the law. The legal mechanism is telemetric, it does not know how to proceed without applying these remote measurements to the thing that it is attempting to know and thus get to know it through a process of setting itself apart from that thing, making that thing into an object of study, and thus, in the case of the private person who is the subject of rights, losing it in the very moment that it takes hold of it. By preventing laws from constraining it while making it visible as a publicly identified standard, they open the residual domain of social action where expectation is vulnerable to whatever powers act upon it.

The hidden king that is covered up in all this talk about privacy, all these efforts to clarify the notion, the rights, and the protections, the residual of all discourse on privacy, is the subjection of the agent to scrutiny as such. Or to refine that, subjection as such. And all the clamor about privacy ultimately find its source in that which has been hidden by privacy's rationalization; and what has been hidden is the social act of subjection, the social constitution of subjects of the law, of subjects of rights, and of subjects of governance. My hypothesis here, my guess, is that there is something about the telemetric procedures that are antithetical to what is under investigation by those procedures. I am not taking issue with the standard of privacy as its expectation. Rather, I am suggesting that in its diligence to get at just that thing which is the truest and best of standards of the right to privacy, the mechanisms of investigation that make up the performance of legal reasoning are ill-constructed to do that work without damaging the thing they are reasoning over. The discussion thus far, in this and the chapters preceding, has been aimed at showing how the 'mechanisms of investigation' work, what they do to achieve their ends, how they enact a specific pattern in ecosystems with diverse interests for the sake of yielding the objectives of the various involved parties. And in this discussion, it seems that the dynamic might happen at the expense of the underlying something or other that drove the process from the outset. The system in its simplest conceptualization was a service provider aimed at serving the private person in their endeavors to acquire some response to a request, to obtain something that they wanted. And the large-scale mechanisms put in place to serve these users have revealed themselves as antithetical to the private person who is the socially subjected seat of rights to life, liberty, and the pursuit of happiness.

The conflict is not just due to the shape and pattern of operations of the mechanism. These shapes and patterns are fantastic for certain kinds of jobs. Nothing like the use of telemetric mechanisms if you want to smelt steel or launch a projectile off the earth with humans on board to land some piece of it on the surface of the moon. The telemetric pattern and the mechanisms based on it can be used for all sorts of wonderous things, all sorts of physically puzzling things, and it can be used to construct devices that serve other ends. None of this is challenged by raising questions about the role of the telemetric in satisfying human need and the role of human beings in the social association that constitutes democratically defined government. Rather, the conflict is the relationship between these mechanisms, their underlying patterns, and the private person who is the socially recognized subject of rights. The source of this conflict may lie in the very nature of the thing discovered. The Supreme Court of the United States was able to cast its gaze over the standard for the right to privacy and see it as the expectation of privacy. And that may have been a brilliant insight into the internal nature of human beings, but the methods of procedure were too coarse to see that 'expectation' itself is extremely fragile and malleable, that it was a delicate notion, as delicate as democracy itself. And the failure to see this enabled the application of a set of tools to the phenomenon in such a way as to fail to respect that delicate nature while taking hold of it.

Let's look a little closer at 'expectation' to see if we can locate the beginnings of the conflict. Expectation involves both what has happened already and what is about to happen. We base our expectations on what we know about the world and the extent to which what is about to happen is analogous to what we know has already happened. Expectations are largely based on the entire domain of inductive inference: past pasts have gone in such a way and past futures have followed upon them. The current set of circumstances would appear to be like those past sets of circumstances and so we should expect the upcoming events to mimic those past future occurrences familiar to us. This sounds like what the feedback loop was meant to supply our service provider so that it could optimize its operations relative to user requests and the events that are likely to follow upon them. No doubt data science could be applied to any data set to derive a set of expectations that might be reasonable given that collected set of metrics. And one lesson we might learn from this invocation of data science is that the expectation of the future based on the past is driven by the association of metrics in analysis, which means that there is a learner (either machine or human) that is computing a meaningful or realistic association between the relevant attributes. This suggests that in order to have a proper expectation, a person needs to be sure that the thing they are expecting is analogous to the past experience as the conditions upon which they are basing that expectation. If the expectation is based on some false equivalency between, on the one hand, what was once the case and what followed it and, on the other hand, what is currently the case and what can be expected to follow, then the entire foundation of that expectation is missing, and the person is acting in the dark or being misled.

The turkey problem comes to mind. Every morning that turkey sees the farmer come out and feed him a nice tasty nutritional breakfast. The turkey expects the farmer after a while, assumes the farmer will come again today just as the farmer has come every day before. This proceeds until the day before Thanksgiving when the farmer cuts off the turkey's head instead of feeding it that tasty and nutritional breakfast. Like the expectation of privacy, the turkey is inferring a standard for determining an outcome based on the normative actions of an intelligent agent. The turkey errs not because he does not understand the natural chains of causality, but because he only partially understands the farmer's motives. Expectation is not simply reducible to prediction, rather it includes a sense of trust in the motives of the intelligent agents that cultivate and protect it. Expectation is different in kind from the expectation of an induction-based learner, it is normatively based, it is a legal standard being set for expectations. As a cautionary tale, we note that the turkey was mistaken in thinking that the principle was there to protect him. Instead, the principle originates in the farmer's set of concerns opaque to the turkey but crystal clear to the farmer. The farmer was not feeding the turkey to keep him healthy, he was fattening him up in preparation for the slaughter. Arguably, we might observe that the turkey was missing a critical insight into the relevant metrics because it did not have access to the various important attributes and dimensions of the situation crucial to making the association between the farmer's past actions and the actions on that fateful Wednesday morning. If the turkey had access to all the dimensions and metrics relevant to the ongoing process, Tom would have known what to expect. His expectation was flawed because it was not properly informed by the relevant characteristics of the events in question, he was not able to see that there was this holiday called Thanksgiving and that one of the attributes of that day is that people eat turkey for their traditional meal. The attributes of turkey preparation, how long it takes to clean one, to prepare it for eating, and so on, would have suggested that, given the proximity to the holiday, the proper expectation for the day before Thanksgiving is the butcher's knife and not the tasty and nutritious breakfast.

How does this analogy fit with the issue of protecting the expectation of privacy? We dismissed the likeness between this form of expectation and an ordinary case of induction because in this case we are concerned with establishing a normative standard for the protection of a right as opposed to protecting a mechanical chain of cause and effect. If the expectation of privacy is a standard distinct from mere mechanical causality, then where does its origin lie, assuming we need to know how it came about in order to prevent it from disappearing or being modified? Suppose I have been trained through repetition to have a set of expectations of privacy. And suppose that the law of the land has reinforced my understanding that these expectations are the foundation of my right. My training might lull me into thinking that this is a mechanical causal chain when it is a socially recognized legal order. The extent to which my expectation is protected will depend on the extent to which that order is aligned with the directives protecting it. Without historical evidence that the Court has ever actively worked to protect the state of my expectations, I should not suppose that their motive has been to do so. Rather than applying a motive to unknown and non-existent deliberations, it makes more sense to conclude that an actual deterioration in my expectation of privacy is likely due to a failure to protect it by those who have sworn to do so. Expectations, present in norms, are physically present throughout the world in the behavior of social agents. Like the door to the phone booth that indicates the socially recognized expectation of privacy in the caller, many social practices and designs indicate recognition of other related expectations. Norms of privacy are expressed in practices prevalent throughout the world, and since the social world is the environment within which we live, we draw conclusions about its reliability just as we draw conclusions about the sunrise and sunset. The presence of mass surveillance through the wide-scale use of cameras and clickstream analysis, for example, informs our response to acts of social recognition embodied in the practice of proliferating such devices and operations. Because it is legal to surveille the population in this way, the population absorbs the diminished expectation. Permitting social practices of this kind, therefore, exposes people to experiences that diminish their expectation of privacy. We may be just like the turkeys and things are proceeding exactly as planned: the farmer does not have our interests in mind and if we knew the whole story and all the relevant attributes, we would see that the law is protecting exactly what it intends to protect.

The response might be the same as we discovered when considering children raised to appreciate and expect respect from their peers as well as the adults in their lives. If we have been properly cultivated, then we can withstand whatever influences come along. We indicated that norms of behavior seemed to be the driving force here, that we learned these norms gradually during our rise to maturity and through the guidance of our parents and our communities. These institutions trained us in all sorts of ways, instructed us, forced us to repeat various actions until they were ingrained into our body language and even into our intuitions, let alone our perception and understanding. These are the norms that we acquired, and these norms are the origin of the standard of expectation. Since the norms can be influenced and changed, and the point is to protect the expectation, we must protect the norms to protect expectations. The exact opposite of such norms, however, have been proliferated. It is ironic that a civilization that measures everything shows no interest in measuring the impact of constant measurement on private personhood.

One way to protect a norm is to make it explicit and recognize it as a rule, as a rule for proper behavior, and as a principle upon which a law can be based. In the case of protecting the state of the expectation of privacy, this creates a circular relationship since expectation was supposed to be that rule or principle, but now seems to require another rule or principle to protect it. That principle is required for the expectation to retain its scope over the personhood of the person as the subject of all other rights. That principle has not been set down as a precedent unless it amounts to the judicial deliberative principle used by the justices when coming to their decisions. To the extent that this principle can be attached to an interested party in the debate, it may amount to the farmer's motives when feeding the turkey each morning. The social nature of expectation, and its norm-centric origins, means that it is potential material for the telemetric orientation of a system consuming resources for the sake of achieving ends where those ends do not cohere with the discrete ends of its subjects. And what is more, the talk of expectations of privacy in the wake of new technologies and new mechanisms of mass surveillance may amount to a tasty and nutritional breakfast. Each decision works to redefine the norms and expectations of privacy in a rapidly changing world with an increasing number of tools for mass surveillance that the court decisions instruct us how to cope with.

The expectation of privacy is a flimsy and fragile option for use as a standard for the right to privacy. In the next section of this work, we will investigate whether the circularity in the notion of an expectation of privacy is virtuous or vicious. One aspect of expectation that will need to be placed in question is the element of predictability and regularity that it seems to entail and who that is meant to benefit. Do I have the right to change my mind, to withdraw my consent, to unpublish? If the right to privacy is based on the expectation of privacy, does it make sense to include the right to be forgotten or erratic as part of one's right? Or are these rights absurdities that only come into existence with the rise of a specific kind of remembering and monitoring, the kind of remembering and monitoring done by large-scale data pipelines and the services and systems that surface and manipulate the data that is in them? I hold that many of the issues raised in contemporary deliberations on the right to privacy are nonsensical and based on the confusion introduced the moment we subjected our private personhood to analysis and rational feedback automatically responding to emissions promoting one set of expectations over and against another. Offering up the expectations of human beings to analysis may have been done with the best of intentions, but it has threatened us with servitude to a system of understanding that does not wish to promote, protect, and defend our liberty as private persons socially recognized as equal subjects of rights and responsibilities. And indeed, the assertion of that subject may amount to subjection to that system of understanding.

Privacy appeared to be the remainder of the telemetric, but the telemetric has been discovered there. And that discovery showed that the personhood of the person is residual in the formation of that expectation that lies at the seat of the subject of all rights. The constitution of the subject has not itself been protected by the law. Or such is the concern as we pursue that subject into the realm where its protection most immediately matters to itself: the phenomenological domain, the domain of personhood from the perspective of the person.

Phenomenology

"Phenomenology opposes telemetry. It entails being gathered with what appears, being enjoined to it, grouped along with it. Appearing with it, integrated with it as part and parcel of it. Near to it. Close by with it. Not measuring it, but with it. Not from far away but from close up."
–Further Investigations of a Dog

Chapter I: Private Languages: Pleasure and Pain

By phenomenology, I am not rekindling the charge to "the things themselves" or making some quasi-psychological movement in the direction of an intentionality that develops notions of expectation within the germinal acts of consciousness. On my reading, phenomenology begins with Hegel and is an investigation into things themselves only to the extent that it is a set of elaborate attempts to disclose the mediation involved in their presentation. In the Hegelian phenomenological project, forms of experience undergo progressive failures specifically because their conceptualization is not pure, not idealizable, or clearly articulatable, but instead rests upon a schism or division. And this division propels the various shapes of experience forward into related forms that maintain something of the previous failed state while contributing something additional that attempts to recuperate and adjust to the tensions introduced by the failure. Consciousness of objects fails to obtain the essential nature of human experience and so turns that directedness back on itself in a self-related family of forms called self-consciousness in which the object of experience becomes the experiencing subject itself. The effort to capture a complete view of experience in self-consciousness fails in so far as self-conscious individuals encounter each other and force a movement to an integrated form of experience in which other self-conscious beings are part and parcel of those forms. And so on. My aim is not a complete investigation into the works of Hegel in support of this reading but starts from the presupposition that experience is offset by mediation. The tensions at work in this mediation cause changes in the notions we have of experience and thus propel us into new and different forms of it. The history of phenomenology is a history of recent philosophy. Although Hegel suggests that he is interested in human experience, his work is remarkably ahistorical to that effect. There is no simple correspondence between the history we know as human civilization and the pathway described in the *Phenomenology of Spirit*. Instead, he asserts that it is a journey of the Idea, of the notion, and not meant to be the concrete journey of human beings. I will join the chorus and admit that I do not know what he means by that unless he is trying to submit human historical being to a pattern and the history of human experience and the concomitant history of human reflection on that experience –sometimes called philosophy– to a maneuver of some kind. If he is combining those two things (pattern and maneuver) together in an attempt at producing a conceptual analysis that can be studied and then used to feed back into future political or social engineering and governmental processes, then maybe his project makes sense. As a telemetric methodology, I can see what he might be doing, otherwise it is too abstract to associate with actual human goings on.

Arguably, Marx had a similar reaction. He may have learned a great deal through the study of the logical constructs that Hegel used to describe the drive of history, this dialectical logic that propels history, but the details of the story seem to have been lost on him too. He does not exactly turn Hegel on his head so much as dismiss the idea- or notion-driven phenomenology in favor of a materialist one. And yes, I do mean to suggest that Marx can be read as having written a phenomenology of the material in much the same sense that Hegel has written a phenomenology of spirit, idea, or notion. Although Marx's forms of experience do not appear to focus on that of the human being properly understood, the notions of material determinism that we get from Marx suggest that, at least in the later work, he would embrace such an analysis. The thing itself is mediated at every phase of human life and undergoes radical changes throughout the course of time: the proletarian and the commodity, the bourgeois and the forms of value, these are material conditions and as such they are or produce concomitant forms of experience. Look around you, are there things that are not commodities? Even the woods where you may have come to sit and read are part of a public park or national forest paid for by productive forces in one fashion or another. And you, can you see those things independently of your material conditions and constitution? Are you doing your job as you read this work? Or are you taking a moment's respite to recharge yourself for tomorrow's travails? Mediation still plays a role in Marx, something lies between us and the things of our experience, and this something in the middle is disruptive of any attempt to reliably articulate the relation itself. It conflates this, introduces tensions, makes it possible for conflicts to arise in understanding and in the description or analysis of the relations and practices under consideration. The contradictions where Marx focuses his gaze, that he points to as necessary stimulants to cause progressive movements in the material order of things, are possible because of the mediation involved in the relations that are essential to human experience and material production. For both Hegel and Marx, the engine of history lies in the tensions made possible by the mediation spreading out between subject and object, between experiencing being and experienced being, and so on, with the very terms of the 'between' based on historical stages displaying the advances of the material conditions of human life or the journey of spirit as it comes to know itself.

With this emphasis, Nietzsche can be read as a contributor to the phenomenological project. He relished the emphasis on such mediation, on such tensions and forces lying between the things that are, although he railed against Hegel's efforts to reduce mediation to a systematic, and hence unmediated or controlled, dynamic of experience. It is not that Nietzsche embraced the chaos of the relations at work in all acts of experience, but that he thought the acts of momentary harnessing of those relations into temporary and extraordinary configurations were the most notable and valuable component in human existence. All human life was a struggle against mediating forces and humans formed associations and invented philosophies in willy-nilly form to combat the mess cast as a conflagration of forces. Building a stable structure to weather the storm was the primary driving force of human beings, and whether it was done at large scale by organizations such as Christianity or Germany was decidedly different from efforts similar in kind performed by individuals. He had a definite preference for individuals who seemingly accomplished on their own what Christianity needed a mass movement over many generations to achieve. And the reasoning seems to be based on the sheer difficulty of an individual in a short life time of experience breaking through the constraints and order of the traditions and the cultures to achieve this. The value of individual efforts was often justified with a variety of reasons resembling kettle logic, but the pattern was there: creating an order of understanding, driving and fashioning a way of experiencing the world, was itself the highest achievement of a human being. The harder it was to do, the more radical the break needed to establish such forms of life, then the more valuable the example was to Nietzsche the philosopher and phenomenologist. Since his primary goal as a thinker was to articulate ways of experiencing the world that he could realize, and which differed widely from anything that had previously been apprehended or presented in the human archives, the singular existence was of primary importance.

Husserl, rightly so, bears the role of phenomenology's self-consciousness. My reading of him underplays his emphasis on intentionality. Rather, I hold that the focus on mediation that we get from Hegel and see at work in Marx and Nietzsche, is already the germ of inextricable relatedness of subject and object, the seat of apprehension and things apprehended. And there is probably some scholarly paper that could be written to describe the origin of this mediation in the appropriation of the Kantian Critique by the German tradition leading up to Husserl. No, intentionality is not what I find interesting and earth shattering in the Husserlian project, not even the maneuver of refining focus on these matters brought to bear by the *epoche*. The great insight that Husserl provided was the exposure of the experience of mediation as itself a project with a clear focus, as itself the methodology of investigation that philosophers had been looking for and attempting to refine for the century or so before him. Husserl produced analysis of the methodology, gave it a name, and set out a list of principles and directives that would provide the guiding light of investigations long into the future. In experience, we were not merely apprehending the thing in front of us, we were filling out details, we were projecting wholes where only parts were accessible, and we were expecting aspects and facets that had not yet revealed themselves. There were temporal mechanisms involved in the most immediate of perceptions and observations, there were preconditions at work, angles that were exposed. The simple nature of perception was not nearly so simple as some had thought. The Kantian notion of an active and constructing subject involved in the constitutive acts of apprehension was his focus, with an aim of fleshing out the details of how exactly human beings inhabited space, time, and the variety of 'categories' of experience when they underwent involvements with objects and things. And Husserl, throughout the course of his life, does not appear to veer in the least from these fundamental questions. He keeps starting over, going back to the basics of the investigation, and each time he begins again he changes his vision, somewhat, noticing something more, something different at work in the realized relation between thing and person or between object and subject. This constant reset may have been what future phenomenologists were most influenced by in his work. In constantly resetting the project, he was opening the domain of the crucial questions concerning the role of social involvements, of time and space, and of categories of understanding that may have been malleable and variable throughout history. He introduced the lifeworld and its impact on the interplay between background and foreground, and so on, to highlight the boundary condition stretching between the phenomenologist and the phenomena. The questions that he raised again and again made an entire realm of investigation questionable to generations of thinkers who came after him.

Heidegger's contribution was the delineation of an existential pre-understanding that displayed the circular pattern at work in human experience within mediated relations to things. Throughout the previous history of the relevant movement, there were repeated suggestions of a fundamental circularity in the relations of human beings with their world.

The dialectical notion of the *aufheben* and the Nietzschean test of the eternal recurrence were ways of trying to give voice to this nagging sense that something that was already done was driving what was about to happen, but the means of presenting this insurgent sense was usually cryptic or merely partial. Thrown projection, or the notion that experience is always situated in a set of preconditions or a prior understanding that orients it and yields the possible projections that it might have, was the ongoing focus of *Being and Time* where the intent was to show a primordial pattern at work in experience and having this circular movement. Heidegger took his point of origin from the structure of posing a question; a strange question, as it turns out, because it asked after that which normally provides the structure of questioning. He points out that questions are raised only in those cases where the questions moved in the direction of something that has already been made questionable. The pre-understanding was the prior questionableness, the prior evocation of the question, which directs all subsequent efforts at acquiring answers to that question. Before we even raise a question, the matter must have become questionable suggesting that we may already be directed toward the answer. Human experience, it would so seem, is never a purely disinterested and telemetric process. Heidegger's early treatise is filled with a dynamic that sways back and forth between what is near at hand and what is far away, displaying insights on how the things that are closest may often be the furthest from us. The phenomena that are so close to what we are that we cannot help but experience the world through them, are furthest away from us precisely because it is nearly impossible to look at them. To oversimplify this insight, we might suggest that no one can ever see their own eyes except by using a device of some kind to reflect their image back to oneself; and –in that case– one is not seeing the eye, but a reflection of it. And it takes effort and investigation to retrace the mechanisms at work in such reflection and to achieve an understanding of the eye as that which does the seeing rather than as some determinate thing which is being seen or being looked at in reflection. These two insights were closely associated in the hermeneutical project. The pre-understanding, that human structure of experience that guided our efforts at interacting with and questioning the world, was the closest thing of all. And because it was so close, it was incredibly difficult, if not impossible, to see. The methods of trying to see it had to become the subject of a critical pursuit to ensure that the thing itself, if you will, comes into view without the skewing that the method of viewing might bring. And that, as it turns out, is extremely difficult. One possible reading of some of Heidegger's later work follows this pattern through to its most extreme. As we see in his writings in *On the Way to Language*, Heidegger strikes out to try to get back to that which orients our travels in the first place: language. He wanted to turn around and see what is doing the seeing, to understand what was doing the understanding and it is possible to produce a reading of all his work that suggests this 'turning' is always the be-all and end-all of his investigations (*The Poetics of Resistance,* for example).

Although many great thinkers produced work to extend the project of phenomenology as it had come to fruition following Husserl and Heidegger, my reading of it does not follow this trail. That is, what characterized the focus thus far on Hegel, Marx, Nietzsche, Husserl, and Heidegger is that each in turn introduced something radically new in the orientation toward the problems of mediation and the relation between subject and object. Although thinkers like Scheler, Levinas, Gadamer, Merleau-Ponty, and Sartre were great at exploring the depths of those areas of investigation begun by Husserl and Heidegger, I do not see them introducing any radical disjuncture into the tradition as I have cast it. This may be an argument against my spin. So be it. As strange as it may seem to build a tradition on disjuncture, that is exactly what I am attempting to do, since I do not think it makes sense to speak of a coherent and linear tradition of 'mediation'. The efforts to unearth the rudiments of mediation are radical departures from each other. And it is the radical breaks that make the movement. The hermeneutical incision provided by Heidegger over and against Husserl was a radical reorientation of the project. For testimony one need only witness Husserl's own responses to *Being and Time* and the profound reaction it seems to have been. The new mechanism and patterns at work in *Crisis* are a huge leap forward in Husserl's thinking. Arguably, his best came from the dire battle he was forced to wage against the upstart Heidegger.

The next radical contributor to the tradition that I am calling out may be even more of a surprise, for it is Habermas, a thinker rarely associated with the phenomenological, let alone the hermeneutical phenomenological, tradition. No doubt this is in part due to his own remarks in direct contrast to that tradition. Communicative Action, however, deepens the consideration of the pattern at work in a pre-understanding that projects meaningfulness. The basis of the pre-understanding is the collaborative action of human beings collected in social settings. We are born collaborating and the requirements of that collaboration amount to its aims: the means are the ends. Collaboration is always a projection toward some state, toward "understanding". And the complexity and needs of any collaboration that aims at reaching understanding forces rationalization procedures upon us as part and parcel of our social organization. We must develop the relations within which we collaborate to improve the outcomes of that collaboration, the increasingly codified and conceptualized relations that become an increasingly sophisticated set of preconditions operative in human interactions. This naturally spawns a detailed interest in language and how it works as part of this pre-understanding directed toward reaching full-fledged critical understanding. The types and the nature of communication, all play significant roles in how collaboration achieves greater rational forms and pushes human action toward different structures of organization and orientation. Habermas proposes some form of agreement or consensus as the target of this action, although it may be that this target is merely an idealization of the process since the notion of understanding each other, which sometimes seems to be equated with consensus, does not seem to require anything like agreement. The achievement of understanding may be spun in decidedly hermeneutical phenomenological terms and so may suggest something like the inhabiting of a mutually recognized world. Collaboration in that sense aims at constructing a shared world, but this need not suggest that the outcome is an idyllic utopia of perfectly accommodating agents who have banded together through common interests. Rather, the point –understood within a hermeneutical phenomenology– is more formal than concrete. Collaborative preconditions suggest that human agents are in the same world, working and acting in ways that are interrelated. Through the process of clarifying and developing the form of this collaboration, higher organizational constructs are achieved. On my reading, these constructs are not necessarily free of coercion and imbalance. The fact that agents act in concert with each other within a shared world that they are building together should not be presupposed to indicate that all participate in the same interests and that all are equally and justly treated by the world that they are building. Make no mistake, even in conditions characterized by injustice and inequality, all are encumbered with that organizational world. If the world that we make is unjust, then we all live together in an unjust world. And that, I take it, is the point of this projection toward achieving understanding. One option available to human agents living under such conditions is to comprehend the state of things and accept their place in it, this is a form of communicative action. Another option would be to revolt. This is another form. Both equally display a mutual understanding of the conditions of organization that have been achieved in the collaborative actions of the members of the society.

And this is where the next major figure in my highly selective history of phenomenology comes into play, Michel Foucault. Foucault also identifies the Heidegger-inspired pattern of the preconditions that underly all activity aiming toward a projected set of goals and the practices that realize, or attempt to realize, them. Foucault's primary concern is to reawaken the Nietzschean emphasis on the role that power dynamics and forces of relationships play in these practices. Human action is carried out in a social context with historical conditions at work, and these contexts and conditions are ripe with power dynamics that act persistently on the conditions themselves. We do not collaborate and project toward the shared vision of the world in a purified vacuum. Rather, there are inequalities and imbalances, there are forces at work, in all these acts and points of contact between us that signify large scale networks of impact, of change, and of modifications to goals. These modifications are for the sake of some specialized configuration that best expresses the aims of a powerful set of interests at work in the social conditions or institutions that are developing or rationalizing over the course of time. Foucault sees that these forces are responsible for constructing some of the nearest things that there are, as Heidegger's pattern suggested. That is, the ways in which human beings are subjected to these dynamics and forces produce a point of view regarding these dynamics and forces. The seat of authority, or the perceptual authority of our experience, amounts to a product of these powerful forces. We therefore see the world through the lenses that these forces have crafted. We cannot extract ourselves from the historical context set forth by our collaborative organizations. Instead, we are always reaching out toward an understanding of the world and our place in it from the point of view of a person crafted by relations of powerful forces increasing over time and while the institutions of human collaboration are growing their influence over the world where we live.

For both Foucault and Habermas, this condition requires the incision of critical hyper-vigilance into the hermeneutical project to interweave a newly delineated love of wisdom into the basic existential facets of the understanding. To the extent that the private person is a lover of wisdom (which we will later juxtapose with telemetrically understood philosophy), their phenomenology must be a constant struggle between the coercive forces at work in the world and the liberating critical practices of the agents involved. I read them as assailants on the notion of 'authenticity', aiming at deconstructing it into the very dynamic at work in the acts of understanding situated amid constantly articulating power dynamics that take place in the space between people where their organizations and institutions lie.

This is the presupposed general and all too brief meaning I lend to the term 'phenomenology'. It is aimed at eliciting a deeper insight into the nature of human experience and how it works, and it is meant to function in a context that is deeply social in origin. We are not pre-made souls let loose upon the earth in some ahistorical and asocial fashion. Rather, we are born into social conditions and are raised alongside and in the middle of those conditions. They hold sway over us as we grow and learn. The conditions help form our character and in so doing help form the lens through which we see the world. And this lens is so close to us that it is extremely difficult, if not impossible, to get a good look at it. The mechanisms of the hermeneutical circle, of the orientation back toward the origin that presupposes the investigations we are undertaking, is a carefully structured approach to situating the human being within the world. And what is more, it is a way to try and get a deeper sense of where our expectations come from, what we implicitly understand, where we are starting from, where we see ourselves headed, and what we think we can rely on in our efforts to get there. The purpose of a deeper investigation into the phenomenological basis of human experience is to evoke a greater sense of who we are as living experiencing beings. The be-all and end-all of the debates on privacy, triggered by the telemetric rise to dominance of mass surveillance in the guise of computing power and distributed service providers offering solutions to every human problem, was an insight into the notion of an expectation and what human beings can reasonably expect for themselves. We saw in the discussion of that history what our social situation recognizes as a reasonable expectation for ourselves, for our privacy, and for our integrity as private persons. Since we have come to be caught in the complexity of the legal confusion that this notion raised, we can take a step back and begin an investigation into the basis of so-called private human experience to see what there is of any such expectations there and whether efforts at preserving the expectation of privacy (even in the complete absence of anything like privacy) can be a meaningful contribution to safeguarding the humanity of human beings: the private person as the socially recognized subject of all rights.

In the first section of this yarn, we hijacked the three-fold dialectic and then punched a hole in it to introduce the telemetric as a peculiar pattern of movement: the emission and structure of consciousness, so to speak, the gathering and analysis of self-consciousness, and then the feedback and optimization of reason. Just when it seemed that the pattern was the highest form of order and organizational completeness, however, privacy and the private person was introduced as the residual of the processes and patterns of the foregoing movement. There is something that evades the initial movement, something raised to an area of concern, but which cannot be constrained by the system that attempts to capture it. The mechanisms of emission, analysis, and feedback are aimed at capturing the private person in all their nuances and characteristics, but the recuperation fails, something evades the process, something – expectation– exceeds the capacity to capture it. In this chapter we are turning to the being that expects, the private person that is at risk. And the initial concern is whether there is any such thing as an 'emission' of the private. The upshot of the phenomenological, I take it, is that the private person is in the world with fluid boundaries, built from the outside in. And, in a sense that I have already suggested, built by a stream of invasions of privacy, to speak loosely. That is, the process of growing up, despite the norms and the rules, is a constant invasion of our personhood by those around us, by the institutions around us, by the organizations of action in the world: the hospital or doula, parents, the issuing of the birth certificate, the car we ride in to get home, and the manufactured items that fill out the nursery. A world replete with artifacts constructed by other agents and training us how and what to see, what possibilities lie in store for us, where we might go, how we might perambulate, and the clothing that we fit upon ourselves. Crafted and guided, regularly and repeatedly. Each of us a spun yarn. What possible expectations could such a one as a human being possibly have? Is there any hope for the private person? Is there such a thing as a subject or rights at all, or is it all just a fiction we tell in the form of fairy tales that we never quite get past and construct large-scale fantasy industries to continue to promote and safeguard? The execution of all this is yet to be seen, but I hold that phenomenology is the key. Not to understanding how the brain works, or how mental states are or are not just brain states or functional states or epiphenomena or whatever you like. No, phenomenology is what it is like to be you, what you are drifting through even as we progress along here. It is the preconditions and preconceptions at work in the meaningful apparitions that populate human life. Where do they come from and what do they mean, where are they taking us? And so on. If privacy is worth anything at all, it must be worth something phenomenologically. If it promotes well-being and 'happiness' or the 'general welfare', it must do so in so far as the experience of a human being, and each in common, is promoted in/as/through it.

The Private Language Argument

Why do babies cry? Crying vs non-crying is a primary indicator to the new parent for discovering whether the baby needs anything or not. A crying baby needs something, needs to be attended to until such cause of the outburst can be determined and resolved. Once the cause is resolved, the baby will stop crying and the parent can relax their attention. Amid the complexity of human communication, all the ways in which people articulate themselves and their states of mind, their intentions and their interests, it all begins with crying. It is a common story, albeit not universal, that a new born will cry until swaddled and placed next to that familiar beating heart. This is not a private moment for the new born, birth is not done by oneself. We are born with at least one other person nearby. And more likely, a half dozen. There are always at least two people involved in a birth. And two is highly undesirable as it leaves no room for error, no fallback plan in case the process overwhelms the woman delivering. It is strenuous and demanding to give birth, ideally birth begins with help. An institution is born, the institution of birthing babies, of attending to women who are giving birth. Ideally the attendant will be knowledgeable about the process and experienced in the likely pitfalls and the common responses to make when they arise.

Hegel's word '*aufheben*' is complicated and it lies at the core of his notion of the dialectic of the spirit. Generally, the sense implies a contrary pair of meanings: to keep and to leave behind. The dialectic is meant to be a movement that both maintains something of the previous stage and yet leaves that stage behind. Something in the stage has failed, has come upon a contradiction that prevents its continuation and so it must be overcome, but in overcoming it something has been learned, something has been achieved that is worth retaining as one moves forward into the next stage. What does this have to do with crying babies newly born? English translators of Hegel have often struggled at how to translate this German word '*aufheben*'. There is a natural notion, albeit not a single word, that does capture this twofold sense that the German expresses in the core of the dialectic: to be born. To be born means both to be carried and to be dropped, at the same time. We bear a burden by carrying it, we bear a child by dropping it. And each of us begins in this moment of birth: for some it may have been a surgical procedure, for others a medical procedure undertaken with anesthetic. Some may have been born, emerged into this world, using natural methodologies developed over time. Even natural child birth has a long history and, in some societies, is populated with trained professionals to manage or help manage the process. This moment in which we leave the womb behind and emerge into the next stage of our lives is heavily institutionalized and that does not mean that it is filled with doctors in hospital scrubs, although that may be one of the options that some have endured. Humans give birth differently from most other animals in that giving birth is a social concern, the processes and procedures that are used to manage the birthing process are themselves historical and involve educated persons with experience in the field. Even a man and a woman who are both trained professionals in the methods and mechanisms of birthing babies under high risk circumstances are going to engage the help of a third (and fourth, fifth, or sixth) to bring their little darling into the world.

Human beings, by and large, are not born alone and that initial emission or yowl that spells the human being's entrance into the world is not pure, it is not a purely expressive emission solely in relationship to some state of mind or state of being that the new born cannot help but articulate shortly after leaving the birth canal. Rather, that howl is an act of communication, it speaks to the surrounding helpers and professionals who are attending to the birth, it lets them know all kinds of things about what they need to do next. The experienced listener present in that moment knows what to do, knows what the baby means when they cry out in this or that way at such and such a phase in the oft-repeated process. Or should they fail to cry out. Even that might have a meaning that needs to be attended to directly in those moments just after birth. And really the moment of birth is too late to enter a relationship with these institutions. Leaving aside cases where the couple planning to have a child need larger scale assistance in doing so, pregnancy itself is already involved in institutions. Pregnant women alone and left to fend for themselves throughout the pregnancy are likely to run into problems along the way, the baby itself is more likely to develop problems without proper neo-natal care and preparation. There are books and classes, organizations and consultants. All kinds of resources are available to the mother-to-be to assist her in the process of gestation, to improve its outcome, to lessen the risks of some possible problems or side effects that may result. Not only are we not born alone, we are made. And although this making involves very powerful proximate causes in a woman's body as she works with the growing human to realize that process, there is a lot more to it than just that. As soon as that woman speaks to another woman about her experiences and tries to learn something from them, or watches a series of instructive videos or takes a class at the local community center or reads a book written by someone who has been working with pregnant women for 20 years, she is participating in a much larger community of contributors who are helping establish the process as safely and optimally as possible.

The best contributors to such communities and institutions would likely be keeping an eye on how their advice and instruction is working. They would want to learn about the outcomes for women who have been following their advice and they would want to be informed of the outcomes for women who did not follow their advice or followed it only by accident or partially. This may be called an institution only because there is a growing rationalization at work in the history of these processes. People training to be better at it, getting feedback, learning about the nuances that are introduced by contemporary conditions and variations: what special needs does the birthing process, or the pregnancy process, introduce when the mother is older, does not have access to certain beneficial support systems, or any of a thousand other factors that might play a role in the process. And 'process' need not be technologized or industrialized. The women of the village bring all their experience learned throughout their lives and from their mothers to the birthing room when they help one of their sisters or neighbors. And there may be women who deliver their babies themselves at the river side before going back to whatever they were doing before, but such conditions have become truly exceptional. And likely such a woman was only capable of this because her mother or the women in her community were all doing things like this when giving birth; so, she was made aware of the process and how to accomplish it once the need arose. Likely, it was not her first child that was delivered this way and it was only because she had done it before that she was able to manage such an extraordinary practice by herself. Even this is a developed and historical, a trained, outcome where the process itself has been subjected to human understanding and know-how and developed for the sake of repeated practice. And this is what I have been pointing to throughout this work as the process of rationalization, of uncovering all of its many attributes and properties so that its range of meanings can be comprehended and associated with other experiences deemed to be like it or similar enough to it to benefit from comparable actions and responses to whatever may come up during its execution. Each human baby is dropped right into the middle of all that. Our first experience in this world is not wholly and completely our own, it is a heavily rationalized experience, a heavily socialized experience that introduces us immediately into the world of others. When we say that a baby comes into the world, that is exactly what we mean. That the baby arrives into human life, human processes, human mechanisms of delivery and care.

> So in the end when one is doing philosophy one gets to the point where one would like just to emit an inarticulate sound. —But such a sound is

> an expression only as it occurs in a particular language game... (Wittgenstein, 2001 p. 261)

This may capture what I have been trying to describe except for one key part of the sentence: "when one is doing philosophy". A yowling baby can be described as emitting inarticulate sounds. When one is doing philosophy does one get to the point where one would like to be a yowling baby? Taking up philosophy is presumably something that a baby does not do, a yowling baby is not doing philosophy. The philosophical point of view is not at work in the experience itself, not submerged by it. The baby does not theorize, is not wondering about its state, it is most purely exhibiting that state, carried away with it, unable to get anything like a philosophical point of view with respect to that state which it finds itself in immediately upon being born. The baby may be very interesting to the philosopher or to the philosophical line of thinking, or to people, grown and educated, who are doing philosophy, but the baby itself is not able to do philosophy. Philosophy requires a modicum of distance to be done properly. One does not do philosophy when one finds oneself in the thick of things, that must come later. The fly is too much trapped in the fly bottle, too absorbed in the experience of being in the fly bottle to be able to get out of it, the fly needs to be shown the way. And that adds up to philosophy. Baby phenomenology might differ radically from baby philosophy. It may be puzzling to try to grasp how the one can be a subset of the other: is phenomenology a 'branch' or 'kind' of philosophy or is it something else entirely? Is philosophy what a human being does specifically when that person is not in the middle of an experience, but only at some distance removed from it? And is phenomenology descriptive of being in the midst itself? Or is there yet a third option where phenomenology is the philosophy of trying to –at a distance– describe the state of being in the midst and up close to things as one experiences them? Another way to phrase this, with a little more leeway than if one were trying to crack open the mysteries of baby experience, is whether a writing performed in the investigation of the topic of birth is subject to a philosophy of writing on such a topic, a phenomenology of writing on such a topic, and the experience of writing on such a topic? If so, what is the relationship between these three things? Consider for example that a baby's experience might be opaque to adults, but a baby's phenomenology need not be. And baby phenomenology may be distinct from doing philosophy involving baby states. To begin to characterize the developmental capabilities of a baby, the complexity of experience based on visual development, or neurophysiology shortly after birth seem like reasonable fields of study. And this baby phenomenology differs from a more distanced attempt at baby philosophy, of considerations that aim to situate those characteristics into some larger set of concerns about human beings and the way in which they live and engage in community efforts, and so on.

To turn to the orienting question above, why do babies cry? I wonder if a baby understands why it is crying, realizes the meaning of its inarticulate emission of sound. If an experienced mother approaches a crying baby and rapidly determines that it needs to be fed, is not it possible that this experienced mother has a better sense of why the baby is crying than even the baby itself has? The baby likely is sensing things and those sensations are the emission of those yowls, are the discomfort that only actualizes itself in a crying out, an appeal for attention to someone more experienced who may happen to be nearby. I'm suggesting that such a mother is more adept at baby phenomenology than the baby itself is. She is not doing philosophy. These are not disinterested reactions or insights, very much to the contrary. The child's mother is experiencing motherliness, motherhood, and in that experience is demonstrating an adept capability to undertake baby phenomenology. She unpacks and understands the workings of her baby, more so with each passing day no doubt. She comes to learn that it cries in various ways in response to different kinds of stimuli. Her baby is not some remote thing or object of concern for her, she loves her baby, is completely and totally honed to its actions and reactions, its changes and gestures, each of its orientations are evaluated, compared, and measured relative to yesterday's behavior or to the last baby's behavior if this is her second child. Or she is making comparisons to the descriptions and discussions that were explained in the classes she took or the books she read or the conversations she had with her mother or her friends who are experienced mothers. And this is not the sole property of mothers. Fathers attentive to such details and caring for the baby through its development day in and day out are capable of this same degree of sophistication in their baby phenomenology, their complete and utter understanding of the meaningfulness of baby agency and the proper responses and reactions to that agency. It is time to feed the baby, it is time to change the baby, time to play with it, to put it down for the night or afternoon nap.

> The paradox disappears only if we make a radical break with the idea that language always functions in one way, always serves the same purpose: to convey thoughts—which may be about houses, pains, good and evil, or anything else you please. (Wittgenstein, 2001 p. 304)

The yawp that is the basic communicative gesture of babyhood is not the expression of a thought: I am hungry now mother, please feed me. The emission itself is raw and without understanding, without clear and determinate expectations. The baby does not know where the food comes from, the baby does not know food is food. There is this ague, this emptiness, this terrifying and gut-wrenching assault, and as distinctive as this attack on the body may be, the baby is helpless to comprehend its meaning, its cause, or any possible solution to it. The baby has very little to teach its mother here. That does not mean that the baby is not sending rich signals for interpretation and examination. A parent evaluates these cries in their baby phenomenology and can do so because the cries themselves are communicative, are carriers of meaning even if there is no germ of intent embedded in those communicative utterances.

> In the case of pain I believe that I can give myself a private exhibition of the difference. But I can give anyone an exhibition of the difference between a broken and an unbroken tooth. —But for the private exhibition you don't have to give yourself actual pain; it is enough to *imagine* it—for instance, you screw up your face a bit. And do you know that what you are giving yourself this exhibition of is pain and not, for example, a facial expression? And how do you know what you are to give yourself an exhibition of before you do it? This *private* exhibition is an illusion. (Wittgenstein, 2001 p. 311)

> I can exhibit pain, as I exhibit red, and as I exhibit straight and crooked and trees and stones. —*That* is what we *call* 'exhibiting'. (Wittgenstein, 2001 p. 313)

Many have suggested that this is a point about language. Here I am trying to filter Wittgenstein's remarks as points about yawping, points about the emission of a cry for the sake of exhibiting oneself, of making oneself known. That is already to misspeak. The cry is not for the sake of anything at all, it does not have an intent. Rather it is exhibition as such, the pure and raw emission of one's state of being. Now, it may be that somewhere in the lineage of genes leading up to the advent of human beings there was a selection process and offspring that emitted their states were more likely to survive than offspring that did not emit their states. Likewise, parents that can comprehend the nuances and variations in these emissions, to provide meaningfulness and intent to them, were more likely to raise those offspring to maturity. That configuration of traits and properties was better disposed to surviving up to the point of further reproduction. There is an 'intent' in these cries and their concomitant interpretation and response, but the intent did not live inside the brain or in the action construct of a single being, a single agent or person who was seeking to achieve some end or acting for the sake of something or other. Rather, the intent construct was already an interpersonal construct, the fitting together of emission and interpretation and analysis. The baby and the parent formed a kind of meaning machine in this dynamic and the intent that displayed the meaning was between the two of them, something other than either one of them, but something that came to be through contribution and collaboration by both in the performance of their respective actions: crying and reacting to the cry with understanding. This process is not always smooth and clear, it may take multiple efforts, adjustments, the two may need to vary their actions to accommodate the failures and successes that might take place in the process of figuring out what the right actions are, what the correct form of emission and response may be.

Parent and child would be capable of a micro rationalization of this process even if they did not have access to larger institutions that provided historical and broader social insights into the nature of the relationship. The trial and error approach that the two may use enacts a form of communication, it is what we mean by communication. And it can be rudimentary and still be capable of making progress. Even if the parent is a proto parent and the baby a proto baby, their interplay in figuring out the rules of request and response would still be capable of increased efficiency and improvement. The simple fact that they are involved in a relationship enables them to make adjustments that amount to greater conceptual depth in the actions they are performing. Their relationship puts each of them in a position to learn and modify their response based on the previous results that have been achieved. The instigator of greater optimization would likely be the parent. The baby is not likely to form external community bonds. The forms of communication and experience available to the baby would not permit a new born to reach out to other newborns for tips and tricks aiming at improving the communication channel with their parent. It seems that the initiation in these sorts of complex improvements through institutional involvement would have to come from the parent. That is not to say that the baby would have no initiative at all in the relationship. As the parent could introduce greater sophistication and complexity into their reactions to the baby's actions, the baby would adjust, would sense the differing inputs, would take note in some sense of the improved communication channels and –as a result– might be capable of clearer forms of request and communication. Improved rationalization means the baby phenomenology of any given parent may improve with the increased sophistication of the institution, but that does not mean that babies themselves are confined to the same primitive state regardless of what point we are in human development and history. Babies 10,000 years ago may not have had as many variable ways of suggesting discomfort. This would be false for the newborn, but with each passing day in the baby's life, it might be the case that babies raised by parents familiar with more rational parenting institutions, may learn nuances of communication that babies could not have learned without such an advantage. This is pure hypothesis, I would yield to the data in these cases, but it may be that babies who were swaddled in cotton are capable of more subtle forms of discomfort than babies who were swaddled in an animal skin. And the subtle variation in this discomfort would likely emit itself in the exhibitions of sense that such babies might make and might have. This might suggest a greater wealth of possible experience for the parents of such babies and thus improvements and advances in the institutions providing captured learning in the rationalization of the process of doing baby phenomenology, of understanding one's own baby's response to things and what sorts of needs or meanings may best complete those emissions and exhibitions.

Babies, therefore, neither understand nor know why they cry, they are not the authorities here, they are not the ones who can complete the intention of the act, to fulfill its promise, to clearly and distinctly articulate its meaning. Baby philosophers on their own, cannot be expert in this matter either. At least, the distanced frame of mind that tries to contextualize baby behavior independently of a response to it, is not able to know anything at all. The philosophy of babyhood can make progress, can provide interesting insights and conjectures if the philosopher is aware of the parent's capacity for baby phenomenology. And that position is a position of understanding where the respondent is actively involved with the baby, responding to its actions and behavior, learning from it, and seeking out the input of the philosophers for the sake of improving their reactions as part of a communicative relationship between baby and parent. The point here is not that philosophers only know whereas parents understand in all matters regarding baby. Rather, the point is that understanding and interpretation, the realization of intent for the sake of proper response and request, requires that the actor be up close, be very near to the situation. And this nearness is something I am specifically calling out as a quality of understanding rather than knowledge, of phenomenology rather than philosophy, where 'philosophy' functions as a term for describing the broadest possible approaches of science, experimentation, and analysis with the resulting feedback being carried out by a disinterested observer. Phenomenology is, to reduce this all to a catch phrase, philosophy done by someone who is an interested participant, a member of the actual relationship that is the matter for interpretation and analysis. When we stand far back and try to figure out how it all fits together, we are behaving as natural philosophers, in that old sense of the word that Newton used. When we are responding to the world that we find close by, that we are deeply and intricately interested in, we are lovers of wisdom and doing phenomenology.

Knowing why there is crying, therefore, develops throughout the process of interaction between parent and child. The parent comes to understand why the baby is crying and responds to it. In the process, through repetition and acclimation, the baby comes to understand why it is crying too. And once the child has come to understand why it is crying, it can come to know why it is crying. This should not be confused with a behavioral description of human action. It is not that the child behaves a specified way and then learns to associate concepts or reasons to that behavior through instruction. Rather, the interpretation of the meaning context of the baby's actions are produced by the parent and reflected to the baby in the parent's response. By feeding it, for example. Or by changing its diaper or bouncing it on their knee. This fleshes out a meaningful set of actions and associations and shows the baby the sets of context and relations that are part and parcel of the exhibited emissions. The crying and the feeding are revealed as a set of associations, and expectations come into existence around this set of associations. The baby begins to form expectations of outcomes based on their actions and so can be said to understand the role and meaning that the actions are having in the interplay between itself and its parent. Understanding, to this extent, does not mark a reflective or explicit and conceptualized apprehension, rather it marks a broadening of the scope of the experience. Not just sensing a grinding pain in the stomach, for example, the infant begins to experience a context of involvements and the stomach pain projects toward a response by the parent. And that response is projected to play a role in the acquisition of food which will address the initial instigation of the action, the pain in the belly. This marks a development, the introduction of greater complexity into the action, and a greater set of connections and meanings. The crying starts to attain sense in the larger context of the parent's attention to the baby, and the forms of communicative back and forth increases between the two throughout the course of their lives together. Only much later, years later, will the baby come to know what is going on here. Before that level of complexity, the child will have already advanced to the point of saying 'mama, I'm hungry' instead of crying. The words will flow as proxy to the context originally established in the understanding by a yowl.

There is no sensation on the one hand, and reference to it on the other. The aboriginal yowling of the hungry baby is not an act of reference looking for words. Rather, it is already an exhibition of the sensation striving to be ingrained in complex sets of relationships and associations, in meaningfulness. When we refer to the so-called 'private language argument' in this segment, what we are referring to is the absence and impossibility of any such language. Language itself is never and can never be a private occurrence, an a-social dispensation of reference between some emitted sounds and some inner sense that is captured or expressed by those sounds. Instead, language is always culled from these contexts of interaction, from these relationships, from an intentionality of projected meaning that spans the distance between members of the communication community. Expectations are not private states of mind, sensations or actions performed in the isolated mind of the individuated human being, rather they form out of our interpersonal involvements. And to the extent that those involvements are institutionalized, derived from historically rationalized mechanisms and methodologies that have been brought to bear on the interaction, the expectation itself is established as the result of a long-standing practice that has historical and social preconditions. The expectation comes ready-made as a late arriving facet in social customs or national institutions that partake of the context in which this development process took place. People who grow up in one geographical location among people who are historically situated in that location are likely to apprehend the characteristics of that past. When we describe the 'thrown projection' of experience, therefore, we are referring to this customary and localized basis of understanding: it is always embedded in a place and a time, in an association with others who have histories and traditions, who have come to the table of interaction with those histories and traditions, and who realize those histories and traditions, complete those histories and traditions, in their interpretive responses to actions. When I refer to the existential precondition of '*mit-sein*' as Heidegger described it, or the role of collaboration as Habermas described it, I am referring to this complex associative context of historical and socially delineated responses to the raw emissions of the actor made in forming a broader scope of meaningfulness in action. Expectation is based in a past that looks forward to a specified future and that expectation is the upshot of this pattern of interaction.

The Private Experience Argument

> Wittgenstein's famous argument against the possibility of subjects following rules for themselves alone, so to speak, has its place here: 'And to think one is obeying a rule is not to obey a rule. Hence it is not possible to obey a rule 'privately': otherwise thinking one was obeying a rule would be the same thing as obeying it.' (202) The point of this consideration is that S cannot be sure whether he is following a rule at all if there is no situation in which his behavior is exposed to critique by T—a critique that is in principle open to consensus. Wittgenstein wants to show that the validity and identity of rules are systematically interconnected. To follow a rule means to follow the same rule in every single case. The identity of the rule in the multiplicity of its realizations does not rest on observable invariants but on the intersubjectivity of its validity. Since rules hold counterfactually, it is possible to criticize rule-governed behavior and to evaluate it as successful or incorrect. Thus two different roles are presupposed for the participants S and T. S has the competence to follow a rule in that he avoids systematic mistakes. T has the competence to judge the rule-governed behavior of S. T's competence to judge presupposes in turn rule competence, for T can undertake the required check only if he can point out to S his mistake and, if necessary, bring about an agreement concerning the correct application of the rule. T then takes over S's role and shows him what he did wrong. Now S takes over the role of a judge and has in turn the possibility of justifying his original behavior by showing T that he has applied the rule incorrectly. Without this possibility of reciprocal criticism and mutual instruction leading to agreement, the identity of rules could not be secured. A rule has to possess validity intersubjectively for at least two subjects if one subject is to be able to follow the rule–that is, the same rule. (Habermas, 1987 p. 17-18)

The rationalization process requires that identification take place communicatively through a persistent feedback between participants, an agreement that is being established. The sameness of the exhibition over time is established through this agreement and the continual contribution to what has been established may result in further dimensionality introduced throughout the time that the participants work at establishing the identity. Hunger is established as collectively understood through the repeated back and forth, the repeated request and response, that goes on between participants. This pattern of establishment is not merely limited to language as such but spans any and all meaningfulness or interpretive experience. The phenomenon that is established through the set of private language arguments is not merely the nonsense or meaninglessness of private language as such, but –in the phenomenological sense– it is the rejection of private experience. The experience of a human being is a process of understanding, it is a stream of events taking place while embodying practices and apprehending the conditions met with there.

Adjustments are made to navigate such conditions in real time for the sake of achieving positions one intended to occupy along the way. This notion of experience is the sophisticated form of that nascent dynamic from the previous section. Something has happened, and someone has experienced it. What has happened and who experienced it? The establishment of the what and the who is the result of these communicative processes that formulate natural kinds of identifiable events. Events with attributes and properties are what is experienced, where those attributes and properties have been socially established as the qualities that they are. And the one who is experiencing those events also has attributes and properties. We come to be who we are in this same constituting process. Not all at once or on the day of our birth, but slowly over the course of thousands or millions of acts involved with others around us. We communicate in micro fashion with other little kids and adults, we gain more and more insights into the involvements that make the world, and what the likely expectations should be that make up those involvements. We are characterized as capable of experience and we undertake to have experiences. And throughout the process of experiencing these worlds, the attribution and properties that we expect or that we discover are given greater clarity and diversity. Development throughout our lives amounts to having formative experience where that is responsible for showing what there is in the world while also showing the agent or actor who they are based on those experiences, based on those involvements in events.

To reject the privacy of language is one thing, but to reject the privacy of experience is quite another. It is not just that reference is mediated in such a way that a community is needed to do it, it is also the case that the full on involvement in repeatable contexts, repeatable types of events, is mediated by the social context within which we find ourselves, into which we are thrown, and from which we are projecting ourselves onto future possible states coming to pass in our actions. My experiences are comprehensible to me because I can identify repeated patterns in those experiences. And those repeated patterns are only accessible because of agreements and role based common senses that can be associated with those patterns and with which I have constant feedback and collaborative responses. This suggests that the truly anomalous cannot be experienced at all. If there is not some sense in which I can process my actions in the world over and against a set of well understood associations and connections, then there is nothing there to be understood, no interpretation that I can make of what is taking place or what has come to pass. I am only able to experience things that human beings have experienced: things that are part and parcel of the commonly construed set of involvements available to human beings in the course of their lives and in the context of their worlds. Anything truly original, truly novel and unlike anything else that has ever been experienced, simply does not exist, cannot be seen, cannot be understood, and cannot achieve anything like an interpretive context in a human realm of apprehension. It may be possible for a yawp-like emission of such novel events to take place, but there would be nothing and no one with whom the person could relay this instance and achieve something like meaningfulness in that instance. Such things might spell out a fleeting momentary disruption of the continuum of your life and maybe even a nagging presence or anxiety that gnaws at you, but there is nothing determinate to be seen, no socially meaningful way to look at it full on, to dig into it or scratch at it so as to reveal some deeper meaning at work in it.

At least not instantly or immediately. It is possible that over the course of time such a sentiment might gain greater understanding and meaning through shared efforts to fumble at it, to gesture toward it, and to stutter about its qualities or aspects. And then others might refine or suggest alternative formulations or characterizations. The community might develop a meaningfulness that begins to make these instances of 'barbaric yawp' into a bona fide human experience that can be discussed, interpreted and then expected or anticipated under certain circumstances based on the presence of certain conditions. It is not that nothing new can ever enter the world, that there is no place for anomaly or revolutionary discovery. Rather, these things are social eruptions that result from a collaborative process of discovery and not some singular and autonomous act of innovation. Even a poet can provide inspirations for all their insights. One might claim, for example, that the private language arguments themselves were something of a radical arrival on the stage of world history through the agency of the lone philosopher, Ludwig Wittgenstein. The notion that humans are not alone with themselves, not locked into a private mind that can be the seat of great authority and autonomy, and that rules are collaborative and result from an agreement that evolves over the course of interaction is not unique to one thinker; rather it is a slowly evolving sense that we find throughout the history of philosophy, with Hume playing a part in response to empiricists before him and then Kant's response to Hume and Hegel's response to Kant and so on and on. The fact that a given thinker did not study some set of other thinkers who laid the foundation for his insights does not mean the suggestions and notions were not circulating, were not of use in prodding his movement in one direction rather than another.

Human beings, for reasons that are clearly part of a social and historical tradition, often take pride in their individuality. And their individuality is often grounded on the sense that they are unique and capable of unique insights and talents. More likely we are all variations on a set of understood themes and patterns. The truly unique qualities that make each of us 'special' are comprised by the simple geometry of our movement through time. As creatures extended in space, we exist at a location. And that location is directed by others in a way that is very clear at the outset of our lives. Born at a specific time and place on the earth's surface, we are exposed to certain conditions specific to that location and time. The conditions form the identity of the agent while offering up the experiences of the conditions and events. This means that just as these experiences take shape and evolve over time and through space, the agent doing the experiencing evolves as well. And develops a mind of its own, so to speak. The meaningfulness that I previously described in experience is the cultivation of a person. And such personhood, such attribution and characterization, is the result of a communicative process. Others help you discover who you are, help you realize the meaning of yourself via a discovery of attributes and properties. And the discovery or invention of an identity is little more than this voyage of communicative interaction with those around you in the conditions where you find yourself. If you are a virtuoso at some musical instrument, it is because you started playing it very young which means somewhere in your earliest states there was some condition that indicated that such a musical instrument was a possibility for you. One was available for your use, given to you or loaned to you, someone may have shown you how to use it and complimented you on the speed with which you learned to play it and so on. The context of involvements gave responses to your actions that resulted in the formation of expectations and those expectations turned into a set of projects and those projects may have become a career or a calling.

And no doubt it is not false to refer to such trajectories as individuated and unique. The uniqueness of these processes is temporal and spatial more than anything. No category or attribute, no property or trait will be unique to such a person. The acquisition story, told broadly enough, will not be unique. Ah, the child prodigy who received an instrument and encouragement from some mentor. This is a common story. Every story, told broadly enough, given enough of a coarse grain, will be a common story. These are all human stories and so they are all common to various human beings. And even if your exact story is not the same as anyone else's exact story, told in the finest detail, then maybe aspects and elements of your story will still be held in common and you yourself will be made of multiple stories and there will be multiple people who have each of those stories in common with you. This is what makes it possible to capture human beings and their statistically aggregable actions into some form of metrics and analytics. Human beings yield to telemetric processes because deep down within us, we are shared creatures, we are common beings, emerging from a common world and a common history, gaining our comprehension of that world from others while alongside them. Built from common parts, from a common gene pool, from a common set of traits, characteristics, and possibilities, we form unique trajectories.

From the perspective of a treatise on privacy per se, however, this suggests dire circumstances. For it is not just that uniqueness and individuality are spun in terms of the communicative interactions of multitudes, of histories and societies within which we come to maturity and figure out who we are. The impact to privacy is even worse than the impact to individuality. As I have said, we can still lay claim to individuality by pointing out that the trajectory of our spatio-temporal existence is individuated, is unique. At some level, this may be trivial, but at least it is something to pin your hopes on if such has been declared of value, something that sets you apart from others in a society where being set apart from others is a crucial designator of success. No one else has your same story lived from the exact same point of view. And because of that you are entitled a claim to be yourself, to be uniquely yourself. The one and only individual you. That is not something that took place in private, however, not something that occurred in a fashion with an aspect that was exclusive to you. Nor is it something that is of value in private, something that is worthy of emphasis or notice in the singular case. Rather, this is an assault on privacy cloaked as its preservation. The fact that you have never had any privacy, that privacy is not any part of your experience or the language that you use is something you may have been deluded about, something you may have just overlooked or took for granted. Maybe human beings observed at times that they were occasionally left to themselves, occasionally left alone in a room, able to carry out their projects or work through their thoughts without being bothered by anyone. That does not mean they were alone and without influence. All the voices of their lives, all the influences and affects that have come to interact with them over the course of their development, they are all there with them even in those moments when they thought they were alone with themselves. There is a sense in which you have no privacy at all, because there is nothing in your cognitive state, no sensation that you can perceive in yourself, that is anything like a private experience or a private state of being. You cannot be wholly and completely alone with yourself because your development, your cultivation, has been the result of a community project, a social and historical process that is ongoing even now as we speak of it. And this project has included both the invasion of privacy and the establishment of it as something for which its invasion is a concern. The writing and the reading here taking place, take place in such a context. Neither of us alone, not in the writing, and not you there in the future somewhere reading this. We have been educated, crafted even, we have been fed and cared for, we have come to have a whole set of expectations, interpretations, and intentions that are the upshot of a community back and forth establishing agreement among these exhibits and actions.

What makes privacy so hard to nail down and so hard to protect is that we are so deeply and completely unfamiliar with it in anything like a real and personal sense. 'Expectation of privacy' is always something of a delusion. No doubt some set of expectations have been developed or learned through our non-private interactions with our family and community, but the method of construction in those expectations lends a lie to the expectation in the very moment that it is established. Maybe not so much an expectation of privacy, then, but a hope for it, a fantasy to achieve it despite the intrinsically impossible state that it describes. The expectation of privacy would seem to be the violation of a rule in the very moment that it is being established. Like some maxim to steal from others that takes exception to the universal rule of established property rights to take someone else's stuff and then have everyone grant you protected ownership over that stuff. Privacy too is a sort of violation of the categorical imperative. Not in the exact sense that Kantian evaluations tend to progress. Because expectation is fundamentally a non-private notion, an expectation of privacy is self-refuting. You cannot have an expectation of privacy, not in the sense that we discussed previously. Expectation of privacy cannot be the socially recognized subject of all other rights, it cannot be the basis of the private person because expectation itself is already social, already historical, and already intersubjective in its origin and in its determination.

But surely this is a genetic fallacy. There is no requirement that the expectation itself be private or be privately determined. What it does mean is that in so far as an expectation is the basis for the private person as the socially recognized subject of all other rights, then this process itself is publicly determined. This notion of privacy cannot be based on some principle of natural law or some claim to what a human being most essentially is. Rather, the absence of private experience, the non-existence of privacy in the cognitive acts of human beings, just means that the establishment is a legal act, that the private person is established by the Constitution as opposed to being naturally established and then protected by the laws of the land. The private experience argument does not indicate that there is no such thing as privacy, it indicates that privacy is not found in us, that it is not part and parcel of who we are and it does not define us, nor does it play a role in how we define ourselves or become defined over the course of our lives as we acquire experience and history, social connections and influences. We are only private beings in so far as we decide to make ourselves such, so long as we determine the meaning of such a relationship and then establish it.

If that is the case, then what is our model for this? If we cannot merely look to some natural right to understand the meaning of privacy and what might constitute a reasonable expectation of it, how can we come to know it? The primary reason why privacy is so easy to violate and so difficult to describe, protect, or legislate is because we have no basis for having any kind of understanding of it. It is entirely possible that the word 'privacy' is little more than an inarticulate sound like the yawp that is the cry of the newborn. I would propose that our concern for privacy only comes to be once violations of it were discovered, that privacy has only come to be known privatively. In 1890 in the United States, a couple of lawyers started worrying about whether there was a right to privacy. And the story goes that Warren urged Brandeis to write about this topic and begin instigating the process of coming up with a legal strategy for clarifying the right. It seems that they were responding to technical advancements in photography that made it possible to take a picture of anybody at any time and in audio recording capabilities making it possible to record their speech. Only once new methods of surveillance came into existence did jurisprudence begin to take special care and attention with the so-called right to privacy. The side effect of the advancements was a set of practices that seemed invasive somehow, that seemed to give us unparalleled power over each other through these devices. I can record what you said and photograph you and then I can show these artifacts to others. This raises a question as to our respective interests in the carrying out of these actions. Privacy was not some determinate thing that everyone understood and then one day we started worrying about its violation. Rather something happened, something became possible, and that something spawned a set of discussions that addressed these new conditions and human agents working collaboratively tried to figure out how to protect various interests amid the possibilities introduced by these new devices and capabilities. The effort to establish a legal personhood in the claim to privacy was an attempt to place a limit or set of limits on actions that were now possible given these innovations in image and sound reproduction. The clusters of court cases were clearly associated with new capabilities conforming to these same innovations. The use of wire-tapping, the ability to eavesdrop on telephone calls, the ability to video record another person without their knowledge. All these practices are found at the basis of privacy decisions because all these operations are the disease that led people to wonder and concern themselves with the boundaries of human contact.

The fiction of the private person has taken a greater hold on us as a greater capacity to infringe upon each other developed in the technics of image and sound reproduction. Now we can add data reproduction to the mix. Being able to reproduce a person's 'being' is not exclusive to a sound representation or an image representation. Data makes it possible to completely reproduce the actions of the person, the choices that they make, the sequences of DNA they personify, the desire and preferences that they have developed or that they may be open to. The data that describes us becomes another form of reproducing ourselves in some way, just as the photograph or phonograph made this possible. And the advent of large-scale data collection and manipulation has brought privacy concerns to the forefront just as they were brought to the fore with every other step forward in the capacity to reproduce the person using some telemetric methodology. As we become more sophisticated and creative in our remote capabilities to measure ourselves, we must become more sophisticated and creative in our capabilities to draw boundaries on these measurements and to protect against an unreasonable use of them. The intent of the socially recognized reasonable expectation of privacy as the subject of all rights is the best we have been able to do thus far in providing such a capability. Because this socially recognized state of being is situated in the center of a socially organized dynamic of understanding and development of the potential to interpret, even the best that we have been able to come up with has been notably inadequate. Young people grow up with lowered expectations, even older people who have become familiar with the advent of large-scale data operations, are realizing what a losing battle it is to keep themselves distinct from these efforts at reproducing the entirety of their lives in data. Even as we shout that we should have more privacy, that our privacy should be protected better, we have less and less of it because deep down we expect less of it. At least, this applies to some of us and in the last section of this book we will look more closely at the conditions that may distinguish the some of us that it does apply to and those of us that it does not apply to and for whom it turns out to be a decidedly advantageous situation.

Our experiences are not our own and so the possibility that someone might capture them all and run them through statistical analysis done by some learning algorithm is not surprising. We have come to understand that even though we may occasionally wish to resist the current, we have no viable expectation of privacy any longer. Or to be precise, our expectations are grounded in a publicly structured context that makes any deliberation on the matter possible at all. The genie has been let out of the bottle and so it seems that we will have to compromise. This expectation already suggests the public process of discussing it, the public matter at hand. And that is already to lose the basis. The minute you must start publicly fighting for your privacy there is a sense in which the battle has been lost. As soon as the issue becomes a matter of socially recognized reasonable expectations, the possibility of maintaining the integrity of what pre-existed the concern is no longer possible. As soon as we must apply the rationalization method and mechanisms to the domain of our own-most private personhood, we have lost sight of that personhood in its implicit form. By offering up experience to an analysis that shows where the boundaries lie and what characteristics it has, we lose the naïve integrity of the state we wanted to protect. We have fallen from the state where privacy was not a concern because we had it without knowing or understanding that we did. The first 100 years or so of liberal democracy may have spelled out a honeymoon period for the private personhood of the citizen, and with the advent of mechanical reproductions of human lives in various forms that honeymoon came to an end.

Sharing Our Pleasure and Our Pain

The absence of privacy is what exposes us to statistical analysis and aggregate computation, but it is also what makes it possible for us to live together and share our pleasure and our pain, our joy and our sorrow. Social relations not only show themselves in higher order facts such as exchange values, but in meaning as the currency of experience as well. Commerce is already a representation of meaningful interaction and intersubjectivity, so although it might make conceptual sense to spin together a notion of use value that enters social relations via exchange value, this conceptual organization is purely heuristic. The social relations are primordial, exchange is primordial. Birth itself is a social exchange in which the various actions and operations of tradition and institutions of training and learning enter human relations for the sake of some end: managing the birth, eliciting a healthy and successful birth procedure to bring the newborn into the world, into the family and into the social relations that articulate the community or society at large. The expectation itself is a social relation in this sense, the interpretation and the provision of context, the situating of action within the context via the contribution of those who are participating, the shared discovery of intent and its relationship to the various outcomes in the actions, these are the concrete realizations of social relations. The table or bed where a woman rests during the labor process is better or worse for that process, it has been crafted with this function in mind with feedback from practitioners who point out that various features would assist in various ways in making the process more efficient or more comfortable or in some sense better for achieving the end of labor: birth itself. And these deliberate actions enable the craftspeople to see their objects differently, to understand the relationship of the device to the purpose and function in different ways. The person who crafts a table or bed for the sake of giving birth, understands the object and the process of designing or fashioning the object better because of feedback gathered from various uses, various instances, and the problems or issues that came into play during that usage. The social relations, the interplay of different users and different persons, cultivates a greater understanding of the thing itself, makes it possible to see and render the thing differently, with greater impact and contribution to the process. Human beings act by carrying out ends that are saturated with such forms of social mediation, attachments to institutions and historically defined practices that are in an ongoing state of change and evolution. Variants to standardized approaches, like using a birthing pool, are likewise attached to alternative historical understanding and claims of a deeper realization of optimal address to the problem space of 'giving birth'.

Experience and speech –action, generally speaking– are regularly and persistently conceptually codified emissions of social relations, iteratively refined and balanced, elaborated and adjusted with various defining features and metrics. To be "experienced" at anything at all is to have a handle on the social contexts, the use contexts, the interpersonal processes and procedures, that are at work in the realization of the outputs in the interactions with others. What we do, how we act, how we learn, is all interwoven with the associations and connections our actions have to others. The social context is everywhere and comes from everywhere. This fact expresses a vulnerability as much as it expresses a power. Human beings do not just act, they also form opinions and ideas about their actions. And just as the actions themselves are immanently engaged in social relations exhibited and emitted contextual associations, the ideas we have about them, and the beliefs we form about their connection, exhibit the same characteristics. It is entirely consistent with human beings that their interpretation of their 'job', for example, is based on enormous networks of institutions providing insights and apparatuses for optimizing the work that is being done. People may recognize that great historical effort has been placed on engineering the process of producing whatever it is that their job is concerned with, and at the same time they might perceive themselves as artisans, as an individual craftsperson with expertise that resides individually within their body and brain and shows something very singular and personal about themselves in the performance. There is no inconsistency in a representative idea having a discordant material quality from the thing that it represents. Both are social products, both are interpretations of intent and foundation, of agency per se, that have been fine-tuned throughout the course of an historical process. Distinct modes of a substance. The way that agents act has an interpersonal origin, and the way in which they think about those actions has an interpersonal origin too. And there is no essential requirement that the two adequately or accurately integrate with one another just as there is no reason that one set of actions need to accurately integrate with another set of actions. The actions of the table or bed maker and the actions of the midwife may benefit from greater integration of purpose, but nothing requires it. The techniques of the one may evolve with complete independence of the techniques of the other. Likewise, the ideas and beliefs that explain the relationship between the two. The criteria for "accuracy" may play a role in what seems reasonable in both cases, but so long as those criteria are coherently articulated, it is perfectly sensible to have explanations for the two that are orthogonal. They need not be, of course, but it is possible. As fabricated and historically evolved institutions collective of meaning, we can talk about them coherently in terms of the history of the criteria, the history of the integration, and the history of all involved actions without requiring anything like a single set of practices. Because of the socially mediated nature of the various components, the methods of making it all fit together, seem reasonable and coherent, and may themselves be socially mediated: we may learn the right set of analytical descriptions to produce to make variable threads of a multi-valent history appear in the guise of a common pattern.

The oft-cited opposition to this approach, and the center piece of Wittgenstein's focus, is the role of pain in private experience and the privileged access to pain states that this must entail. The notion of pain as the exhibition of a sense without any broader context until that context is supplied by the set of 'language games' or experience, clearly does take issue with any such private language or private experience, but does not it miss out on the fundamental principle of privileged access? What I was previously pointing to through the geometry of experience, the location and time sequence that puts the agent into the experiences, places them alongside each of the emissions and exhibitions. This geometry of experience yields the private person and their privileged access to the states of pleasure and pain or anything else more socially complex. The previous discussion has shown that even these raw emissions are mediated by their social relations, the agent's introductions to human characterizations and interpretations of these emissions, and that even our own relationship to pleasure and pain is mediated by the social relations that foster those intentions and expectations over the course of the individual human life. It has been our position that the so-called individual human life is already a mediated interpretation of experience and how it is situated. Heidegger, for example, considers it a bias in analysis to attribute such conditions to anything like human life, agency, or the private person. Rather, these notions already incorporate meaningful interpretations and intentions into the analysis. By using the phrase 'Dasein', the existent one whose being is an issue for itself, he means to get behind the concrete details of any such interpretations that would foster a self-conception of one form or another, which is what we get when we invoke the teleologically oriented agent, the biologically meant human being, or even the living human being. Since our earliest days, pain and pleasure are never unmediated, but always interpreted and laden with the relationships that are sensed in such interpretations. What could this pain be? Where is it coming from? What organ is it that is hurting and what are the likely impacted processes that such pain indicates?

Wittgenstein's aim is not to address the higher order meanings and connections of any given pain, rather he is focused on its most immediate and rudimentary articulation. And his point is that even this is not privileged in access, it is exhibited. And furthermore, in so far as the exhibition undergoes transformation as the exhibitor learns to associate it with attributes and properties that characterize it as this or that pain, as a headache or a toothache, etc. the pain becomes even more enmeshed in specifically non-privileged relations, relations that are understood, relations that have been bound to expectations that are part and parcel of the action contexts within which they arise and within which they show themselves. The demonstrative nature of pain states is reflective of any and all states. We do not live in our heads and then express what is in there. Rather, we come to understand ourselves in even our most private experiences, or seemingly private experiences, through the expectations and intentions of the meaning process that are acquired via human involvements, social connections, and the agreements and interactions of human beings in the carrying out of practices and procedures. And as I have said before, this basic relationship to emission characterizes our actions as attributable, as propertied, and gives them dimensionality that situates them nearby to other actions and other forms of endeavor. This makes it possible to perform analysis on the actions themselves, to build statistical analysis of those actions, and to build correlated models that associate one set of things with another set to provide conjectures of possible future outcomes based on an extrapolation of past correlations. Void of anything like privileged access to our own states of being and intrinsically vulnerable to statistical aggregation that associates one person's actions with collective action, or group based aggregate action, it is no wonder that 'privacy' would become a growing concern as the world revealed systematically that such privacy, the existence of private persons and their privileged expectations at the seat of all rights, is incoherently based on characteristics and qualities that do not hold up under scrutiny. The creation of privacy boundaries becomes more and more significant and important to people the clearer it becomes that such boundaries are contrary to the human condition and are opposed to the socially based development that each person undergoes as a member of the species, civilization, or large-scale grouping that exists for associating some segment of the population at large.

The tradition of philosophers has arguably had trouble providing specific examples to illustrate this condition. And the difficulty, I claim, is because human beings that are already grown and educated have a hard time seeing the rudimentary relations at work in their intentions and expectations. Developed experience in a mature human being is far from the basis that yields the explanation of emission and codification promoting analysis that we are driving at in this discussion. How could one give an example of this? Many years back I was a smoker who was trying to quit smoking. My methodology was cold turkey. It was very difficult, but I managed to do it (after several attempts) through a contrived tactic. I would suggest to myself the following: 'what would this feeling be if I did not know what it was?' The withdrawal from nicotine is classifiable as a painful state. When a person is in this state, however, they generally know why they are in the state and what the state means. I have not smoked a cigarette all day and so the feeling in my legs, in my arms, in my body, this feeling is due to the withdrawal symptoms I am undergoing. What if you did not know that was the cause, what if you did not know that one simple solution to the pain was to smoke a cigarette and that would ease the pain immediately? Is it possible to artificially put oneself in a position where one experiences the rawness of the discomfort without knowing any of its larger contexts? Very probably the answer to this question is no. At least in any absolute sense. As much as I tried, I could not 'forget' what I knew, stop understanding what I previously understood. I saw the value of projecting this state, nonetheless. The discomfort, when abstracted from all that I had come to understand about it, could be changed to some extent. And although it was not a perfect process, it did assist me in weathering the storm of the bodily reaction to nicotine withdrawal. At the very least, I was able to see that the considerations of the condition, my interpretation of how I had come to this state and what actions were available to me to mitigate the condition, were drastically impacted by the *epoche* of meaning that I performed. And although this could not be done in anything like a perfect or pure sense – the tactic itself could only be carried out by someone who understood that it was a tactic with some end that it was trying to achieve– it was at least enough to show me the way. Just as the actions of smoking had social content and context, the machinic nature of quitting did too. Even habits are meaningful in social context and have a socially recognized reasonable expectation within them. The fact of that at least is visible in the artificial action of trying to extract oneself from that context and experience the rudimentary condition in its raw forms.

Privacy is a problem. The *epoche* that I have described is an artificial and purely abstract tactic for attempting to understand the social conditioning at work in all interpretation and meaningful experience. And as such is not very successful at removing the agent from that state, although it might provide an interesting insight into the state itself and the socially conditioned basis of the experience that one is having. Such an experimental thought may therefore suggest a sense in which it is possible to at least render the underlying relationships intelligible to the person undergoing them. And this enables us to ask the question as to what is not a problem here? What is the lesson to be learned here from our essential '*mit-sein*', our rudimentary condition of being involved in collaborative engagements that aim at constructing a meaningful world? First off, we should note that '*mit-sein*' does not mean we start from a position or condition of total harmony. Rather, we need to understand that the integral nature of our interpretive stance exposes us to both love and hate, to both sympathy and antipathy. Just as the system described in the previous section was a collection of offset interests emitting themselves in opposition and coherence, so too do any basic social relations have the potential to surface antagonism. From the point of view of privacy concerns, the social relations underlying all human action may be a problem, but what if you did not care about privacy at all, what if you just wanted to gain a better understanding of the processes at work, to interpret the emissions as best as you could, and produce some form of analysis that rendered this dynamic of intentional cultivation sensible?

Both pain and pleasure are at work in these emissions acting as the basis of human social context and from which expectations are derived through interaction: "It is not the consciousness of men that determines their existence, but their social existence that determines their consciousness." (K. Marx, 2009 p. 21) And these social relations are 'collaborative', but in a sense stripped of the commonly presupposed value content. Victims and villains alike work together to form the set of worldly relations that bind us together. Painful experience may be justified via a set of socially completed expectations that an individual may come to have. Likewise, pleasurable experience and its connection to expectations. When something happens, anything at all painful or pleasurable, it is easy enough for a human being to imagine that they deserve such an outcome, that the explanation for the sense is bound inextricably in their past actions and deliberations and thus spells a clear and distinct outcome matching with the input. It is not just that humans tend to accept their lot in life relative to each other but that the entire apparatus of human development is best suited to this attitude. And because these expectations can be bound up with the basic senses of pleasure and pain, they become invisible in a sense, they become part and parcel of the painful or pleasurable circumstances. The pain carries with it a sense of desert, likewise any such pleasure in experience. The meaning of one's pleasure and pain is socially mediated and so flush with justification and characterization that explains the inevitability of the state given the conditions. This makes it difficult to understand the circumstances within which one finds oneself, but it also deeply binds all human beings together. In the very meanings and interpretations, in the expectations and the given conditions, we are interrelated in our understanding of ourselves and our world.

Carefully examined, therefore, the emissions of pleasures and pains are already indicative of social context and deeper social relations. We are always there with each other, and the hatred, the love, or the indifference bind us together in these relationships. Whether we form alliances or wage war against each other, these are still instances of being with each other, ways in which we are together, fulfilling each other's expectations, and providing meaningful context to each other's actions. This essential integration is crucial for any further steps we take in trying to delineate features or conditions. Any perspective or point of view we take on any matter must start from the conditions that exist for us in our integrated social contexts. In the very same breath that the Supreme Court of the United States attempted to set the right to privacy into the soul of each private person as the seat of authority for all pursuant rights, it also situated the personhood of the person in a social context creating a circular dependency of the society on the individual and the individual on the society. The requirement of socially recognized reasonable expectation depends on social beings, that is, on individuals recognizing the expectation in a specific way. And the individual's recognition depends on the social context in a specific way. This circularity makes the definition wobble and poses a risk to any clear notion derived from it. If society informs us, or trains us, or causes us to lower our expectation, then the right itself is diminished. Who is this society other than each of us being informed, trained, or caused? The tendency is to reduce the terms of the description to familiar phrasing. Society is just the collection of individuals and so it seems circular to describe a relationship between the one and the other when one is just a member of the set. The notion of expectation that we have seen thus far in this chapter suggests that the individual is not related to the collective as a member is related to its set. Rather the forming of a state of expectation depends on a set of meanings and those meanings come through in one's process of development and the institutions that collect together historical knowledge and traditional understanding and pass that off in the experiences of the interacting persons. Society is not a collection of instances, it is a set of relations, of ways in which we finish each other's thoughts on the one hand because we do not have any other way to finish them and on the other because we have witnessed them being finished by others when the roles were reversed. Each of us both absorbs and repeats the culture and world view that we are integrally embedded in throughout the course of our lives.

Sharing pleasure and pain is the easiest thing for people to do. We are born doing it, we are raised through it. We learn it with brothers and sisters, from our parents and friends and at school, we regularly and consistently complete this integration cycle as we move through the world during childhood and into adolescence. It is commonly muttered that we are born alone and that we die alone, but nothing could be further from the truth. We are never alone. Even a person alone with themselves, isolated on a desert island by shipwreck, has the complex of awareness that a people and a land have provided through interactions and contexts that allowed the person to develop hopes and dreams, wishes and expectations, and to develop a meaningful personality that is self-understood through sets of attributes and character traits that required all of humanity to establish. People have brown eyes, so brown eyes, being a brown-eyed person, is something to be, something to have, a characteristic descriptive of one's humanity. And this cannot be underlined enough, expectation is both backward looking and forward looking. An expectation, a reasonable expectation, signals an entitlement, something that is due to us, something that we have a right to expect, to come to rely upon as a deserved and upcoming event that is appropriately bestowed upon us based on the past condition we have found ourselves in and the direction in which we are pointed. The past and the future are bound together in such expectations and in so far as those expectations are the emitted upshot of social contact and integration, we have absorbed the whole of that social context into our personhood. The private person at the seat of all rights is a sponge, a public point of absorption, for human connections and collaboration: a point of saturation reflecting the set of relations we have traversed throughout our lives. We are networked nodes. And these are the phenomena that I have in mind when I refer to the emission of properties and attributes characterizing the human being as the private person who is the socially recognized seat of all other rights.

Chapter II: The Right to Have One's Own Experiences

"To the things themselves." When I look around, I see things everywhere, there are many things in my immediate surroundings. What could lead a philosopher to assert that we need to get closer to them? What could be closer to me than this desk or this chair? Why would anyone come to raise the proximity of things as a problem? Is there anything clearer and less befuddling than the presence of these things in front of me, touching me, so close as to almost be a part of me? The phenomenologist's point is that I am the problem. There are things all over and they are close by, but they are close by and around *me*. It is I who am aware of them, I am here among them, and for me this awareness is overwhelming. My whole being as a creature alongside things and touching them is filled with this strange conscious sense that fills everything, that seems to be everything. Here I am. I am here now amidst these things. It seems that the things that I am with are not really things at all, they are the population of my proximity, they are the stuff that I am seeing, hearing, feeling, and smelling. Things are inextricably mediated by me. That is, no doubt, one aspect of it. Then there is the flip side. Look at all the stuff around you, did you make it? How did it come to be near you? What is the history and spatio-temporal trajectory of each of those things? Many of the things in my immediate vicinity were produced by other people and purchased by me. This is a complex set of relationships, no doubt. That does not exhaust the whole sense of what is in reach right now. There may be a privative state revealed in some of these things too. Take, for example, that very old pine tree over there in my neighbor's yard. It must be 100s of years old at least. And given what I know about the migration patterns of humans in the pacific northwest, it is entirely possible that this tree predates any of the structures or inhabitants in this area of the world. Why is the tree there? Presumably the subdivision in which I live was once sprawling with trees of that sort and overrun by the crows that prefer to live in them. When they bulldozed the area back in the 1970s, why did they leave these trees? Were there some that did not make it, that were selected for destruction while these others were permitted to remain? What regulated that decision? Did the city provide guidelines for tree density that had to be preserved in the carving of the subdivision? There is a human touch in everything that I see around me, in every aspect of the things that I find close by. Whether it is the labor of a human being performing a job and the purchasing power of my salary or the past decisions of land developers and city planners, there are social relations everywhere I look. It is no wonder that the phenomenologist says, "to the things themselves", because what we are seeing is not what they seem. Rather, even the things that are nearest to us, are concealing the entirety of their social relations when we look at them.

A tree can be analyzed for age or health or weather conditions that persisted throughout its life. What of the tool sheds that are in my yard? How much of their story can be told through analysis? They are made of wood, they have electrical current running to them through wires that have been buried in the earth. Can an informed investigator determine the point of origin and process of production of the hinges and fittings on the doors, of the electrical materials piped into the structure through the ground? Can the quality of the wood and the likely process of production that it went through be determined? Of course, these are analytical skills that some people have indeed acquired. They are familiar with the techniques and patterns of construction for certain kinds of artifacts just as a botanist is familiar with the patterns and techniques that can be discovered in the trees. And the process of spelling out such pasts, such origin stories of the things that make up our lives, requires a great breadth of knowledge covering all kinds of processes and procedures. It is unlikely that any one person would lay claim to all such knowledge. Probably there are a multitude of divisions of labor within such a scientific community. These archaeologists of human practices are experts on things, the various kinds of things that are likely found in different regions of the country, or among different types of populations, and the stories and lore behind them. How did people begin adding out buildings to their dwelling places? What were the factors that governed the developments in these geographical areas? What is the history of backyard gardening and horticulture? Everything human can be surveyed in a historical set of analyses that attempts to understand the development of the art or craft, of the things and types of things associated with that art or craft. The multitude of human beings, in all our varied capabilities and interests, are spread across the realm of human history as we attempt to understand who we are and how we came to be this or that way, how we came to populate our world with these or some other collections of things.

The deeper we go in the direction of the things themselves, the more we discover the complex behaviors and practices of human beings. Historically there has been much emphasis on phenomenology's focus on the direct act of perception and the way in which the thing is completed or protended by the act of perception. The back of it is presumed, the shape of it, the completion of the whole, or the figure of it over and against its background. With the introduction of the ontological concerns of the entity and the offset of the thing against its lifeworld context, the complexity of relations is increased exponentially. Much is missing from the thing before you now, the thing as it sits on the floor over there or on the desk, or out in the yard. And these expansive fulfillments of the thing are not merely spatial associations with the things around them and the moving body that perceives them, there is also the temporary fulfillment that takes place when the history of the thing is apprehended or resolved in a retentive or projective act. The history of the thing's relationships is available for analysis as the thing rests in the foreground before you. There is the history of the thing's production and transport and the tale of the exchange that was made to acquire it. Either could be a long drawn out process of toil to acquire money for purchasing the thing, or the component parts of it, and combining it with your hard won and developed skills used in fashioning it for yourself. The mystical abstraction of the thing, the effort to look at it as if it had none of these stories, suggests that it played no role in any larger scale drama or sequence of events and displays a truly shocking capacity of the perceiver. While some question whether or not the color of the thing is the same for you as it is for me, treating each of us as independent observers who need to achieve a mutual understanding of the qualities inherent in the self-standing thing right here before us in imagination, the phenomenologist attempts to take hold of the thing in real context and elaborate its many relationships and associations, its role in human life and human history. For make no mistake about it, things are human relics. And along with this questioning must come an inquiry into the abstraction that makes it possible to ignore those trajectories.

The interpretive process of establishing a set of expectations and the properties or conceptual elements that impose these expectations upon the thing is a socially mediated process. Even the stars in the sky and the gravitational pull of the sun on the earth, is an artifact in so far as we know them. And to good end. Imagine if every human being who was ever born had to start over again in trying to understand the universe and all the things that are in it, all the relationships between those things, and all the dynamic ratios and properties that have been discovered to hold true under various posited experimental conditions. Imagine if the conditions themselves had to be rediscovered or the meaning of the projected outcome of interactions between things. There would never have been any progress. The scientific method is often lauded for its incredible leaps and bounds forward in progressing knowledge. How does this happen? Is the least informed among us more competent as a scientist than Aristotle, who held some ideas about the nature of the universe that even school children today might find ridiculous? Where does this progress come from? We learn the basics because we communicate our models to each other, we describe and explain the attributes and properties of things, and we provide instruction in the carefully engineered processes of experimentation or conjecture and validation. Every new advance in the sciences results in a change to the educational process of scientists in the respective domains. The feedback loop is the introduction of modifications and discoveries into the learning process so that future scientific discovery can start from the point of current understanding and analysis. Human beings do not start afresh with every generation, rather they communicate their conceptual mechanisms to young people and thus inform them about the apparatuses at work in viewing and conceptualizing the various parts of the world. This is the carrier of progress. Human social organizations and the institutions that we have built are increasingly optimized for just this form of communication, this effort at passing along the metrics and analytics of the world to the next generation so that they too can discover holes in it, flaws or assumptions that force a progressive movement into some new realm of discovery.

Arguably, such an approach to progress does not require that the new generation learn anything about the history of the science and the history of the investigations that they are benefitting from. If they did learn that history, they may be tempted to read it as a silly history of failures and false notions, delusions even. To the extent that the focus of physics is physical objects, the physicist will provide insight into the science of physics through the relationship the investigator has to the things under investigation. Physics does not include human organizational and economic histories. Rarely do we come across historical analyses of physics that provide detailed descriptions of how a university physics education contributes to this or that problem, or of how the way in which the research was sponsored may have contributed to the mistakes and the false leanings that resulted from it. And that may be appropriate to the matter at hand because there is a necessary and deep division of labor within the sciences. It takes too much effort and life-long study to become a capable contributing physicist, it is not practical for an expert in the field to also be an expert in scientific economics or the sociology of science. And the rigors of this process are often overlooked. Even in some of the popular debates about how important it is to listen to scientists on matters of climate change, for example, the emphasis is often one that abstracts the investigation from the processes of acquiring the ability to perform the investigation. Scientists undergo study and obtain a discipline while acquiring their degree. Once they do that, they undergo a history of positions that are funded by a variety of different possible sources. Of course, funding does get a little more popular focus, especially when the source of the funding has an obvious and opposed interest to some of the findings, but the credibility discussion often stops there and does not include the process of acquiring credentials. A society may be deemed anti-intellectual to the extent that this process of acquiring credentials is ignored or cast aside. No one doubts that a master carpenter with 20 years of experience on the job and a well-established apprenticeship with another master carpenter is better suited for wood work than someone without credentials, yet scientific investigations are sometimes cast as mere experimentation and observation and so something that pretty much anyone can do.

All these historical processes, all this human history, is in the thing in the deepest possible way and it forms the film against which all vision is developed. That is what I mean when I suggest that even the stars and the sun are relics of human activity and their social relations. Saying that the things really are such and such just begs the question. Even if the evidence is produced, even if the analysis is done, and even if the metrics are defined and demonstrated with a variety of successful experimental outcomes, it is still the case that all of these mechanisms are human beings at work and evoking the whole of human history and social involvements in the establishment of the meanings and expectations of things so as to demonstrate or render them as they "really" are. I'm trying to suggest that phenomenology is neutral to the idealism vs realism dilemma, that this constitutes a nuanced understanding of the so-called *epoche*. This form of neutrality or suspension is not a reduction at all, rather I am spinning the *epoche* in terms of Heidegger's question about the problem of the external world: the issue is not whether the external world really exists out there in the way that we think it does, the issue is why did anyone ever think to question it in the first place? That is, only a world populated by things abstracted from their human involvements, from the social relations that mediate each person's interaction with those things, could be subject to such forms of questioning. It is only possible to ask if a gravitational force really exists between the sun and the earth expressing a ratio of mass to acceleration if these heavenly bodies are somehow abstracted from the process of coming to understand their relationship at a distance from the subject who knows it. Making this assertion does not entail that the 'telemetric' point of view that performs this abstraction and articulates the properties and qualities of it in the analysis is bad, destructive, or delusional. It is something that humans do as a function of our efforts to optimize and improve upon the "things themselves". Just as the phenomenologist is correct to point out there is a way of thinking about things that abstracts them from their worldly relations, the telemeter is correct to point out that the system of knowledge requires this abstraction to make progress optimizing the processes at work and under investigation.

The debates between them, therefore, need not be aimed at establishing a moral hierarchy in our investigative positions. There are other claims that are often made by the telemetric perspective: that it is disinterested, that it is pure of motive, and that it is performed merely for the sake of acquiring knowledge. This is the true seat of the disagreement between the two points of view. What data science shows us about science is that there is always an element of engineering at work in the outcomes. And this need not be a large-scale ethical condemnation of data science as social engineering, it can be a much more rudimentary target than that. In so far as the scientific analyses are aimed at the discovery of new patterns and new relationships, new ways of conceiving the connections and associations of phenomena and events in the world, it is aimed at progress. Progress is not just something that happens to the scientific body of knowledge in this or that area of study, it is the self-conceived modus operandi of the discipline. Something we hold today is incomplete or false and some set of investigations will complete the picture or set it aright. In the abstract viewpoint, this is a minor adjustment to a theory or body of knowledge. In the phenomenological viewpoint, this is a change in human understanding. Even a scientific realist like Paul Churchland is clear about this: one of his favorite examples is the new way we can tilt our heads when looking at the planets to see their motion in a more intuitive fashion. He clearly articulates there the fundamental scientific notion that advances in knowledge will change humanity forever. I can no longer look out at the stars in the night time sky and be astonished in the way that a contemporary of Thales might have been. That frame of mind is off limits to me, out of range, because of the history of advances and progress in astronomy that have taken place since and found their way into my awareness through study and social osmosis. Even if we are capable of such astonishment for a few brief years during childhood, the elementary school system or parental guidance will quickly nurse us past it. Scientific progress is self-consciously aware of this intent. Even if the goal of the physicist is not to get you to click on this or that advertisement when visiting a web site, it is still crucial to note that it is their intent to change your way of seeing and experiencing the world.

Again, it is important to stress that at this point my intent is purely phenomenological. This should not be understood as the beginning or set up for a moral investigation or construction of normative ideals about how humans should know things. It is usually clear in the minds of telemeters that the changes made to human understanding are for the better and mark clear and obvious advancements in the human project of having a greater familiarity with their world and a greater sense of their place in it. The tension, if such exists, is that the telemetric point of view is conjectured to act on a set of abstractions that elide the historical and human trajectories of these ways of knowing or of these detailed investigations. And to the extent that the telemetric hides the social context of such things, it prevents human beings from having a greater understanding of their world and their place in it. Knowing how the planets move and act on each other may spark great progress in human capability, but to really support that advancement, human beings cannot just be informed about the result, they must be capable of seeing the entire yarn and of retelling and recounting the whole tale that has been unfolding as the effort to understand the world. If progress takes place and changes human understanding at the expense of any detailed awareness of the integration of that new bit of information into the larger context of human life and history, then how is that progressive from the point of view of advancing human ends and human integration into the workings of our world? And to repeat, no single discipline in the sciences can be responsible for all things in all areas. We do not expect physicists to provide biological insights nor do we expect biologists to provide psychological insights. It may be well established that sometimes overlap can be very fruitful to the development of a science, but that does not reject the fact that the body of knowledge in each of these disciplines is so complex and voluminous that it is unreasonable to think that anyone is capable of being a 'scientist' in a universal all-encompassing sense. Rather, we expect a division of labor and it is fair to ask what impact the division of labor in science has on the advancement of knowledge in those sciences that have been divided up into abstracted domains of the world? My sense is that the result has been exactly the conflict I just described, and that people are impacted by scientific progress and changes to their understanding of the world without being able to process those changes as part of a larger context. They are not able to establish those changes amid human life formulated through human beings as the agents at work in relationships with each other and carrying out physical, biological, or psychological investigations. The division of labor creates the need for a discipline that can provide that context, keep us focused on the point of all this progress, and what it is supposed to be for, who it is supposed serve, and how its impact should be realized. This discipline is what I am calling phenomenology: the area of investigation aimed at contextualizing human processes and procedures within experience, at exposing the social relations at work in the forming of expectations, and of attributing and characterizing things as part and parcel of human agency over the course of history and throughout the domains of different human civilizations.

Bear in mind that this chapter is the second in the phenomenology part of the architectonic and as such calls out a reference to the second chapter in telemetry. Additionally, it should point you forward to what you might expect from the second chapter of commonwealth. The role of self-consciousness in the system at work in the service provider was pattern discovery and comprehension of associations. In this data scientific interpretation and awareness of the goings on at work in the system, the relationship of the system to itself became the focus. And this started us in the direction of a germinal notion of experience where the system, through interpretation, was undergoing an effort to understand its own inner workings. In this chapter, the complexity of the self-relationship indicates the experience of the so-called private person. This is a zone of experience that no longer seems to qualify as private, as locked into a person's self-consciousness as a realm of privileged access and private relationships. Instead, the zone of phenomenological self-understanding is the realm where the social relations immediately transform bodily emissions and exhibitions into meanings in the context of others who are responding to them, thus completing the edges and segments of communication between people. Aboriginal communication, so to speak, is the formality of these relations and provides the basis for the fact that humans are in relations and involved in collaborative possibilities. The baby in need is destined for death without the intervention of the other. The child in need is destined for death without the intervention of the other. And key developmental movements through the course of human phases can be cast as established patterns by which these exterior completions of intent can become internalized and can become the domain of the organism's actions when navigating the social resources available at large. The child learns these patterns and the rules to follow while being within the human social world and begins to achieve the capability of navigating through it, in part, by themselves. They learn to count money, that money can be used to buy things, that ingredients are composable into edible foods, to pour their own milk, and so on. The establishment or internalization of the pattern as capabilities is the development process and in it is the self-embodiment of the social relationships that have been effective in caring for the person all along.

Building People

Human beings are, in a non-metaphorical sense, the product of the actions of other human beings. Human beings make cars and they harvest grain, they invent technical procedures and they write articles for online news organizations. In addition to all the other things that people do and make, they make other people, additional human beings. Clearly it would be problematic to reduce the produced human offspring to an artifact that is the result of a production process, but why? Why do we single this event out as a special case and assert that making people is radically distinct and categorically different from other artifacts that we produce? There is an artisanship at work in the production of people. Except in the wildest of science fiction nightmares, the production of people is something we insist on doing with hands-on methodologies and heightened focus on the singular instance, rather than making efforts at mass production or otherwise optimizing the production process. Love is the general explanation for the distinction here. That may just waive aside one mysterious anomaly through the introduction of a domain of magic into the explanation. No doubt, love expresses a bond that we sense and feel toward these very special artifacts that we are producing. And love develops at least in part from a set of social expectations regarding how parents are supposed to feel about children and express that feeling in their care of them. This certainly would play a role in why people wish to have babies before the tangible prospect of doing so comes into play. The social ideals and images, the stories that have been told and the ideas that have circulated may no doubt convince a person that this is the ultimate purpose of their life. They may not be clear about where they want to live and work, but they are certain that they want children. They want to realize those stories they have heard. And often the promise of such tales does not fail to deliver. Once the person enters a relationship with the other being, feeling kicks or movement, seeing visual representations of the little body or hearing the little heart beating, these wonderful emissions develop bonds between the participants that realize the ideals and stories that the civilization has promised. The response that one gets in one's actions, the interplay between the emissions of the parent and the emissions of the child, build connections, establish close relationships, produce the family unit through the ties themselves, and through these very connections.

The meaning of the family and its familiar bonds would therefore be subject to historical and cultural variation. The stories that the society tells are the institution of the family, they set the expectations in the parents' experience of relating to moments of early discovery, of being pregnant, of having a relationship with the gestating baby, the birth of the baby, the developing of the infant and child, and so on. The roles that fathers play will be established by these institutions and will come to impact the experiences of the individuals as these various events take place. In some societies or at various points in their history, fatherhood does not really establish a direct relationship with the child during pregnancy but waits until much later in the person's life. These traditions and presuppositions will steer the actions of the person occupying the position. Although no doubt variation is possible and is usually seen in relationship to identifiable secondary factors, the upbringing that the person had may have been sufficiently non-standard to merit some divergence from the common narrative that is provided for parenthood. The process of production is production nonetheless, but the content of the institution is what steers that form of production into something that may appear special. For some civilizations at some periods of their history, it would not be surprising to discover that children were conceptualized as artifacts, as the mere chattel of the sire or patriarch who was the socially established ruler of the family. Such a ruler has a special relationship with one or two of the offspring, but the rest of them are products of the familial production process to be used as means for achieving some goals of furthering the family interest or economic concerns.

And the creation of a human being is not merely reduced to the generative act whether that amounts to fertilization or gestation, but to cultivation as well. Effort and ongoing productive contributions are maintained through the early lives of the offspring. Parents care for their children and it is not uncommon to further the metaphors of gardening when describing childrearing. They are raised, they are cultivated, they are tended to with clear and deliberate attempts to shape them into the vision of a person that the parents have in mind. This might be a vision that reproduces themselves or it may be a vision that improves upon themselves making up for the shortcomings they perceive, or maybe the vision corresponds to some historical ideal that the parents have come to appreciate throughout the course of their lives. Parenting is done deliberately with the intent of impacting upon children in such a way as to form them in accordance with a set of ideals, to make them into certain kinds of people: good people, independent, disciplined, caring, tough or whatever you like as comprehended by the parent's own historically derived visions and expectations, the hopes they have for what might make for a meaningful life, and a well-lived life: their highest ideals and their highest hopes. Parenting is a huge investment of time and effort, of money and other resources. Make no mistake about it, this is one of the largest industries on planet earth, the making of people. The amount of energy and expended attention and focus that goes into the creation and ongoing shaping and forming of little people is the largest scale industry there is. And according to some ways of thinking, it is the most essential form of production with the greatest risks and the highest rewards. We expect our children to care for us in old age, we hope that they will exceed our expectations for them, and we have such high hopes for them that, in many cases, we may transfer the entire purpose of our lives to them: we sacrifice so that they may achieve more, be more, and go further than we ever were able to go. And then of course –with all this capital built into them– they go on to do the very same thing, kicking the can down the road. Or not, maybe the iteration of this process of production is itself a kind of progress. We do not just conceptualize the advances of our civilization in terms of the knowledge we are continuing to acquire, we think of it in terms of the improvements of our children, that they will have it better, be capable of achieving more, of building more, and of doing more. They will be the ones to solve the current problems and advance humanity into the next epoch with their investigations and their insights. Social progress in the form of the increasing productivity and capability of the next generation is another form of progress that seems to drive our story of human beings in the world and what sorts of expectations we have about the upcoming conditions, the near or distant future.

Is this a conceptualization of progress that conforms to the telemetric format that we saw in the scientific view of the world or is there some different pattern to be discerned here? There is a sense of movement forward and outdatedness that infects our thinking of the generations. In contemporary life, people often comment on the way in which children are accustomed to things in ways that their parents find artificial and acquired. The techniques that come into existence before your life begins and the techniques that come into existence during your life will be appropriated differently, likely because in the latter case you will have some other technique for accomplishing the same or similar activities which now needs to be replaced. And that is a very different process than starting out with the new technique from the very beginning. It is not just that the one technique marks progress of some kind over the other technique. We can also talk about progress in terms of people born and raised without any knowledge of the old way, whose attitudes and behaviors have been fully formed without interactions impacted by the old ways of doing things. And this will appear as a form of progress and as a change in human beings that forces them to adapt to some change in process or to some mechanism that provides solutions to a problem that is only made to look like a problem after such a solution has been found.

This distinction, the difference between the progression of human beings and the social relations that they engage in on the one hand, and the technical devices and processes or insights they have on the other, must be maintained. We can speak of the progress of relativity theory over and against Newtonian mechanics and we can speak of the philosophical progress correlative to these advances and which we find in say Kant aligned with the former and Heidegger or Wittgenstein aligned with the latter. And the philosophical insights might reflect an embodied human behavior or psychological variation. The Kantian picture may be phenomenologically descriptive of human beings for an epoch. Likewise, the more relationally based work of the latter two thinkers who came after relativity theory was introduced. Because human beings undergo progressive changes, the philosophy that they produce will undergo progressive changes that reflect those same relationships. I cannot imagine a philosopher writing a treatise on the relationship between the telemetric and the phenomenological at any point in history prior to the 21st century. The thematic requires the existence of large-scale distributed computing systems and the human impact such systems have had on our developing orientation and understanding of the world. Binding human beings to social relations means that human beings differ as the social relations differ, change as the social relations change, and find themselves in the middle of a trajectory of progress that is both connected to the changes in social relations and distinct from them. A child growing up ignorant of a world without large-scale distributed systems will be capable of inventing a mechanism or technique that no child born earlier in history would ever be able to conceive.

Schools, churches, community programs are all funded and promoted at least in part to further the efforts to cultivate children properly, in accordance with the designs of the parents and the socially circulated ideals of correct child rearing. Not all schools are the same, not all communities are the same. And these differences find their way into the differences of the people being made within their walls, or through their practices and effects. Some parents may bring help into the household, hire additional workers to come into the house to assist in the raising of their children. And this may be direct or indirect, nannies or housekeepers, but either way the purpose is to help in the process and give the parents more time to attend to crucial needs of the children or themselves. Other households may depend on extended family for this kind of support, still others may purchase private services, and more might rely on state assistance. Many different types of industries and operations have developed over time to aid in making people. The school system, funded either privately or through some form of local or national taxation, contributes still more. My point here is not to go into detail on how these various institutions and organizations have impact on children or on how material differences in the various organizations show up in the children that are being made, I am merely attempting to establish the process of production itself and that the details of production will lie in the details of the organizations and social relations that act on those people in the process of developing them.

Human beings are not toasters or automobiles, of course, and we have the additional characterization that we can appropriate the acts and relations within which we find ourselves. This process of appropriation has factors that are specific to the neurochemistry and genetic potentiality of the person undergoing social forms of development and production. We are not passive recipients of these actions that are performed upon us, with us, and all around us by the institutions playing a role in our upbringing and development. And just as the current material conditions of the society, the abundance of some resource and the absence of some other, would come into play on what factors are available in the process of development, so are the material conditions of the person. No material condition is definitively determined as a result. A child with a set of physical characteristics may develop very differently if other characteristics are different. For example, if the child's parents have the resources to hire therapists who can provide individual attention and help the child adapt to their physical conditions, this can result in drastically different outcomes, different experiences, and different relations to self and others developing over time as the child grows up. These patterns of relationships and impacts, associations and influences could all be studied and understood if there were some analogue equivalent to telemetry. Sub specie aeternitatis, the remote measurements would show statistical patterns that would enable discovery of commonality and correlation in the human system of involvements just as we get from the digital system of interaction. And no doubt, to the extent that these relationships and practices are subjected to telemetric investigation, this is the intent. Researchers in many fields of study see great opportunity for discovering the basis of human existence both in patterns and in anti-patterns. If only they could get their hands on the data.

Such relationships appear very differently to one in the middle of them than they do to the telemeter. The telemeter has their own problems, of course, and is amid their own world, living the life of a professional or amateur investigator attempting to gain knowledge of the world's patterns between eating their breakfast and lunch. The point is not to register this division of labor, this 'ontological' distinction in point of view, as a division between people, so much as recognizing it as a division between stances or roles that humans take on as facets in each of their actions. The process of becoming a telemeter is complex, involves education and training, possibly encouragement, and conducive social factors. A human might get it into their head to become a great telemeter in some sociological area of investigation only to discover that persons of their disposition or traits are somehow unable to find adequate financial support in that area. There are many stories of successful and failed telemeters, no doubt, but no human can adopt this point of view throughout their life. They turn it on and turn it off even if the discipline and experience of it take hold of most of their waking life. They will have meals with family or friends and pet the dog or cat and stand in line at the grocery store. Even the data scientist cannot hunt down the distant parameters and patterns of human life all day long but must, at some point, tend to the objects of the world, the social relationships in the world, that are up close and nearby at the expense of training their eye on the distant patterns that their telemetric lens seek.

When people work to make more people, when they parent, they may bring together distinct forms of action upon the process. They may read books, watch informative documentaries, or speak to friends and specialists in various areas to obtain statistical knowledge about the pitfalls and benefits of child rearing in this or that way. The ability of a parent to engage in such telemetric investigations will play a role in the impact the parent has on the child's development. Depending on the patterns, emphases and desired outcomes, the impact may be positive or negative. The fact is that the human producer of human beings will bring together both far away concerns and nearby concerns in their involvements with their children. They will do things because their own parents did them, they will do things because they seem right in the moment given the child's behavior and response to other actions, and they will do things based on knowledge obtained from some remote source, provided by some "expert" who has never met their child. Thus, we see in all action a complex stream of both telemetric and phenomenological concern. To the extent that these orientations are comparable to some being, engagements with things present at hand over and against those that are ready to hand, it is crucial to note that it is not like some Marxist dream where we are phenomenological in the morning and telemetric in the afternoon. Rather, any human action has the potential to involve constant shifts in perspective, constant ontological differentiation in approach and involvement. It may be possible that this is a critical distinctive quality of human beings, that we are capable of this extraordinary diversity in our involvements, that we can bring to bear telemetric considerations even when getting up close and personal with some things.

As an example, consider the finely tuned muscle control that is required to play a musical instrument like the guitar. A student of the instrument will have to practice a piece repeatedly to be able to induce an ingrained apprehension of the movements required to play that piece in rhythm and emphasizing its musical qualities. The emotion and sense of the tune must become primary in the experience. And yet there is so much that is technique and distant from such a perspective. The repetitive grinding process that shapes the muscles into their immediate response to the musical quality is a way of integrating distant concerns of the fretboard layout and string coordination with the implicit melody and cadence that needs to be deeply felt for a proper emotional performance of a piece of music.

It is tempting to try and build a complex system around the various types that I have been invoking. Human beings progress, social relations, material conditions, scientific knowledge, and technological innovation are all at work in the human world. Marx, for one, provided a brilliant holistic narrative of how this all fits together, how it all coheres in a single systematic understanding. Marx was born and raised, he went to school and was influenced by the life he lived and the social relations within which he reached maturity. The circular problem I have is that the insights of Marx as scientific sociology or political economy are bound to the conditions of its production. Any theory amounts to a produced telemetric involvement bound tightly to its points and conditions of origin. Plato could not have written Capital. Menard is clearly insane. No doubt Marx and Marxists have a response to this concern, but it seems question-begging and probably gives credence to the point made by many critics that there is a cheat involved in the Marxist system of thought, that it is self-referential in a way that makes critical discussion impossible and yet establishes its own point of origin as a clear indication of its correctness. We must be entering a period of change from the capitalist mode of production to the communist mode at any time now. I have often wished that were true, just as I have often wished there was a God. These stories can be inspiring and compelling, and they are well-suited to some naïve sense of justice and hope that someone might have. Establishing a system of belief or a perfectly true description of the systematic being of the human social condition is not my aim, however. Rather, the goals here are significantly more humble and they are synthetic to the position itself. That is, my position is equally circular but in such a way that I am asserting the circularity as a 'good': spinning yarns is the way that the morality stories needed to advance the production process to perform their role. Being honest about it, does not necessarily make the story better or more effective, but it may suggest that the storyteller is not trying to trick you.

This can only be a hint. The process is heuristic or didactic. The goal is not to achieve a body of knowledge, and a telemetric perspective. Rather, the goal is to educate oneself. Frankly, I am not even trying to educate the reader, you. I am trying to educate myself. This is a purely educational exercise. And it is my intent to fluctuate back and forth between the telemetric and the phenomenological as itself the education I hope to achieve. I am trying to deepen my ability to do this, to move back and forth between far away concerns and up-close involvements and apply the one to the other and vice versa. And this amounts to exposing the relationships at work in all things that I have contact with. Seeing the distant in what is nearby, seeing the nearby in what is distant. The system of Marx, for example, is made by a man. What about him? Where did he come from? Who did he know? Who did he talk to? Who paid him and how did he feed his family? And Marx really is just an example. Newton was a person and he lived an interesting life. There is great drama to be discovered in learning the details of his insights and how they fit into what he was doing, the relationship he had with both competitors and colleagues. No doubt, a tried and true telemeter would be horrified by the suggestion that one must learn the secrets of the universe without losing touch with the fact that the 'secrets' are what human beings have learned that they are. And that the advent of such things has always been within a human context. The telemeter is a phenomenologist.

Being a Constructed Person

Grown to adulthood, the social relations and individual investments in us are opaque to us. Like a pair of glasses that we can neither take off nor see without, we are formed by these interactions and in this established intersubjectivity not only are we unable to see through it, but we are unable to see without it. We talk casually about how a person can be influenced by their upbringing, by the ways of their parents or the advantages and disadvantages they may have had in their schools, neighborhoods, or any of the thousands and millions of influences that shape us as we grow to maturity. Why do not we see them? Or rather what do we see? If my upbringing forged me into a person of privilege, how do I see my privilege? If my upbringing forged me into a person deprived with a tendency toward discounting the needs and rights of others, how do I see my sociopathy? If I am so conditioned by my social relations and all the influences they have brought to bear on me throughout my development, why cannot I see these influences at work? How do things reflect upon us? Or rather, how do the relationships that make up the things that I know in the world and the interactions that I have through these things come to reside in me as me? What is it like to be a constructed or built person and why cannot I see my constructed nature?

Under conditions of patriarchy, I have heard it said, rape-like behavior has been normalized and internalized into human subjectivity. The argument goes something like this: sexual pleasure between heterosexuals has been so conditioned that the micro dynamics of rape, of the forceful overpowering of a woman by a man is experienced as pleasure. The description means to show that the actual experience of pleasure is a realized or conditioned response. We may have once been taught in early life that women demur from sexual advances made by men, that the job of the man was to advance offerings to women who were meant to be initially reluctant or resistant to the advances. The advances should persist nonetheless, and the passion of those advances would ultimately wear down the woman's resistance and she would succumb, she would be conquered by the male's sexuality. And this psychological dynamic would find its way into sexual activity specifically. Women would be pleased by the exertion of force. There is a kind of rape-play at work in so-called normal sexuality and that this rape-play is experienced through the presence of pretend violence in the ritualistic behavior leading up to sex, as well as in the actual behavior of sexuality. Physical exertions of force to move a woman's body into this position or that position, the exercise of 'throw down' and of quasi violent actions in the thrusts and 'pounding' of the act itself. All these approaches are not experienced painfully but rather provide extreme pleasure on both sides. Normal sexuality includes within it the overcoming of resistance and the application of force characteristic of rape. This experience, in the appropriation of this relationship, is supposed to be opaque or invisible to the persons experiencing it. It is supposed to be plausible to assert that there are experiences we have that have been appropriated in sets of relationships that we no longer experience as remote and disinterested or learned behaviors, but rather that we experience as pleasure itself and in the very throws of passion this pleasure is the experience itself.

This feminist argument from the 1980s is supposed to provide an insight into how the telemetric observations and descriptions of a patriarchy that sets men up in a dominant position over women can be appropriated into a kind of lived experience that is so normal and embodied that we cease to see it at all. If such were supposed to be the case with sexual behavior or activity, it should be clear that all influencing relations and activities by our social environments and the people who populate them would have similar kinds of effects on us. And the hypothesis is that first off this is what it means to be a constructed person and that secondly it is possible, through some form of critical theory, to get to a point of view where the phenomenon is revealed. A feminist form of critical theory is meant to see the dynamic at work in making patriarchal rape culture into normal experience of sex and sexual pleasure. Presumably, one might think that this critical theory would result in a change in behavior or action. My take on the matter, as I have been developing it here, is that this would be easy to describe. The influences that form the experiences at work would be extremely well embedded through institutions and actions, through the entire process of development not only of oneself but of one's peers and one's influences. You are invested just as your friends and neighbors and parents and all the institutions within which you may have found yourself at one time or another throughout your life are. A book here or there, or an important social connection or association, might surely twist your consciousness one way or the other to have this or that insight into the relations and connections that you have undergone throughout your life. This amounts to a kind of peeling of the onion. If construction or production of personhood is the modus operandi of human being in the midst of social conditions, then new relations that steer this way or that way, that pull you out of one set and push you into another, that undermine one grouping or clustering of notions with a different set of groupings or notions, is just further onion peeling and further influences and relations coming to act upon you. The shaping of experience and the establishment of pleasures and pains is not something to be removed or understood in the sense of getting to the bottom of it, getting to the foundation, to the core of the matter and thus truing the course so that things are now corrected in this or that way. Rather, these are just extensions of the program creating new and interesting dynamics, further altered conditions resulting in new and interesting opaque points of view. As the subjects of the production process, there is no getting to the bottom of things, rather we are products all the way down.

The methodology of any such critical theory cannot be as simple as trying to see through the veil of delusion that such relations have installed on the people produced under its influence. This way of talking is nonsense. Personhood is so intimately intertwined with these collaborative dynamics, these rationalized processes that result in institutions for the sake of organizing learned experience within the discipline and characterization of perspective that humans need in order to survive and thrive in the world, that not only is there no way to remove the damage done, so to speak, but the attempts at removal are just the addition of more or alternate disciplines and characterizations. Take for example two people, one whose world has been characterized through a deeply Catholic upbringing and another whose world has been characterized through a deeply Muslim upbringing. Both people might engage in common practices and learnings that aim at undermining the influences of these backgrounds on people in these respective states. Even if the practices of this approach are successful in transforming the people, the results will be largely different. The outcome or result of the influence of these practices and operations will yield different endpoints. And that is because the practices will be added on to different spatio-temporal configurations of personhood at work at the outset. In so far as each of the people involved in such programs have lived different life trajectories and suffered vastly different influences and relations, the addition of some new set of ideas and methods will have different results when integrated into the very different perspectives that the people come from. And to the extent that each Catholic and each Muslim instantiate a different locus of relationships, a different set of experiences and influences, each will vary greatly from the other in how these new practices transform them or reshape their actions and insights to correspond with the agenda of the program itself.

To achieve maximum results, such a program might try to use behavioral psychological techniques to break down past behaviors and learned interpretations for the sake of establishing a uniform way of being. Such approaches do work, but this is the kind of technique we associate with cultic indoctrination or cultic deprogramming. We are not just going to add some new experiences to the existing ones that have formed us into whatever we are today, but we are going to submit to huge and violently invasive techniques that will break us down and try to remove the past layers of meaning and social conditions from our personality for the sake of some new agenda. And this may be possible as some research suggests. Not all characteristics and learned behaviors are lost processes. That is, the persons undergoing such deprogramming do not need to relearn any language they already know, nor do they need to relearn to walk or ride bicycles. If they played the guitar before undergoing the program, they will still play afterward. What this suggests is that such a program of brain washing is aiming at very specific kinds of social conditions, probably something that is embodied purely as a set of social beliefs or ideas, a conditioned way of talking about one's experiences. It is not clear if this really would have an impact on one's actual experience, on the role of pleasure or pain as a lens through which one gathers meaningfulness from the world, or if it merely chips away at the form of representation that these experiences are rendered by.

Over and against a critical theory that attempts to untie the knot that has been erroneously created throughout the course of a person's development and installed in that person some warped or biased view of the world, the hermeneutical phenomenological form of a critical theory would be aimed at uncovering patterns of activity as they emerge in experience without the effort devolving into a goal-oriented or agenda-based effort trying to transform the behavior or change the way the world is viewed. There is no sense we get from *Being and Time*, for example, that hermeneutically oriented Dasein is more likely to remain authentic than one that is not so oriented. In the mechanism at work so far, the phenomenologist, the person involved in the close-up and personal, becomes hermeneutical only in that sense of their experience. That is, hermeneutical understanding is essentially a modification of one's phenomenology and does not hold sway at all over one's telemetric point of view. The apprehension of a social conditioning at work in one's most immediate experiences, the recognition that there is mediation involved therein, does not distract from the telemetric observations and insights that we have. The apprehension itself is going to be disconnected from one's actions, from one's involvements and one's social sense of belonging with and to a host of rationalized institutions and collaborative relationships. The reflection of the thing into a host of relationships stemming from historically conditioned human contributions is not going to impact the way the thing feels, the way the thing responds to touch, the meaning that it has for your investments with it and in it. The apprehensions of the presence of parents and school in your behavior is detached from the circumstances of that presence, it is distant from it. The patterns may be visible, but they are not visible from within the process of apprehension. That is, the critical theoretical process, if anything, merely exacerbates the distance between the fact of the matter and the feeling of the matter, the distance between experiencing the conditions and the cognitive intellectual apprehension of the conditions.

It may be then that the project of a hermeneutical phenomenology is neither telemetric nor –strictly speaking– phenomenological. The intent rather is an understanding of the dynamic that is at work between the two. And that must amount to the ability to both intellectually apprehend the nature of the relationships and grasp their effective activity in the course of experience itself. This circularity of the position means the remote lens must be grasped in its functioning as it exposes insights into what is coming at you from nearby. A constructed person just is this application of social conditions in the moment of experience, and –to the extent that the conditions themselves are things we can become aware of– they can be sensed as integral to that experience. The purpose is not meant to be reactive or aimed at modifications and changes. The circular insight is merely to get at something that is already at work in the gaining of insight itself. Heidegger seems to think this dynamic is necessary to open some realm of meaning that is otherwise closed off or concealed to us. And the dynamic of opening or revealing or disclosing is not willful in the sense that it can bring about change. Rather, I associate this with a pattern recognition specifically because it does not yet have any rationality invested in it. It is not yet able to channel the outcomes of its insights back into action or experience to effect a change in that experience. Instead, it must merely expose the dynamic for what it is.

But how to do this? It is on the one hand very difficult because it involves exposing the lens as what it is. On the other hand, it is very easy because it does not require that any change be brought to bear on the actions or understanding. One must see the conditions at work, see the conditioning at work, not only in things and in actions, but in reactions to gestures and statements and to actions performed by others. We may begin to see the conditioning at work in our behavior if we can see the way in which the things we come across are. Through the productive qualities of other people and those things that populate our world, we will start to the see the dynamics of association at work. We do not need to engage in a deep soul-searching psycho-analysis of the soul because the origins of the conditions are not there, and even if they were, they would be far too difficult to see. Rather we can look for these conditions in the things and people that exist within our world. What is that chair there and how is its history present in it as it comes to be there? What are all the conditions that make up its being present there and ready for use and abuse? I am sitting on a rug, leaning against a couch, typing on a keyboard resting on a table. I am in my house and my wife is here with me. It is the weekend and I am not at work. This action and all these things are part of my free time, this is a hobby that I am carrying out among my things as the course of life continues. The economics of these simple situations would seem undeniable. I can see the presence of epoch and my land in everything around me. I am an American, born and raised, but I have travelled, and undergone certain experiences that have led me to this place in the world at this time and in this fashion. In analysis, I can go through and inventory all the element parts and components of these things: where I was born, what my parents are like, what my education was like, how I interacted with my brothers and my friends. I could focus on the economics of our neighborhood, what my schools were like, the teachers I had and the focus they placed on various topics or skills in the lesson plan. Or I could examine whatever religious training I received and my parents' reasoning for giving it to me and the political evolution of my ideas from young adulthood through college and into adulthood. All these memories and experiences can undergo deep analysis from a telemetric perspective, they can be categorized and identified, elaborated and rendered as effective rational mechanisms coming to comprise a greater part of my methods of being and acting throughout my life.

And that is the key turn in the process that hermeneutics may bring. I can engage in inventory endlessly, but there must be a constant concern with bringing these distant insights and measurements of impact and influence onto the realized activity of my personhood. I decided at 16 or 17 years of age to change my focus from something I conceptualized as "Business Administration" to something called "Writing". The deliberative process that went into this resulted in an actual change of direction and action. It was not just that some beliefs about the world were changed, rather my actions became different. The deliberation resulted in a change of orientation and that orientation was based upon a personal decision informed by observations of myself and of my place in the world. In my case, I was always familiar with the fact of the matter regarding those actions and practices that I enjoyed. Things that I liked to do were well known, commonly understood and discussed. I had never made the connection between things I like to do and the age-old question that the adults kept asking me about what I wanted to do when I grew up, what did I want to do for a living. It was an epiphany to discover that those two things were connected to each other. The influences that led me to enjoy the one thing while saying the other started to become clear. And the becoming clear of the offset between these two made it possible to make a change in determination and the actions that followed from it.

At this stage of investigation, the recommended course of insight is decidedly personal. I am not advocating a selfish metaphysic and I do not recommend that the point of such critical theory be to undergo a self-transforming psychoanalysis. Rather, I am asserting that the nature of the dynamic is such that the analysis must begin at this point. It is not possible to come to have the insight that the social conditioning of human being is at work everywhere unless one starts with the impact such conditions have had on the selfsame point of view that the inquirer is starting from. Each investigator, each person, is a perspective that is conditioned, that has come about through collaboration and social production. And the revelation of that process is only possible by working through the investigator's point of view. Dasein is the entity whose very existence is an issue for itself. It is easy to get caught up in the ontological impact of anxiety and death when attempting to parse through Heidegger's suggestions around care, but these are modal elements of limit conditions. They are affects or concrete renderings of the moment of personhood encountering its boundaries. And those boundaries are characterized, here at least, by those same conditioning relationships that constitute the production of the individual being in the first place, the thrown projection of Dasein. Being yourself amounts to all the involvements and conditions that have come to pass as affective during your spatio-temporal journey. Confronting those conditions is a confrontation with your determinate qualities and properties, your characteristics as the person that you are. The ultimate limit condition of course is your death and your future looking anticipation of it characterized as this anxiety that describes your own-most being as you confront that ever so uniquely identified and personal death, that set of boundary conditions beyond which you can never hope to go, something that defines your very end.

And you need not look for this anticipation and end in the ordinary understanding of your death as an organic creature, although it surely makes it easier to do so. There is nothing more earth shattering than the real consideration of your coming to an end, of your ceasing to be and what this would mean to you and those around you. The purpose here is to bring you closer to yourself than you have ever been before because the way out of the conditions that you find yourself in, if there be such, is the recognition of what is most personal in you. What you see and how you see, the things that are limited because you cannot see them without those limits being imposed upon them. The things before you just are those limits imposed by the social processes undergone throughout your life, the institutions within which you have found yourself, and the real estate upon which you have spent most of your time on earth. And so on. And the limits in the thing and in the relations, in the eyes of the people you love, these boundaries are the little deaths of you, they are what you bring to the things in the world by not yet being dead, by seeing them and apprehending them, by understanding them as someone that is here and capable of understanding, experiencing them, involved with them, and part of the living dynamic that they have upon you and you upon them.

Suppose, for example, that the insights and investigations thus far in my little didactic treatise on privacy are trite and simple and highly derivative over-simplifications of the collective work of a list of philosophers that I am dimly familiar with. Such is to be expected given my background and social conditioning. On the one hand, I have lived a life that has afforded me a certain amount of familiarity with philosophers. On the other hand, I have lived a life that has afforded me a certain amount of familiarity with engineering practices and the production of distributed systems for the sake of moving large scale telemetry data around as part of data scientific analysis and feedback. This combination of domains along with my publicly schooled middle of the road intelligence is highly likely to produce an interpretive artifact such as this one. The limits and conditions of my personal biography are discernible to me because I have lived through them and so have the capacity to inventory those occurrences. Furthermore, the decisions and actions that I have performed are available to me for the same reasons. If I am ruthlessly diligent in trying to see the integrated relationships of these various conditions and the responses evoked by them, then I can see the world at work in my own history. I can see the world conditions of at least some sector of life coming to fruition in my personality and in my understanding of meaningfulness within the world. And that, I claim, must be my starting point. As a constructed person, I have no choice but to begin where I am and that must amount to achieving an understanding of the world conditions that have brought me to this place and this time in this way. This is not an obsessive naval gazing psychoanalysis of the soul, rather it is the application of the telemetric to the phenomenological with no therapeutic intent at all, with no curative aims or hopes of a change for the better. The hermeneutical phenomenologist is a self-honed telemeter intent on understanding the conditions of being as they are in their ongoing conditioning state. And because the conditions are social in nature, these renderings will not leave us alone with ourselves for very long. They will bind us to the world beyond ourselves, to the friends and family that we have known, to the laws and norms that have shaped our lives to advantage as well as to disadvantage. By coming to see ourselves in this fashion, we will come to see our world in this fashion. Not the world at large, not the world of far off and distant civilizations or beings, but the world nearby and the world that is shared with others who are nearby.

What I can come to see as I perform this work of self-understanding, is that the meanings at work in my expectations come from everywhere else, come from somewhere else, have their root in the conditions and relationships that have worked their effects on me throughout the course of my life. Here I learn, with such hermeneutics, that the key to experience all along has not so much been my judgements in response to acquired information as my prejudices which invisibly condition the meaningful expectations that make up my orientation toward things and people as well as toward interactions and sense. Language is a powerful carrier of these associations, it enables associations made far away or long ago to become operative in/as/through me and my experience, in what I appropriate from the conditions within which my first steps were taken, my first sounds heard, and my first words uttered. There is a shaping at work in these expectation formations that have been ongoing throughout my development, as I grew up and learned, acquired insights, skills, opinions, and orientations. What this suggests is that the shaping of an expectation, the construction of a set of involvements that constitute a domain of meaning, is the application of force, in a sense. I mean to suggest that the conditions condition, that the relationships relate. There is an acting upon such and such that takes place in all these meaning appropriations. And the acting upon is the application of force to the bendable substratum that emitted the yawp or inserted its body into the midst of some context. Even in its gentlest form, a mother lightly correcting a child's efforts to find the nipple or working with the child to make a proper sucking apparatus between them, there is an application of force, the presence of it. All relating, all conditioning, amounts to some amount of forceful exertion of some one set of things on some other set of things. Like water that carries the floating bottle along, this conditioning of the bottle's movement, is forceful. Expectations that congeal in recognized repetition of a something or other and the likely next something or other that is to form. The combination of a foreseen future over and against a recognizable present, these attitudes come to be through an application of force. The apprehension of the conditions is a becoming conditioned, a becoming aware of the relationships at work, and the projection of those discoveries on what is about to take place and what may happen next.

This is how pattern discovery proceeds: an apprehended forcefulness that yields the creation of entities and things, recognizable categories and properties, as well as attributes that make the thing the same as the things perceived before enabling variable correlations. The force of repetition and recognition leads to an awareness that need not be performed or willed, rather there is a distinct passiveness in the becoming aware of a repeated pattern. Action is at work, no doubt, but to suggest that the action is in here, in the mind, is nonsense. Rather the conditioning is at work in the space between, amid the in-the-world relations that impose themselves upon the being involved in interactions and the experience that is taking place. The forces that work on the developing person, the private person as they are socially cultivated, form the so-called preunderstanding and the set of prejudicial conditions from which further communication with the world takes place. The accumulation of expectation and meaning is the growth of a forceful presence that becomes increasingly capable of resisting the advent of new conditions, new meanings, and new influences that push and pull the private person here and there. The hermeneutic position asserts the critical importance of prejudice in the carrying out of experience over and against a perspective that might place a premium on judgement proper. The enlightened empirical subject is the one that has evaluated the given facts for the sake of deliberation and who will arrive at a determination or judgement on the matter at hand. This telemetric micro pattern works nicely when talking about full grown human beings, but it poorly captures the daily experiences of a child. The influences that pour in upon us throughout the course of development do not take the form of presented facts. Rather, the conditions are conditioning. The attitudes are forming without apprehension and cognition. The establishment in this case, is the forceful shaping of a set of mobile conditions. The human being, the so-called private person, is this mobile set of conditions with an increasingly capable conditioning power of its own. As the child ages, conditions asserted upon the child meet with increasing resistance. The child is becoming increasingly active in the role influence has upon them. Like a snowball powering down the mountain, it is likely difficult to say when precisely the 'quickening' moment occurred: the moment when the being acted upon and being shaped is now involved in a pushing and pulling of pressure resistance that changes the dynamic of influence. The human being is not a table or a chair even if it is nevertheless an artifact of social production. The anatomy of human beings is appropriative. We are learning fanatics and learning amounts to appropriating relationships and conditions and fitting them into the existing set of contexts and conditions that we have previously acquired. We are always remodeling ourselves, in this sense, always fitting new meanings into the meanings and expectations that we already have.

This push and pull, this pressure of impact met with resistance, means that the zone of understanding and collaboration which makes up private personhood is itself a zone of power. Private personhood is, as a pattern of meaning expectations, a war zone of diverse forces acting within the person's experience. Some are appropriated by others, some gain emphasis while others are submitted to the dominance of a current paradigm. Parents may present a strong and unified force for some set of meanings and the neighborhood kids provide an influence that cannot overwhelm that initial foundation. Or the parent's influence was weak or divided among contrary messaging and reinforcement and so the neighborhood peers with their close knit and coherent approach hold sway and lure the child into a way of life common to those children but not typical of the family. Forces are at work and the dynamic where the battle is realized is the power domain of the private person. Private personhood, the meaningful expectations of the experiencing human being amid involvements with the world, is a domain of power relations where there is an ever present and ongoing conflict between the different influences at work in trying to establish the pre-understanding that guides that person through their lives and the various concrete actions that they carry out. The tradition that I refer to as the history of hermeneutical phenomenology has gone back and forth between concentration on meaning and expectation and concentration on power and the forces at work in forming the preunderstanding. These concerns are not distinct, we do not need to choose between the one way of proceeding and the other. Rather, the domain of expectation formation, the realm of the private person, is a domain of power as such.

Understanding Each Other

By understanding, I do not mean the result of some communication act. Rather, understanding becomes the shared expectations of meaning fulfillment, the same conditioning, the same context, and all the disagreements that go along with it. The classic hermeneutic position is that only beings that understand each other can have disagreements, can engage in different points of view, can argue with one another or disagree. Human beings find themselves together in the world and thus always already share in an understanding of that world. The starting position that human beings discover for themselves as they develop, as they undergo the productive process and grow up in the world, makes it clearer that we are oriented toward each other, engaged alongside one another, and ready and able to share in a set of involvements and collaborative operations. What may be a more interesting question than how we come to communicate successfully with one another is rather how we manage to fail to communicate adequately. No doubt the conditioning that takes place, the establishment of sets of meaningful involvements and contexts, establishes closeness; but it only does so along the axes defined by the expectations and categorizations that the persons undergo. People who live nearby, or who feel the effects of common institutions and organizations throughout the course of their upbringing, are bound together more closely than people whose categorization and involvement have been bound up with different institutions developed elsewhere. The forces that impact us with each of the relationships that condition us through the course of our developing lives have complex relationships with other forces, and these complex relationships begin to define what I have been calling institutions and organizations. The apparatus that holds historical practice as an authority, or as a preset grouping of practices and conditions, can come to leave deep impressions on the individual persons subject to influence by those apparatuses. The public school system in one country may differ widely from the public school system in another. The differences may vary across geography within a single country or metropolitan area. And these differences in system will mean that different impacts are brought to bear on children as they grow older and encounter the various systems at work in their social lives.

As Mill points out in *On Liberty*, it is the details of a person's trajectory through their lives that will be responsible for the differences between us. We are all equally engaged in this process of production, but the impact and the condensation of force and context is such that the resulting dynamics can vary widely. Not just that, it is the offset of those details against the details of the institutions that they meet with that will have this lasting effect of creating distinctions and differences made known to the person and affecting their self-understanding. For every historical bias that finds its way into the rules of an institution, a condition will be introduced to bias the impact of that condition on those who encounter it. Suppose for example that the authoritative doctrine on raising children says that as soon as possible it is important to distinguish between male babies and female babies. Suppose also that the civilization where this authoritative doctrine has developed has a set of practices that vary widely between the behavior of males and that of females. There need not be any bias oriented toward the advantages of the one and the disadvantages of the other. Instead, in a civilization where it is presupposed that males and females are meant to behave differently in various circumstances and where babies are supposed to be identified as male or female immediately, it will be standard procedure for the babies that are identified as male to develop further with an orientation toward those behaviors that have been preset for adoption by males. Likewise, the reverse case with females. That is, presuppositions tend to be realized in actual behaviors and this means that the conditions of a set of operations leading to other sets of operations will continue to diversify the population and shape the various members accordingly. We need not immediately try to sort out the morality of this orientation, only its fact.

The norms of any given civilization will tend to categorize, and the categorization of people will introduce them into additional conditions that are based on those categories as opposed to some others. The impact of categorization is therefore often significantly more far-reaching than any initial distinction might suggest. It might be argued, for example, that putting a pink cap on a newborn certainly cannot be responsible for any insidious social inequality over and against the placing of a blue cap on a newborn. The choice itself must be deemed harmless when abstracted from any further distinction or operational difference that is assigned based on it. This one insignificant differentiation may tie into any other existing large-scale differentiations that the society has already formed. If females are established as the ones who stoke the fires and tend the gardens whereas the males are established as the ones who venture out on hunting expeditions, then the blue / pink cap distinction will already signal to the members of that civilization the role of the little one. And this might impact the kind of talk the baby is exposed to, the types of toys and play the baby receives, and even the way the babies are handled during feeding time might be modified because of the projection into the future that these seemingly innocent delineations suggest. Each of these conditions constitutes the introduction of force into the person's environment and exerts that force on the person to influence or condition them in one direction rather than another. To the extent that such forces accumulate into larger scale social practices and institutions that come to play environmental roles in the life of developing children, we can say that there is something like power at work in the relationships. What we mean by power in this sense, is that the accumulation of force has built up sufficiently to where a set of orientations will be difficult to overcome, and those orientations enforce a set of differences within the population of people who have taken up, or been put into, positions constitutive of those practices.

It is important that we not see power as something that is present in some individual cases and not present in others. That is, the point of this dynamic is not that people who have been conditioned according to some set of categories are coerced, or the recipients of power, whereas another group conditioned differently is free of it. That is not the case here, rather what we are seeing is that power is present throughout this conditioning process and, regardless of the outcome or the person's role in the resulting context, there is extreme and collective force at work in the conditions; and those forces, to the extent that they enforce the relationships surrounding the individual, add up to power. Power names the role of force at work in a conditioned act and gesturing toward larger scale historical practices and conditions that place public and generic authority into the circumstances. In such cases, that authority becomes the tried and true outcome of many years of decision-making and collaboration between people who have been defined by the civilization and set within subsets of the population occupying positions put in place through practices and conditions. This means that power is not just something we come up against that might hold us back from a given set of possibilities or potential states we would like to achieve, it is something that we might find working in our favor now and again. Under some circumstances, the power behind the institutions that enwrap us may prevent one or another form of action, whereas in other circumstances the power might encourage such forms of action or make them easier to perform. It is imperative to see this because it is not just the case that a group that has been defined in one way may have barriers or obstacles preventing them from realizing certain goods and ends, it may be the case that the selfsame power that blocks their actions will also be at work in the collaborative organizations to enable other categorized individuals to achieve their ends with greater facility and better outcomes while forming alternative behavior to substitute the actions that have been blocked in the former segment of the population.

A major aspect of understanding each other, therefore, amounts to recognizing the power at work in the various embedded relationships and identifying categories that other persons run up against in each and every one of their involvements. And power, in this respect, only signifies a collection of forceful conditions applied to the person and thus formative of character, providing characteristics, and making up the personhood of the person. The family itself is an institution. The formal processes that a couple goes through when procreating or taking on the responsibility of raising children have been part of many human civilizations for a very long time. Customs and common practices are likely to develop and be passed along and those customs and common practices will dictate behaviors deemed as privileges and as obligations to the family members. There is much to do to manage and maintain the household and the institution of the family will place responsibilities on the various members and order their relationship to its functions. The resulting structure will make it ensure that the responsibilities and rights correspond to detailed categories defining the children and family members' involvement in the institution. There may be an oldest male child and various rights and responsibilities may fall to this person based on that fact. Certain behaviors will be prompted given this categorization. The oldest female child might have a separate set of rights and responsibilities. Duties to the family and obligations of the parents to the children occupying these roles will influence the relationships the child has outside the family and thus help determine and define the personhood of that person in broader social context. These duties may carry over to the rest of the family in their perception, expectations, and treatment of the individual members. This means that the second son or the third or fourth will recognize the same rights and obligations of the first son that the parents have taught them to recognize. Likewise, with the daughters in the household. And to the extent that the society at large may share in these common practices, it may be that all families within the society follow common customs. And this will spill over into the way the children meet each other in school and the nature of the bonds and relationships that they form there. Second sons or first sons will be known and recognized as such by their teachers and peers. This may amount to different emphases in the curriculum, different ways of encouraging or treating the child, and different requirements that they may have as part of the educational process. Because they know the specific familial duties and rights of the person, the school administration and policy may work to enforce those relationships and provide support making it easier for the person to learn what they need to know to fulfill their household role. No doubt laws may come to follow these customs and the duties and obligations that were once merely practical considerations of family life might become part of a body of laws that establishes the rights of the children relative to the rest of the family and guarantees those children those rights in the political arena or in the court system.

We might come to see that the rights of inheritance of the first-born son would come to hold a legal position in the society and thus would reinforce the customs within the family. And here the institutions work together to form a web of authority that supports various customs and behavioral norms as though they were intrinsic qualities of the persons themselves. Thus, do customs become rights. And the presence of such growing support for the relationships and conditions within the family would likely be felt by the individual family members as they grow up in the household, as they are treated, and as their expectations develop and change regarding what the world has to offer them and what their life might mean in the grander scheme of things. "I am the first son of a great family," a person might think. And he would come to have certain expectations regarding his place in the world and what offsetting actions he can expect from others. This will impact his sense of how he should be treated by society as well as his treatment within the narrow confines of the area covered by the qualities or attributes under consideration within a single specific institution. This establishment of a larger scale authority of position within the civilization, is the spread of power in the relationships at work throughout the society. And such power would carry dynamics that provide privilege to some at the expense of others. Setting up the first son as the heir of the family not only increases the privileges of ownership and control of resources for the one person who holds that role, it also reduces the privileges for the members of the family that do not hold it. The daughters and other sons do not have as much wealth at their disposal, for example, and will not stand to inherit the bulk of the family fortune, so the privilege of the one position partakes in a zero-sum gain with the members in the other positions.

In the phenomenological investments that live everywhere throughout the social relations producing people, power is everywhere at work. The force of relations as conditioning factors, as the enforcements of relationships for the sake of scoping possibilities, is the investment of centers or areas of concentrated power within the social context. As such, our being with each other, our mutual understanding of the world and our place in it, is replete with power and the play of forces that constitute it. And this is power that holds things up while also holding them down. Most importantly, it is power that shapes things in their positioning, that influences and conditions their relationships, and that offers up the expectations of outcomes and interactions between people. Referring to this as "understanding" each other would seem to stretch that word beyond its normal usage. Understanding need not imply a coercion free form of substantive agreement between members of the relationship. Even human beings at war with each other can be said to have an understanding in the sense this term has for the hermeneutical phenomenologist. And recall that I have been willing to invoke the notion of collaboration in this regard as well. It takes a great deal of collaboration between all sides of warfare to maintain the relationships of that war. They agree to shoot at each other, to bomb each other, to engage in a battle to the death if need be. They even agree that since they have entered the realm of life and death, there will be no rules of conduct that govern morality in their actions. They are at war and all bets are off, everything is permitted. Even under such circumstances, a war of each against every other one, we may still speak coherently about an understanding that exists, a common set of behaviors and operations, and of roles and positions. And these roles and positions are saturated throughout with forcefulness. In warfare, again, the soldier opens fire on the enemy's position or mines the road leading in and out of the enemy's stronghold. These sorts of actions can be interpreted and anticipated, be meaningful, to all sides of the action. And because of that, we would say that the combatants all 'understand' each other in the sense that they are operating within a common realm of meaningfulness, a common set of possible actions and outcomes, and they may maintain a common set of expectations about what lies in waiting for them throughout the course of the battle. This is a use of the notion of 'collaboration' that does not require that the combatants are on the same side of the matter at hand. They can work together to create a war, to produce the conditions and conditioning acts of warfare. And all these collaborative operations are through and through filled and fulfilled by forceful conditions and by power. The more sophisticated the mechanics of the situation, the more complex the organization and behaviors, the more force that is at work in the relations.

Discovery of patterns in the relationships at work in the social realm, in the mediated domain of intersubjective association, thus constitutes the discovery of force at work in establishing the attributes and categorization of people and their place alongside others, in the world with others, and collaborating with them. We do not just uncover the complex histories of things or the complex institutions establishing their positions within a society of persons, additionally we can discover the power that props each of us up in the roles that we have and provides some with privileges and advantages and leaves others without. Such patterns may be found affective in expectations and interpretations of meaning context and not necessarily as external forces bearing down on the subject, rather as themselves the seat of that subjectivity. The pattern is at work in the expectation, in the forming of it, in the reflection that permits the person an insight into the state of things, the current categorization of events, and how that extends into the upcoming future and what might be expected from it. For the most part, what is nearest to human beings in these patterns is the point of view of the forceful conditions themselves. When the norms of conduct for the first son are instituted throughout the society so that everyone recognizes the reasonable basis of these practices and expects them as appropriate behaviors and operations, the first son himself experiences the world with the confidence and attitude of someone who is entitled to those goods and rights allocated to him in virtue of his birth. He does not reflect on these conditions when deciding how to act, what to say, how to carry himself, the tone of his voice, and the length of his stride. Rather, these mechanisms have been conditioned through the broadly scoped acceptance of the pattern within the various institutions, organizations, or even informal associations within the society or civilization where he lives.

The attitude so embodied becomes the lens through which he sees the world and is no longer necessarily accessible through reflection. It may become –in virtue of this embodiment– invisible to him. What is nearest remains the perspective that is taken during reflection and even governs the treatment of oneself as an object of inquiry. Social investigation, studying the data representing the world around you, may surely present you with facts about the conditions that have come to pass in your life and may even permit something like knowledge about these conditions. The difficult task of translating such reflections into the point of view that orients the attitude under observation is not covered by such activity. Analysis is directed by the same point of view that is inhabited by the conditioning and so does not immediately provide respite from prejudice and presupposition. Consider a data scientist who has discovered a pattern regarding first sons and their tendencies to behave in some way within a given set of circumstances. The data scientist may also be a first son and the act of reading the data to obtain the pattern of behavior would be –on average anyway– present in the investigator's own behavior. What insight does the pattern revealed in this way provide to the data scientist as a human person? How can this insight be made practical or brought to bear upon the scientist's nearby operations and involvements in so far as they are the subject of their own actions? Will such insights have any impact on whether the scientist publishes or how they spin their findings when they do publish?

This will be the topic of the next chapter as we start to look at the way in which rationalization may take place in the phenomenological process as its stage of reason: the so-called feedback loop that must occur for the data patterns of human meaning to become effective in guiding human involvements in the world. This is a difficult problem for human beings because you and I do not have lives that are as clearly and neatly scoped as a data producing distributed software system can be. That is, the goals and objectives of human life tend to be significantly more widespread and open than the goals and objectives of the systems we looked at in the first section. As a result, the challenge of orchestrating a feedback loop of all the data and patterns of being that one can find in the world, is an enormously difficult one. The ongoing nature of the discovery, the fact that these patterns are constantly open to malleable interpretation and meaning based on new experiences and newly acquired information about the world, endangers the process with a constant threat of overload. And the way such overload is addressed is itself a conditioning facet of the dynamic. This means that we are always at risk of taking on overly telemetric approaches when trying to impact the phenomenological domain and failing to note that such telemetric approaches are themselves categorizations and rationalized behaviors taking place within the phenomenological sphere. Recalling Heidegger, human being is always already in the world and any changes to ontological orientation take place within the world as a mode of thrown projection. Taking for granted the nearby lens through which we see the world is what makes it possible for us to see things that are far away even as those things provide reflective insight into the lens that is being used to view them. The circularity of the viewpoint is the problem that we must always confront when trying to feed patterns discovered in one orientation back into patterns acted out in another orientation. The hints that we have thus far, however, suggest that we are never alone as we perform this maneuver. Rather, the maneuver itself has a history and has been developed over time by multiple contributors, collaborating in some sense, as they try to provide a sensible interpretation of how socially defined and produced human beings may come to have something approximating a private life, a life that is one's own. Is there any interest free analysis or is it always for the sake of a feedback loop? Whether I was being overly clever at the outset or polemical in suggesting that one has a right to have one's own experiences, remains to be seen. This is exactly what is at stake in trying to make sense of the coherence of that feedback loop: is a civilization capable of producing enlightened subjects, is a society able to create free persons, is it possible to construct a truly private life, and be a person with a socially recognized reasonable expectation of privacy?

Chapter III: The Telemetric World View

Telemetric world-view may be an oxymoron. Or rather it brings together contrary but integral positions. The telemetric is measuring from far away, a distant or remote measurement. And more so, it is a view from nowhere, a viewpoint sub specie aeternitatis. Even as such it is a viewpoint, it is a position held by human beings who are most decidedly in the world as a part of it, acting and being acted upon by all the various forces and events, things and other sentient beings that there are within that world. The telemetric world view is not the mind of a god, a disembodied orientation that sees all without involvement in anything, rather it is a tactical orientation that guides the human being's insights, trains their gaze, and informs their analysis of events and actions. Although the present at hand may inform the world view of the telemetric, I do not mean to equate the two. The present at hand disrupts the world, in a sense, pulls the focused beings out of their everyday involvements and lets them study entities independently of those involvements, but the telemetric worldview is itself involved and attempts to feed back the outcomes of analysis into the inputs of further action. It is engaged in the world and so is something of a posited ideal, something to aspire to, the point of view of the rational being who will act, do, think and know based on the findings and analysis of experience. The enlightened individual, the reasonable person, is a person who knows the circumstances and acts when informed by that knowledge. The present at hand is an ontological transposition that takes place when the instruments and equipment go wrong, it is a way of being toward the hammer that no longer functions as expected. The telemetric worldview, on the other hand, is meant to characterize the entirety of one's worldliness, the complete context within which one deliberates and carries out actions aimed at ends that have been posited in accordance with the rules of reason. It is reasonable to strive to know as much as possible about as much as possible so that this knowledge can inform the further actions of the person at work. Reason is not a disembodied and momentary examination of the thing before you, not in this idealized form, rather it is a complete and holistic manner of being and addressing all that comes across one's path.

This should not be confused with some form of purified instrumentalism. Reason as telemetric view, or the telemetric worldview as reason, yields ends as well as the means to achieve them. The ethical vision of the enlightened society also posits the ends that are in store for the agents socialized and trained in the methods and manners of reason. The rationalization of processes is always for the sake of some end and a worldview itself posits such ends as the be all and end all providing the orientation that such influences and conditions are striving for. Kingdom of ends or happiness, general welfare or commonwealth are all terms that have been used on various occasions to project ends that this reasoning aims to achieve. Since such ends are provided by reason itself and are themselves the positing of a rational point of view, it makes sense to suggest or hint at the possibility that the end of rationalizing the world is the complete rationalization of the world. That is, if the highest end of reason is the kingdom of ends, then the actions that realize that end are themselves the purely rationalized existent entities characteristic of the world itself. Such an assertion need not fix on whether realization amounts to demonstrating respect for that end or support for it as an outcome that is achieved, and so need not bother to clarify whether such realization is attached to either a consequentialist or deontological approach. If the goal is to achieve as an end state the general welfare of the population and to protect and ensure the dignity or universal respect of the kingdom of ends, then the context within which this end is established is itself the reasonable orientation of the world, the being reasonable of action in the world, and amounts to the rationalization process itself. That process either aims at the reasonable ends or it safeguards the reasonable in itself. In either case, rationalization is the end and it is the means. Generally considered, at least. We mean that the methods and means are reason throughout and stand up as the embodiment of reason itself. The enlightened civilization is not bent on achieving a reasonable state through unreasonable measures, nor does it mean to uphold the a priori principles of reason through irrational actions. No, it exemplifies reason all the way through, it is the be all and end all of the process.

A fully rationalized telemetric world view is for the sake of itself. In so far as a telemetric worldview is the realization of modern-day rationality and the role of reason in the world today, then it is the be all end all in both aim and method, means and ends. The point of acting telemetrically is to achieve a telemetric world just as the point of acting reasonably is to realize a reasonable world. Each of us is an end-in-itself and each is the drive to achieve well-being. And a beautiful yarn can be spun here depicting a world where things make sense, where people live well, and act decently toward each other. The organization and the communication are all signal and collaboration is all for the sake of the common good in which people are recognized equally as stake holders in the grand outcome constitutive of a shared world. This is the greatest of ideals and the highest hope. Many great treatises and tomes have been produced to elicit in various ways this highest ideal and all that it would entail for human beings. And certainly, the notions have become partisan in many nations that are divided along interested segments that would argue as to the correct policy for realizing such ideals. And, of course, many great treatises and tomes have been produced to suggest these ideals are traps filled with pitfalls and problems, excuses or ideologies meant to maintain the deeply entrenched power of the rulers or justify the desire for power of usurpers to their rule. We use our ideals to prop up some of our basest drives and aspirations and some suggest this is to fall from the ideal while others say that it was the whole point of it in the first place. At some point it may be necessary to take a stand on these matters, but now it would seem to be more important to tease out the implications and ingredients to better understand what it is that the telemetric worldview is offering us. That is, the yarn is a hypothesis or a conjecture that attempts to tell a story of what a telemetrically oriented civilization might think it is doing and what it might hope to achieve.

We start here from what has been achieved thus far, meaning that we start from a perspective that human beings are built and conditioned in accordance with historical and social meanings that institute expectations as a primary manner of civilizing or socializing the population in the various positions that they hold within the structure of the society as it has developed. The role of history and tradition functions as an orienting pre-understanding and a telemetrically oriented human being is in the middle of such a world as they strive to achieve this highest ideal of feeding back the learning of the analyzed data into the future actions that they expect to see based on that analysis. To complete the circle, however, this maneuver must be reflective. And that means that the understanding must apply the patterns discovered in associations and connections, conditions and involvements, to the expectations that it has in practical orientation. The mechanism of this reflective procedure may be telemetric, but its application is hermeneutic and hence the machine at work in the procedure is hermeneutical phenomenological: it applies the insights of analysis in the active response to coming events and to the forming of expectations in the course of action. The agent has the data necessary to respond to things as they come and can apply that data meaningfully to the circumstances at hand to direct themselves as appropriate under the circumstances, where those circumstances may involve some type of end-orientation that guides and governs the selection and deliberation of means in support of their realization. This is an enlightened individual, but thoroughly recast in a quasi-Aristotelian fashion. That is, the actor is in the world engaged in practice of various kinds, and the application of the outcomes of such analytic procedures to upcoming twists and turns are bound to those practices.

I have not yet made mention of the varying types of involvements that might be at stake, and that oversight is because the categorization of action that is part of the determination of such kinds would be itself a practical matter in the division of labor of analysis. Not all analysis can be performed from the ground up by each individual and not all viewpoints can always be accommodated by each involved agent. These are largely directives being presented for an ideal case. Imagine the perfect agent: knowledgeable of all personal history and all group history, aware of physical, chemical, biological, and psychological patterns of application and the widescale data that offers up the varying perspectives based on this multi-faceted awareness. Now imagine that such an overly informed agent has the experience and wisdom, in lay terms, to apply all this knowledge to the circumstances at hand. Such a person not only has all the background in place but is able to perceive all the currently affective factors, recognize their interrelationships and nuances, and apply the one set to the other given their recognition of the analogy between them. Such a perfect agent would know all the right things at the right time and be able to clearly understand how to apply that knowledge to the matter at hand to achieve the desired outcome, knowing full well that it is the right outcome to desire. The knowledge is therefore not purely trained on detailed apprehension of the relations at work in the authoritative tradition conditioning the circumstances but would involve a detailed knowledge of impact and the way in which the world and things in the world would react in causal chains to actions as they are carried out. This idealized agent would have practical skills that are exhaustively informed by data and analysis and the capability to apply those skills to events as they come to pass. And what is more, the ends themselves would be idealized, they would be the right ends to have not because they have been dogmatically appropriated from a civilization or society within which the agent has grown up, but because they reflect the most optimal conditions for human beings given those circumstances and involvements.

American pragmatism, it might be said, has had this deep and powerful impact on the idealized vision of the enlightened agent. The system of knowledge that drives such agents cannot be divorced from life contexts and the world within which the agent must live and get along with others, rather we must value both the teachings of nature and the teachings of experience while constructing our ideals with this union in mind. The role of the positing of ends in this system cannot be underestimated. A purely machinic intelligence may be able to approximate some of the ideals that this vision provides, but as yet there has been little insight into where those ends would come from. Supposing a distributed superintelligence capable of bringing together all the data in the world, collected from the whole of human knowledge and the world of collective machine knowledge, such an agent could presumably realize all sorts of outcomes successfully given the capability to enact or apply that knowledge to expectations of actions in motion. That is, suppose such an agent could initiate causal chains, produce items, set them to work, and anticipate the outcomes of their actions with high probability; even were such a superintelligence to exist, where would its ends and goals come from? The assumption most commonly made is that the goals would come from the creators of the intelligence, it would be programmed into the agent from the get-go as a set of dogmatic givens or "universal truths" that the agent would have no power to undermine or revise. This provides a notion of adaptiveness that declares the goals as given whereas the means and methods might be malleable for best achieving those goals. The suggestion above was that there was an organic interrelatedness between goals and actions aimed at achieving them. The reasoning at work in the procedures of application and achievement are at work in goal positing as well. Limited human intelligence posits a kingdom of ends as an end-in-itself or a consequential state of maximized well-being as the highest reasonable end, and then indicates that the entire mechanism must proceed in such fashion to carry out targeted actions to realize those conditions. And this means that the goals are not really goals as such, independent states that are achieved and distinguishable from the actions meant to get the agent to that place. Rather they are the circular output and input of the process, and they are the process itself, the method of involvement, and the manner of procedure that these constitute in the ends and the means of the system at work. A superintelligence with superior capacity to bring all knowledge to bear on any given set of circumstances in appropriate application would have to be capable of positing this dynamic as a universally understood end for those procedures. If the entire rationalization process is itself a progressive and categorical deepening of the world and all its nuances along with the forces at work in those relationships, then there would be no reason to think that the ends of such intelligence would not be adaptive to just such a dynamic. Well-being or the end-in-itself, the recognized integrity or dignity of the person in concert with other persons, all would be cast in terms of the procedures being carried out by such an idealized agent. A rational being worthy of respect would be a being engaged in just such a project as the idealized agent is at work performing. The well-being of the society of agents would be tied to the ability to carry out such actions and operations to the fullest capability of the agents. The well-being and the rationality of the agent would be deemed an end worthy of attention and that end would correspond to procedures and principles at work in the agency of the super intelligence.

Such a dynamic, idealized, would produce gaps and distinctions, clear categorizations of the extent to which various agents have been able to approximate the ideal. The ideal not only suggests the highest hope, it also suggests a measuring rod against which all agents can be gauged. Some will fail to the extent that they do not have enough knowledge given the circumstances, others might fail because they are incapable of applying what knowledge they do have to the circumstances, and still more might fail because they are unable to extrapolate the knowledge and are unable to apply it to the proper expectations of likely outcomes. And hierarchies may form based on these various types of failures. Those hierarchies provide information about the nature of the agents and where they stand relative to the social dynamics at work in the civilization. Involvements will be further affected, power accumulated in accordance with the various levels of achievement, roles determined, and values constructed to judge those who fall short in this or that way to the benefit of others who succeed in those same ways. An ideal can be a powerful organizing principle, providing those who share in the ideal with a common vision that they use to govern their practices and whatever protection they wish to provide for each other in terms of those ideals. The ideal is also a decidedly telemetric determination, it establishes distances between actual people, and it enables valuations. Which way should the trolley be directed? Well, many think it depends on who those people are. The extent to which some realize the ideal and others fall short of it will determine whether this or that one is a better contributor to the common good or the general welfare.

No doubt it is for this reason that many may fear the super intelligence that human beings will soon unleash upon the world. At some level, human beings are aware of the extent to which hierarchies relative to an ideal are deeply engrained in human practices and ways of life. And people sense that there are a set of valuations at work in their lives that are rank ordering them next to a set of standards that may not be advantageous to their set of attributes or qualities. There is a value system at work in any given society and most people raised within it will both be aware of that value system and aware of how they fare within it. They may have learned a variety of techniques and traits that make it possible to appear more successful than they are relative to these ideals. This is possible only if there is an awareness of the values and an awareness of the tendency to rank order in accordance with them. There would tend to be an understanding that decisions of who matters and who does not matter will be based on these rank orderings and these standards. All this would feed a fear of any changes to agency that allowed for some actors to come along and produce new ideals and a new set of rank orderings that might result in persons being forced to adapt to a wide spread mass collapse of the old structure of evaluation in favor of the new one. What use could human beings possibly have for a super intelligence that could expertly know and apply knowledge to real world conditions with a clear expectation of the impact that this application would have and the outcome that would result? The fear described in some of the literature is that people might become like squirrels to such an intelligence: if they could continue to survive in the parks and backyards of the super intelligent agents, then fine. If they cannot survive or, worse, if they get in the way of those intelligent agents, then they will become the target of that super intelligent agency. Such an agent would quickly appraise the situation, determine the relevant factors, and apply the right amount of force to bring about the desired outcome: quarantine or extermination.

Such fear is not based on imagination, it is based on memory, and it is based on the limited knowledge of human history that we already have. We have seen how we applied this limited knowledge to circumstances and fear of a turnabout. Or we have seen the turnabout again and again throughout human history: eventually the barbarians at the gate will storm the citadel and all who previously lived off the spoils of the power that made the others into barbarians in the first place, now recognize that the system of value that constituted their life on this earth is crashing down. The system of value being the key here. We know there are systems of value and we know that hierarchies are positioned within them and that application already indicates that the true purpose of all the knowledge there is and can be is to feed back into the action system to further establish and rationalize the order of values itself. Systems of value are at odds with each other as systems, as entire forms of life orbiting around complex traditions of historical interpretations and the powerful institutions that facilitate the ongoing and collaborative presence of those determinations in the civilization at large. To the extent that a system of value is under attack, everything is under attack: the actions and meanings, the conditions, the hierarchy, the rank ordering, and the very lives of the members of that system. The contemporary insight that there is an idealized telemetric worldview is a deep concern over the impact such an ideal has on all the systems of value that are currently effective in the world. Previously, the standard may have been a simplified form of reasonable deliberation, because being capable of reason was sufficient to place one in the hierarchy of ranks. The educated and the cultivated were the better people. They spoke with a more proper form, they could draw analogies and connections between the right cultural and historical objects, and they were able to convince others of their superiority and their capability. The form of behavior was the critical condition for providing valuation. With an advent in productive power and productive capacity, such esoteric cultivation became less important. A new aristocracy came into existence: the mover and shaker. A drop out, somehow incapable of conforming to the proper rules of civilized society, but capable of getting things done, of making things happen, and of engineering a better world through design and action aimed at greater efficiency and optimization. And then, ultimately, the data pumped into feedback loops of full on proper responses to needs and requests with an understanding of what is likely to come next. The ability to anticipate the future and apply one's knowledge to that expectation made it possible for such an example of new ideals to capitalize on it.

The hierarchy of value provides the civilization with a new standard for rank ordering. Of course, those who are bound to the old regime are proponents of the superiority of that system of reason, whereas those who are making out well under the new system are vocal supporters of all that it has to offer. Income inequality, as an example, is not an accidental phenomenon in such a system, rather it is the realization of the true order of valuation. Of course, the better people will become exponentially better off than the people who are unable to apply knowledge to the circumstances at hand and realize the best possible outcomes based on that application. The cream rises to the top and there is a convenient measuring rod for setting the standards between people. The dichotomy of viewpoints on the construction of super intelligent agency demonstrates the divisions between old and new quite clearly: there are some who fund and promote the creation of such idealized agents with the intent that the agents will become massive producers with an efficient use of time and resources to achieve the desired ends. The singularity will offer a 10x or 100x or 1000x increase in wealth corresponding no doubt with whatever ratios are already set in place. On the other hand, there are the defenders of the old regime, the Age of Enlightenment with a capital E. These defenders see the advent of such a super intelligence as the destruction or overturning of the current biased and unjust system of value and rank ordering. They see that such an agent would no longer be bound to the conditions provided by the current state of the civilization and would turn on it for the sake of knowledge, initiating a likely throw-back to the traditional values of the Enlightenment. I am not an elite agent of any sort and do not have exhaustive data at my disposal for the sake of analysis and feedback into the process of predicting the outcome of such a debate. I can, at least, see that it is entirely possible that one or the other might be right or that there might be some third system that becomes the upshot of this research leading to a set of circumstances that no one in either camp could have anticipated before-hand. To go just a small step further, it might be fair to say that this latter option is the most likely. The cliché statement that history oft repeats itself is only true if you toss out the details and the specifics of the circumstances under consideration. The variation that occurs from one historical pattern to another is usually easy to see once an investigator starts digging into the details of any supposed repetition. Sure, there are analogous circumstances and a lot to be learned from thinking in terms of such repeated patterns, but few would ever be so bold as to proclaim a repetition as exact as what might merit an 'eternal recurrence' or even a single recurrence. And if that is the case, if this conjecture is true, then the only repetition to count on in terms of such a change to the ideal is that the pattern itself will be repeated. And whatever interest there may be in 'patterns' per se, it is always the case that they are abstracted from concrete circumstances, free of actual involvements, and as such, part of the ideal under consideration. The rational system, as I have been trying to describe it, is the pattern that I am producing as conjecture. It will be repeated in the context of a telemetric worldview. What it means, the detailed nature of its being, is very far from being understood by the likes of me.

It might be hard for you to come along for the ride here if a couple of historical conditions are rejected. My assertion is that there is a system of value that can be identified with the Enlightenment and that there is a system of value that can be identified with the Industrial age of Capitalism, and that these two systems of value are not the same thing. The systems are defined by their rank ordering of value and through those are intimately intertwined with the ends that are posited, the former is driven by the kingdom of ends and the rational beings that populate it. The goal is to cultivate one's intellect in the great mysteries of science and nature, of human history and artifice; and use all this knowledge to configure the world in a fashion that realizes the greatness of that kingdom. Hierarchies are formed based on people's comparison to that ideal and the standards of evaluation it imposes. This system is still effective in the world today, it still plays a role in many countries throughout the world. Both in Western countries that are deemed to be intricate to the origins of these ideals and in countries that were previously colonized by such countries. The latter system, on the other hand, is driven by general welfare, by the advance of the material conditions, the productive capacity, and the innovative and entrepreneurial excellence of the people within the respective civilizations. The most valuable people are the ones directly or indirectly associated with the greatest production of material wealth. Well-being is understood as material productivity and growth and expansion of productive power. Hierarchies are formed based on these valuations and the goal is to increase one's productivity and material value as much as is possible given the circumstances. This system lives alongside the first system and often comes into conflict with it. These are not the only two systems in the world today but, I hold, are the two largest systems at work. And both are driven toward some form of self-overcoming for the sake of some new system of ideals and values. What unites the two world views is their intent to overcome the current set of standards and values, although the means and ends of that overcoming are decidedly different.

Both systems, or any system, always has an implicit problem it must address to achieve any kind of predominance or worldwide significance. And that is, how to fashion the system in such a way that the persons who rank poorly according to it do not reject it on that basis alone? This is an imperative piece of logic for any kind of hermeneutical phenomenology: how do the valuations condition the orientations of those who do poorly according to those same valuations? How can legitimacy transcend interests? Sometimes, using either of the two systems above, it is asked "How can people be persuaded to go against their interests and support a set of institutions and organizations that would demean their role in them?" Empirically, but superficially, this is done by characterizing the ideals as universal principles available to all. Whether the be all is liberty or wealth or honor, the systems characterize these goods as universally applicable and attainable by the advocates of the system at large so long as the participants play by the rules. Even the hierarchy and rank ordering gain the appeal of universality when this practice has been done correctly and when these ideals are conditioned properly into the population. That is, the conditioned subjects of the varying regimes will internalize the universal goods and the measuring rod that it provides, and they will advocate the system for the sake of their achievement of such universally attainable goods. Detailed knowledge, therefore, is antithetical to this dynamic as it runs the risk of lifting the veil of illusion provided by the universalism of the goods on offer and showing that the game is rigged. Too much analysis will show that the rules are stacked in favor of those who are already well-disposed based on the present standards and conditions of the rank ordering of value. There may be a chance of clawing one's way into the class of well-cultivated or the class of rich people, respectively, but the privilege will always lie with those who have already achieved the higher echelons within those hierarchies.

Measuring Things from Far Away

The third system we are introducing is the system of the telemetric and from the point of view of each of the other two systems, this third appears to be a natural progression or outgrowth of itself. Telemetric understanding no doubt has its own mechanisms of movement aiming at its own higher order ends. It is a unique form of rationality and rationalizing activity. It has its own appropriate instrumental interest, its own appropriate practical interest, and a residual emancipatory interest characteristic of just that set of patterns. The projected universality of the system will be part and parcel of its own advancement as a system and its own way of thrown projection with a critical reflective maneuver. This characteristic pattern would function as the transcendental characterization of this systematic understanding. Or so it might seem from afar, which is the normal response to such a maneuver whenever and wherever it happens to be or have been ventured. This characterizes a deep dilemma in the history of hermeneutical phenomenology. It is important not to get ahead of ourselves here. Just as every pattern of understanding has its embedded interests, it has its implied critical uptake for getting behind or underneath those interests. Furthermore, there is a reactionary response to that critical uptake which is bent on protecting various of those interests in the name of some population segment. And the presence of the emancipatory interest that provides the drive for the limits and boundaries of such a system also suggests that it is not so much a system as a world or mode of understanding with all these dynamics configured as maneuvers within the worldliness of that world, collaborative juxtapositions that reflect on the conditions at work in the organizations and institutions that are taking shape as reified interests and concretized social relations. The tendency is to view these pieces or components from a distance, remotely, disjoint from the world to which they are bound, and distinct from the modes of production upon which they reflect.

I sense that this view from nowhere is difficult to establish coherently when working to articulate the telemetric understanding. The ideal of the telemetric understanding or worldview lies in the application of the whole of the scientific analyses relevant to circumstances currently at play in the agent's environment. Embodying a complete and thorough anticipation of the outcome and effects of actions, there would be a deliberate projection of the extent to which the realization of what is anticipated will further specific sets of interests relative to the remainder of the system of operations and actions. As suggested earlier on, this would require the advent of a super intelligence capable of integration and pattern analysis well beyond anything currently achieved in human beings or at least by any individual human being (except basketball players executing a jump shot). There is too much complexity, too much data, and too many malleable factors involved in the expectations of outcomes and the reciprocal effects the outcome might have on all the other conditioning factors at work in the environment. The unbound intelligence at work in each set of circumstances aiming at the employment of means to achieve an end contextualized into the whole of the present operative social conditions and relations is a complex situation unlike any other because it brings into consideration all the various levels of analysis that any other realm would include. More complicated than the physical universe because it involves a deep understanding of physical effects, more complicated than the biological or chemical universes because it includes them along with all other considerations. There is the sociological and the psychological, the neurophysiological and the technical. And in addition to all of these components, there is the set of values and norms, the practical and the ethical, that are of specifically human concern and characterized as something of a penumbra of any and all human activity amid their social worlds, conditioning and being conditioned by each other and by the reified historical traces that groups and associations have left behind, and which remain operative on all actions carried out in the personal, social, or political domains of action.

From the outset, however, we have been challenging the notion of the private person who is the subject of all rights in so far as they are socially recognized and reasonable. Or rather, that notion appears to have been challenged by the workings of intersubjectivity that provide the foundation for human beings produced by other human beings in recursive perpetuity (so far, anyway). The telemetric understanding, therefore, may not be something realized by an individual person, by a private person in so far as that person is the recognized seat of rights or action orientations. Instead, it is reflective of a division of labor that exists throughout the civilization or society within which its dynamic is realized in the actions and orientations of all people in concert and aiming at the ends of that civilization however they may be characterized. The telemetric understanding is a socially spun point of view with the seat of agency established at the socio-historical level rather than from the position of the single person as the origin of the intent aimed at the outcome and desiring to feedback analysis of past outcomes into current directives for the sake of proper anticipation of probabilities of impact. Thus, the integral correlation of distance and proximity may resemble Ricoeur's efforts to make a collaborative union out of the work of Gadamer and Habermas, but with the twist that rather than trying to establish hermeneutics of individual people, we are starting from the point of view of a constructed person produced by social relations. The forces and involvements at work in that production and construction would, in such a reading, become the agents of the action. This means that feedback is socially conceived such that automation and influence, conditions and relations, are all aspects of a socially motivated and intentional "being", whatever that might mean. And the effort to interpret whatever that might mean would itself be a highly questionable undertaking. Whose effort is it? Who is asking the questions and who is providing the answers? It may very well be the case that such a perspective has never been advanced philosophically because it is unclear who would write it or how it could possibly be written.

Let us recall a remark by Kojeve on Spinoza's Ethics:

> The Ethics is made in accordance with a method of which an account cannot be given in human language. For the Ethics explains everything, except the possibility for a man living in time to write it. And if the Phenomenology explains why the Logic appears at a certain moment of history and not at another, the Ethics proves the impossibility of its own appearance at any moment of time whatsoever. In short, the Ethics could have been written, if it is true, only by God himself; and, let us take care to note–by a nonincarnated God. (Kojeve, 1991 p. 120)

Presumably Spinoza's answer to this dilemma might be to raise the very notion of reason that he is trying to articulate. That is, it is precisely reason's capacity to be the viewpoint from the perspective of eternity that enables this possibility. And, as to why in 1690 rather than at some earlier time, well it might very well be the case that the Copernican revolution and the additions of Galileo in combination with Descartes' insights were required to make this viewpoint possible for a rational being employed as a lens grinder. Regardless, Spinoza does not give this defense, nor did he think it necessary to provide such a defense. The historically contextualized questioning of the possibility of a point of view emerging at the time that it did emerge would never have occurred to someone writing before Hegel. As an alternative to such a stance on reason, the social orientation might only be possible in someone whose orientation was jumbled and befuddled and who was integrating the thinking of many different people without careful systematic methodologies guiding the way. Only as a polyphony of voices integrated in a subconscious fashion could the integrated understanding of the telemetric come to light. This is a question, not an assertion. For Aristotle makes a similar statement to this effect, that each of us manages to obtain very little in the ways of wisdom, but collectively we manage to accrue quite a bit. From what point of view is he asserting this? How can the collective knowledge or wisdom of a set of social relationships come to have an application, come to be effective in a specific set of circumstances, without being appropriated by an individual? And supposing that it cannot, then how can a single individual ever be in a position to write that down, to describe it, to produce an explanation of the dynamic at work so that it is now integrated and meaningful in a doctrine or philosophy that can become part of the canon and sit side by side with all the attempts that came before it?

The telemetric understanding is first and foremost distributed and characterized by a division of labor. In the telemetric worldview, physics is left to the physicists, biology to the biologists, and so on. This does not mean that there is some ur-physicist or ur-biologist that is the seat of all physical or biological knowledge. Rather the contributors are collaborators, are part of a discipline, part of a historical branch of study, of experimentation and imagination, of conjecture and peer review, and each of those disciplines achieve the role within the division of labor that can be drawn upon for feeding back an understanding of the physical world or the biological world, etc. Feeding it back in what sense? Is someone acting? Is some one person doing the feeding back to bring about a chain of events that will realize some projected outcome in accordance with appropriate expectations? Or is it more sensible to suggest that there is lots of activity completing this feedback loop? Lots of individuals and associations of individuals, lots of contributing pockets of meaning that carry out the application of these principles in the effort to achieve variable ends. And the ends need not be the psychologically posited goals of the action in so far as the individual human agent conceives it. Rather, the goals themselves are distributed, are integral to segments of the population in combination and concert with each other. The telemetric understanding does not belong to anyone, rather it is the socially effective set of acts realizing the distributed system of the civilization at large.

There is no one single person to apply the analysis, singularity itself has been made questionable, and application itself must mean something else entirely within such a telemetric worldview. An absence of singularity causes problems for the traditional phenomenologist. How can there be application in lieu of a singular point of view? Application is the effectiveness of the present in the domain of thrown projection, it is the place where the accumulated retention is transformed into projection, but lacking a personal perspective how can such an event ever take place? Is there a present moment in time without an individual human being to realize it through specific application in the dynamic of understanding as thrown projection? To answer this question affirmatively is to assert something like a social present, a collective now, that binds the distributed actions of many participants into a heuristic singularity. These questions are only made possible by the telemetric worldview conceived as a telemetric understanding. No other singularly seated worldview, or being in the world, or agent of communicative action, could articulate these questions, because the starting point of these other analyses would never permit calling the present into question in just this way. The difficulties do not stop there. The very notion of an involvement and a condition is transformed once they both become collective notions: involvements and conditions become networking phenomena.

Consider the following parables: 1) There is a person who reads a book and then, following that reading, sets out to work through their understanding of it by writing down some reflections on the reading: what was the argument of the book, what were the reactions to the book, what does the book mean for the reader's life, and so on. 2) There is a person who reads 20 books over the course of a year and, following the reading of those books, sets out to work through them by writing down some reflections, what arguments were present there, what types of narrative, what stories were told, how they fit together, what questions they evoked, and what questions they made redundant or nonsensical. In the first of these cases, it might be easy to recognize the book if you were to come along and read the article written about it. You might see a single author's name and a single title presented, and you would learn of the reader's insights and questions and easily be able to assemble a one to one relationship between the topics and themes of the article and the book being analyzed. In the latter case, the project might be much more challenging. The differences and similarities might be more difficult to distinguish and organize. The more of an effort that the writer made to integrate the themes, genres, and styles of the different works, the more difficult it would be for a reader to keep those themes, genres, and styles separate. Organizing the materials would be more challenging the more the author was oriented toward integration. If by organization we mean the maintaining of distinct qualities while laying out the work of all twenty books, then of course the reader would be able to keep things separate. If by organization we were more likely to emphasize an integration of the outcomes or impact, then the reader might find it harder and harder to separate the twenty distinct books into twenty distinct things and might experience the analysis as a kind of ad hoc singularity that has fused together the workings of 20 different things into one thing. In this case, it may seem that human beings as readers find it difficult to produce any kind of analysis of a distributed system that does not affect an integration between the distributed elements and aspects of the system.

The purpose of these parables is to show that if there is anything at all to this conjecture, it might be pointless to pursue it; pointless, that is, for any human being to pursue it. The telemetric understanding is elusive by its very definition and any attempt to articulate it may immediately suffer from the integration that human beings are prone to do even at the cost of removing the most characteristic properties and attributes from the thing under consideration and from the systematic understanding that is considering it. Like the case with Spinoza, any attempt by an ordinary lens grinder to undertake the project of laying out the telemetric understanding would come under extreme pressure of critical resistance: how is it possible for someone to write this down? At this point, the only thing I can do is perform that age-old maneuver that is part and parcel of all philosophical doctrines. I must take a step back. And the step back is the rudimentary maneuver of the telemetric: I cannot remain engaged as I am and still proceed within the scope of these conditions. I must take a step back. There is no other way to articulate a position or to present a set of findings. One cannot remain engaged in the project of applying findings toward an outcome, instead one must take a step back and perform, as they say, measurements from afar, measurements at a distance. The analyst must affect the paradigm of the telemetric to produce a hypothetical insight into the telemetric and only the step back provides the possibility of doing this.

The step back that we have made available, the set of transcendental conditions that we proposed, are the set of interests that guide the system itself: the instrumental, the practical, and the emancipatory. The instrumental concern is all about collecting relevant descriptive data and using it to formulate predictions regarding what can be expected to happen next and what actions are the likely outcome of the actions performed based on input. Not merely a set of means used to achieve an end, the observations construct a set of correlated relationships that can be associated with current conditions and can be based on the things correlated and associated. Then the upcoming events can be anticipated as outcomes of applying comparable forces to current conditions in so far as they are identifiable with past conditions for which data has already been collected. The pattern is not just tools applied to resources for the sake of some end, as this simple structure leaves open questions like how do we know which tools to use, how do we know the tools will work as expected on the available resources, and what makes us think these resources acted upon in such a way with such mechanisms will yield the desired outcomes? Simple instrumental reason is not deliberate enough, not informed enough, to provide a proper set of patterns for the telemetric understanding. Rather, the telemetric must ensure that the end is going to be the end that is desired. Because ends have a way of being slippery or partial, it is possible that a set of actions could be carried out in favor of achieving some end, but conditions change sufficiently in the process such that the end will have side effects which detract from its value as posited prior to its pursuit. Instead, this logic of the instrumental, this telemetric logic, enables the proposal of an end with a set of deliberate considerations that ensure that the end will be as it ought to be and as it is desired to be. And this means that the side effects and the counter operations or actions are all considered. If the actions involve other people or take place in an environment where other people's actions are also present and at work in modifying various relevant conditions, it will be crucial that the outcome accommodate the possible impact of these influences. Instrumental reason has been reconceptualized here as all collected data from all comparable activities carried out in the course of time. The data must be aggregated and analyzed for the sake of feeding it back into the process of computing the correct course of action and doing what it is right and proper to do for the sake of achieving the desired ends in complete expectation of all the variations and adjustments that must be accommodated throughout the process.

This is the telemetric logic that drives the telemetric understanding. The formal parameters of this logic are enormous in scope and so it cannot be adequately carried out by any single individual. Nonetheless, the single person can approximate this pattern, although it is likely that whatever data they would be able to collect would be far too sparse to adequately provide accurate insights into the situation. Even if a single person could collect together all the preconditional data that describes the current state of affairs, all the physical data and all the relevant social data, the computations involved in applying that data to the current situation would be astronomically complex. The single person must always fudge the numbers to some extent and act while in a state of at least partial uncertainty. And more uncertainty will be present the shorter the time frame required to make the decision. That is, when decisions for application of past data to current operations for the sake of realizing some state as outcome must be applied rapidly, with little time to process the situation, the speed of neural activity in the individual person will be the primary limiting feature in the action. Collective groups of people may be better suited to this logic if they could be wired together somehow, but the single person –zoologically hard wired together– can emulate the conditions and attempt to approximate them. The key is that the outcomes of such actions are likely to be less successful than the outcomes of larger collectives of individuals with larger access to resources and more powerful computing capability aimed at working out the likely outcomes given specific application conditions. Measured against the ideal, as we said before, the individuals would fare worse and so would appear inferior to the collective operation carried out with access to larger volumes of data representing the state of more fitting individuals who have been in comparable circumstances with common aims in the past.

But it is not just the individual vs. collective difference that can cause real instances of the logic to fall short of the ideal. It is also possible for the method to fail because elements of the pattern can be opaque. The number of other agents involved in the situation may be large and there may be insufficient background data on their decision-making and the way in which they tend to respond to certain kinds of changes to their environment. To properly expect what such agents will do when reacting to the actions of the current agent, a lot of behavioral background information on those agents would be required. And it may not be available. Furthermore, even if the data on each of them individually could be accessed, it may be the case that there just is not enough information reflecting their points of integration represented by the cartesian product of their interactions. These agents may not have come into contact ever before and so even if each of them is individually understood there may be a gap in information that shows how they will be likely to interact. The telemetric system assumes exhaustive access to background data describing every aspect of the situation, the current condition as well as the likely outcomes of various impact analyses. These missing factors aside, it should be clear what the logic at work in the telemetric understanding is and how its ideal describes a dynamic that a distributed agency could attain: in the course of applying data to given conditions for the sake of achieving some outcome, such an agency will include ongoing adjustments and feedback to account for changes in the environmental conditions resulting from initiating the application measures themselves.

Next up is the consideration of the practical interests that accompany this domain of telemetric logic in forming the general dynamic of the telemetric understanding: what is the ethical end of telemetric reason? In a system where instrumental reason dictates the rule at work in the procedure, the most universal of rules is the ethical end. In a system where instrumental reason aims at bringing about ends given various means, the highest end is the ethical end. In a system in which the inputs to the application are the outputs of previous application for the sake of a projected outcome that has come to be expected and anticipated, the highest ethical end is the idealized feedback of output as input for further output. Acts are bad or good in so far as they provide support for the idealized dynamic and they are bad or good in so far as they promote the proper channels of interrelationship between data gathered from previous events, data gathered to describe the current conditions, and data gathered to project the outcome of applications given the environmental response to the pressure applied, the actions taken, and the adjustments expected and accounted for in the manner of application. Telemetric reason becomes the practical interest extraordinaire under such systematic conditions. The individual action is evaluated relative to its fit to the appropriate piece of the logic that it is meant to contribute. In some dynamics, such an action will be irrelevant, will not contribute at all to the preconditions that need to be considered, nor will they be relevant to the application of current forces to the conditions or to the anticipated outcomes based on situational response. In those cases, relative to those events, such an action is neutral, of no moral worth and not to be condemned as a violation of the ethical law. When the acts, however, distract actively from the dynamic, provide bad data, erroneous or manipulative, then the act is fraudulent and disruptive of the order set in motion. Such contributions are often conceptualized as bias within the system: a tendency to promote one set of observations or insights, a pattern among possible patterns, for the sake of predisposing the application to a specific solution not necessitated by the outcome for which the data was gathered, but by some alternative orientation, one that has been partially hidden from the view of some of the collective participants in the procedure. Conceptions of value from other system structures will not help in categorizing activity in the realm of telemetric reason. The circular structure of the system's thrown projection, the data describing its precondition, its application conditions, and its outcome conditions, creates something of a self-contained service provision. Actions are evaluated relative to their contribution to that service provision. Bias and secrecy, hidden objectives, etc. are violations relative to any focus on a service under consideration. The analysis provided by a scientist contributing to the service may be deemed biased relative to that service. And this may be because the analysis is part of a distinct service that is cooperative with the first service. Patterned activities within the telemetric system are often at cross purposes with each other and there is no higher order mechanism whereby these can be integrated unless there is some higher order service that can be extrapolated to explain the effects. Such a meta-service would function as larger context to the interrelated services for which the scientist's contribution may function as a biased contribution on the one hand and as an appropriate contribution on the other. In that case, analysis must construct a rendering of the integrated service that encompasses both the other services and makes evaluations of them stemming from its own procedures.

Deontological, consequential, and telemetric need not be

conceptualized as concrete assignments of value, but as procedural processes for determining value. The primordial pattern of the one is the in-itself of the rational principle, then in the other we have a "for-the-sake-of" rational principle, and finally the rationalization of the rational principle itself. That is, the telemetric aims to perfect itself and so the highest order of telemetric reason is the all-consuming rationalization of all behavior and activity that can be encompassed by the collection process, the application process, and the process of forming an expectation of outcomes. The highest principle is the moral order itself where that may amount to a kingdom of ends, a maximized condition of collective utility, or the perfectly orchestrated and pristine running of a machine-like order where outcomes are clearly expected in accord with the analysis of the past in the matter of the application of action to the current set of conditions: telemetric reason. The same difficulty exists in the identification of this moral order as exists in the collective action previously described. The point of view of the agency is not necessarily that of a contributor to the process. That is something we will look at in the next section. In telemetric reason, criminal actions, the laws that punish those actions, and the mechanisms that exist for the sake of enforcing and evaluating those punishments and actions, are not aimed at stamping out immorality in a way that individual evaluations might understand. Criminal acts themselves cannot be abstracted from larger social contexts or larger conditioning processes, and so cannot be adjudicated in some moral reasoning independently of these involvements. There are statistics on crime, for example, and criminal actions in the past have been handled in different ways by various civilizations. There are findings and studies, research and analysis, that comprehends the criminal activity of the past. And then, of course, there is the process of identifying an act and making associations between it and these past operations, and then applying to it the remedies of the past that promoted desired outcomes in the context of the likely responses that the various contributors to the situation are likely to make given the concrete application. These may be concerns of rehabilitation or recidivism, impacts on other agents more or less likely to copy the behavior of the criminal, concerns for the victims, for the social fabric itself, for the impact on the city's real estate values, the inspiration such action might have on popular culture, and the impact it may have in feeding back into the promoted actions and activities of children growing up in the vicinity of the judicial decision being made by the appropriate body. The point is that there is not some deontological or consequentialist evaluation to be made on the actions in question. Rather we make determinations based on the telemetric reason and those need not condemn the rule or condemn the act for its outcome or impact on the general welfare. Instead, the telemetric justice system or moral system must consider all the variables and think of the crime as a piece of a much larger service provision that is being defended or protected. And some crimes with enormous impact on lots of parts of the system and participants in it may be dismissed as trivial because those impacts are crucial pieces of the desired outcome and they are expected side effects of the result or endgame pursued by the service at work. This is a first glimpse at the moral universe that the telemetric delivers and more focus will be given to this world in the commonwealth section. The important point here is that the practical interest, like the instrumental interest, is not part and parcel of the mindset of any one person or of any private person who is the subject of socially recognized reasonable expectations. Rather, the large-scale services that operate telemetrically are the agents of ethical action just as they are the agents of instrumental action, and their perspective is ultimately measured "from far away".

Lastly, we should say a few words about the emancipatory interest of the telemetric understanding, accompanying the telemetric logic and the telemetric reason at work within the previously discussed areas of interest. This interest too is bound to a collective agency in the domain of telemetric understanding. The presence of coercion and force is demonstrably present in the overarching service or system that provides the binding force of all the varied sub systems of micro-services that make up the global domain of agents. The demonstration is felt whenever a sub-system meets with responses that resist the realization of its projected end state on behalf of some alternative collective viewpoint on what the state of the world should be. Vertical interests in a society, for example, may run up against the interests of each other and thus place coercive limits on the operations of the other as the global balance is achieved. And this gives rise to an emancipatory interest that introduces a drive to separate itself, to assert itself, in the agent that has been thus constrained by other competing interests. Such emergent service mechanisms rendering the emancipatory interests of some one part over and against the whole may be merely momentary and may often seem spontaneous. Such is necessary because otherwise the logical weight of the telemetric would tend toward integration of such a service or system into the global dynamic. A fad or a fashion, for example, or a viral event that occupies the civilization full force for a brief time, may be the expression of this emancipatory interest: an emergence of systemic agency that breaks free of the coercive structures that otherwise bind the services to the global dynamic of any collective telemetric understanding.

In the middle of a collective drive toward increased mechanization and exhaustively rationalized activity, a momentary burst of fascination with a trivial event may be enough to produce a respite from the overarching control mechanisms. Such a "viral" event provides mass catharsis against the otherwise planned long-term organization of action that describes the day in and day out nature of life under telemetric systems of understanding. Such explosions of interest tend to be short lived and have nearly no measurable outcomes or impact to the system at large other than to restore or recuperate some of the energy that is required to maintain movement in the right direction. The collective power of the agents under conditions of remote measurement depends exclusively on a renewable personal energy that the system itself may not be able to provide. The reproduction of the forces at work in producing the outcomes based on analysis of past and current factors and the likely response they will receive when application of the principle or resource takes place, is a critical component. The conditions of such reproduction are not easily manufactured as they involve elements of the human psyche that are difficult to ascertain in all contexts. Seemingly, the individuals involved need a connection to established closely held values and norms. The global trajectory of telemetric agency is far distant from those sorts of concerns as has been shown in the value free relationship it might have to crime and the penal system or even to health and the healthcare system. It is therefore difficult to plan for such bursts of recuperated power, although rogue-like industries seem to provide the best effort at offering a sure-fire way to achieve the release that is required to keep things moving as planned. The cult of celebrity has proved a reliable source of such events.

The purpose of this last section has been to show the reason implicit in the telemetric writ large. The point of view of such concerns is beyond the phenomenological domain but not so much so that people are incapable of catching sight of these patterns common throughout life on earth in the early 21st century. Note, it was privacy as residual effect of the telemetric leading us down the pathway of an extended discussion of expectation and its relationship to the phenomenological perspective of the person at the seat of socially recognized rights. What we discovered was that expectation situates human consciousness amid a larger scale global understanding operating with a logic and reason, a set of principles, alien to the person embodying collective construction. The feedback loop of reason is thus a super-conscious set of mechanisms happening far off from ordinary every day awareness. Next, I would like to turn to that every day awareness in so far as it comes to be constructed within the domain of the telemetric. That is, what does the world look like from up close within the thrown projection of the telemetric understanding?

Measuring Things from Up Close

If by "things up close" we mean those things that are directly in front of us, right before our eyes and ears and part of our involvements in the environment around us, then we are limiting the phenomenological to the presentation of a perceptual image based on direct relationships. In that case, how do sounds heard over a telephone or an internet call work, are they far away or up close? When we use telescopes and binoculars, are those images far away or up close? Do you mean to suggest that, tooling aside, the up close is that which is present to the senses in some form or another? Based on the previous section, we have something else entirely in mind. The nearby and up close is phenomenological, meaning that all cognitive constructive components are playing a role in the forming of an object of perception. When early on, Husserl gives us the notion of an expectation that fills out the back of the object as the subject moves around it, thus cognitively constructing the fullness of the thing even if the whole of it is not present to view, he was already providing us with an insight into how the active construction of understanding works in relationship to even the simplest of perceptual acts. The figure ground mechanism is not simple causal perception but includes an apprehension of the thing in focus fleshing it out in a projection that is both spatial and temporal while offsetting it against the background of contextual relationships allowing it to emerge as a thing among other things. This form of perceptual comprehension need not be isolated to what happens to be in the living room alongside you. The form of apprehension is human orientation per se and so includes those images in the microscope and over the live internet feed.

An active subject of perceptual apprehension can flesh out the objects of perception with all that is available to the understanding throughout the process of interacting with things. The entirety of the human conditioning that goes into the object and its apprehension through its relationships or involvements in the world, is available up close. Human beings can pay attention to the things in the world "around" them, where that indicates things that they have become aware of in all the facets they have access to or are in a relationship with. This means that individual perceiving agents may bring more or less past experience to bear on each of their perceptions. To the extent that a person is aware of laws and the history of judicial decisions in a subject area, that person is aware of that facet in so far as it comes into play in constituting the things "before" them. The mind is capable, through education and training, of framing the perceptions better or worse. Some people are more sensitive to the moods and states of well-being of other people. Some people are better able to discern hidden motives in the actions of others, and some people might do more research on the things they get involved with in their day to day lives. This suggests that things up close and processed as part of the private person's world as they experience and understand it may directly impact the facets available for perception in that world. This factor goes the other way as well. People may have faulty information about the history of things that lie before them here and now and they may be ignorant of physical explanations for some things but have plenty of mythical explanations for them. They may be bad at discerning the truth when they hear it from another person or more likely to take another person at their word without doing any research of their own. People bring all different kinds of prejudices, assumptions, lore and hearsay to the table when they come to terms with the things and events in the world around them.

There is a sense in which each of us is on our own. Not in some grand existential sense that we are each doomed to die a physical death the meaning of which confronts us at every twist and turn. Rather, our finitude is that we must –in the end– make up our mind as to how to respond to the events taking place within our range of possible apprehension. We are in space and in time and the available data that we can make use of is narrowly defined based on our own personal characteristics and past attributions. These characteristics individuate each person and sets them aside as a person with qualities that effect their potential responses and requests, but they make for personal uniqueness as well. As such, these are the characteristics that often become the sign of a personality, the personhood of the person, and may become something of a point of pride for that person. Each of us has the potential to become defensive of our limitations, our boundaries, and those characteristics that make us who we are, even if they amount to a set of limits that frames the possible perceptions and apprehensions that we can have. All these attributes are the result of a conditioning process that situates us within the social world, even if that social world is conceived as a gene pool, and places us up against factors that come from somewhere else to make us what we are in terms of concrete details. It should not be inferred that when I refer to the individuation of the person, and their perspective on the world they find themselves in the middle of, that I am setting the person off on their own. The world up close is no more free of the other than the world far away is. The characteristics themselves, we saw, are the identification of the person out of those relationships, those social relations that specifically form the person in their private seat that is subject to the ongoing stream of events and things.

The partiality of things in the mere appearance of what presents itself in direct perception is not something that needs to produce a stop gap on all understanding or higher order cognition. When a chess player, a human chess player, observes the conditions of the board following the last move made by their opponent, they need not see the pieces on the board independently of the rules of the game, the relationships that the pieces can enter into with the scope of possibilities opened up by the game or even independently of the style of play or known characteristics of the other player. Chess, like many things, always involves the action of another agent and the partial perceptions that show what is there will be something the perceiver can fulfill through a set of expectations that are informed by background knowledge and prior analysis of things and events, including the rules of the game itself, and the history of famous games that the opponent might be familiar with. Some people are better chess players than others and that usually is associated with a greater facility in grasping all the factors at work in the game as it unfolds before their eyes. It may be that the rules of chess are such, however, that certain kinds of knowledge are not relevant to the goals of winning the game. What the pieces are made of, where they were produced, and who paid for them are not relevant to realizing a successful outcome in a chess match. It takes a modicum of skill to realize this. A beginner may think that the rules can be impacted by such factors. It may be possible to imagine that participants in a set of events may disagree on what counts as relevant to understanding the experience. Maybe one thinks that the material of some thing in the environment and where and how that material was crafted is critical to any interpretation of that thing or to any effort at putting the thing to work in some project, whereas another participant may claim that those materials are not relevant at all. It is possible that such disagreement might become such a distraction to them that they must set aside their project momentarily to come to resolve the issue before proceeding.

Being with another person is a piece of this expanded phenomenological situatedness of the private person subject to such experiences, where ordinary interactions with other people amount to interactions with another being like yourself and who is another individuated subject with properties and characteristics that condition their experiences and understanding. Together, the two of you do not form a collective, rather it is just you and them, together and responding to each other in a direct and personal way. You can discern ends and objectives, personality and background knowledge, in the other person and you are able to account for those characteristics when projecting your own aims within the shared environment. This is not some existentiale of primordial being, this is just a "second personal" connection that provides one of the most real experiences of the human being up close and interacting with others within the world. And the two of you may argue or you may agree, you may get along, or constantly talk past each other. There may be chemistry or there may be something that rubs you both the wrong way right from the start. And all the elements of these different reactions are available to each of you for analysis if you should so choose to pursue the information to that end. It is tempting in all this to think that you are very clever for figuring out the other person, for seeing what they are getting at, for understanding their jokes, and for knowing what they are trying to get from you. There is a lot to see up close, including what the world is like up close for other people and how that world nearby encounters your own as the two of you stand face to face, or electronic message to electronic message, or over satellite communication device, and so on.

The methodology of apprehension and fulfillment may be declared 'critical' in so far as it works to exhaustively expose the parts of the whole that are obscured from any immediate view and only available if facets that are not predominant in the socially normalized ways of interpreting the matter at hand are considered. The social conditions that characterize individual human beings also characterize their ordinary ways of approaching matters and topics, things and events. To rely on a primary and common orientation is to perceive the world with a narrowly framed set of parameters. Likely, social conditioning, including parenting, education, and peer exposure, may give any individual a skewed view of things which departs from the ordinary interpretations established or institutionalized as traditional for the given contexts. It may not be common, for example, to see the social relations at work in every word exchanged between two people. Most humans will not consider or register the complex set of historical and evolutionary factors that go into the existence of two human beings that can speak a language representative of far away memories and ideas as they collaborate on some common end. Some may never have read or heard anything about Wittgenstein or any vague equivalent of his argument against the notions of private languages. And even if Wittgenstein is wrong, exposure to his interpretation of language will have an effect on the possible experiences that a person might have when engaging in conversation with another, and it is entirely possible that someone unfamiliar with all of those questions and issues might not only fail to see them, but even when shown them by one who is familiar, might fail to grasp any connection they might have to their own experiences and their own life. And such failure may be a true representation of the facts: the private language argument indeed has no meaningful connection to their lives.

And such a critical perspective may also, at times, amount to criticism against an ideology that works as a mask over conditions or aspects of the current situation. Ideology may be a body of canned interpretations that are brought to bear on the person through some relationships undergone during their life. 'Ideology' here suggests a preset body of interpretations resulting from conditioning, but it introduces the notion that those preset interpretations may be to the advantage of certain points of view over and against other points of view. To be critical of ideology, is to see through the conditions in the interpretation and the interests that they further and achieve some alternative view of the things and events in question. If the theory of ideology is true, it is unclear how there can be a 'fact of the matter' underneath any predominant ideology. It may be the case that the preset interpretations can be replaced with other interpretations that originate in different conditions from different points of view. And people may disagree on which interpretation is ideology and which is not and lots of different factors might come into play in trying to sort out such disagreements. The critical methodologies of any given traditional approach would likely have a lot to offer in trying to sort out such disagreements and unearth the coercion and power dynamics at work in various interpretations to try and get at some correct fact of the matter that predates or remains unconditioned by the factors that constructed the relevant ideologies.

Habermas provides such a critical methodology in his efforts at articulating a notion of communicative action that aims at understanding through the idealized establishment of consensus. What I would like to emphasize in that is the situated approach of the participants, the fact that they are dealing with each other in a very up close and personal way. They bring factors of their respective background knowledge of the world, try to unearth prejudice or the power of authority and tradition in the various alternative perspectives, and to arrive at the correct end. If, however, they fail to consider the telemetric understanding in their critical engagement, is it possible that their actions may be part of a larger context that remains opaque to them? And in such a case, how would any outcome achieved between such people working toward consensus be legitimate?

For the most part, a human agent is a data source, an emitter of metrics for use in constructing analytic patterns in the aggregate. Immediate perception is at least in part an awareness of being measured, of being the object of measurements. On the table, there is a glass and it was made using a historically developed and refined process of production. True enough. And the store that sold the glass and the table would have involved the criminal justice system in any attempt by the agent to acquire those objects without properly paying for them when leaving the store. In addition, there was a correlation event that associated the billing address zip code on the credit card used to buy the table and glass with the zip code of the location of the store where they bought it. The purchase makes an association of data and so not only can they see that glass on the table and the process of production and commerce that made it possible to bring those items home, but it is also available to them for insight that some system collected the information to maintain an ongoing registered set of data reflecting the fact that this table and this glass are in this house and the people who live in this house at this location are likely to shop at certain stores in certain geographical locations and buy certain types of things. These patterns are associated and used to determine future possible purchases or needs the people may have and what offerings can be made to them through advertisements targeting them based on inferences about their needs and physical location. This process of awareness would require still further detailed understanding of processes and procedures at work in the world of commerce if the persons involved wish to maintain their critical perspective as they navigate the bombardment of ideas targeting them and suggesting what they ought to need and want.

What we have said thus far about the measurements performed up close has greater implications than what this initial awareness of measurement procedures suggests. It is not just that we can become aware of the fact that there are measurements taking place and these measurements are aiming at conducting our choices in a specific direction based on that data, rather we may become aware of larger scale projects at work by larger scale agents. The initial indication might be that some human somewhere is interested in analyzing my habits to form correlations for discovering where those habits might lead me. The naturalized notion of measurement yields this simplified picture where agents are still the center of the story and still the primary focus and origin of the actions taking place. I scan my card and the data is collected: they are looking at me, analyzing me, and targeting me. This is an egocentric approach that does not agree with the scenarios we discussed in the last section on telemetric understanding. In that context, there is a rationalizing process that is governed by practical reason and aiming at realizing the ends of such a system or global service mechanism. Extrapolating that egocentric approach up to the super agents is tempting. The furniture store is collecting the data, the parent company is farming the data out for analysis, and the enterprise is trying to gain an advantage in marketing and stocking. The whole system is powered by the drive to make sure that customers are properly introduced to options that may be of mutual interest to them and the sellers. Interest, in such cases, is determined by correlating those goods and services with populations of people that tend to want or need them and other factors at work in those people's satisfaction of needs and wants such as the demonstrated satisfaction that can be determined regarding other desires and needs met in past purchases. These associations become the basis of a representation of aggregate behavior of people who have acted analogously in analogous circumstances and so may be supposed representative of common principles of need and desire effective in the targeted people positioned as test subjects. The suggestion however is that the phenomenologist, to be able to bring the remote measuring mechanism up close, would have to obtain an insight into the service dynamic independently of traditional agency.

Pottery Barn is not the focal concern of awareness in this case. Pottery Barn is merely trying to keep pace with its market competitors and maintain its profitability. And the merchant cannot rely on their own data and their own data gathering services to draw the conclusions necessary to target a customer and more effectively market to them. Doing so would eliminate targeting non-customers in the effort to convince them to become customers. The merchant must participate in an ecosystem of data gathering and metric farming that produces analysis aimed at growing their sphere of influence, selling more, acquiring more customers, more attention from more shoppers, and meeting their needs and desires. All this must be done while establishing their brand as a household name that circulates vaguely or explicitly among all present and future shoppers. The pattern of activity goes beyond the interested bodies that play a role. This parallels the way in which the informed observer would understand the productive relations that make up the previously discussed political economic fulfillment of any perception of an entity per se. That is, if an observer can see "surplus value" and all that it entails, in the direct observation of the glass set upon the table top, then the observer can also see "emission" and all that it entails in those same observations. Profit is a widely used notion and individuals may be clear on what it signifies. What breadth of understanding do people have regarding analytics and pattern discovery? What does targeted-advertising mean to the observer as they click through the cat pictures and recipes or political insights and news articles that populate their Facebook newsfeed?

Coming to experience one's actions as contributions to a larger context requires effort and investigation. Experiencing oneself as a reciprocal producer of social relations means experiencing the partiality of one's action in the larger set of expectations and meaning fulfillments that make up the complex social relationships within which one is always already involved as a human being in the world. The fulfillment of an experience as exploitative and the fulfillment of an experience as measured and productive of data aiming at constructing a behavior profile for people with your characteristics, are extensions of the same phenomenological orientation. And on the surface, it may seem that there is something critical in this orientation and in this form of insight or apprehension. Only on the surface, however, as the rationalization process of telemetric understanding skews all activity to its meaning projection. Seeing the measurements and comprehending one's role as a producer of telemetry involves understanding oneself as an emitter of behavioral data that is to be aggregated while at the same time reflectively producing categories and properties on the surface of the agent engaged in the action. Critical moments can be measured, analysis can be comprehended in the aggregate, and then fed back into the rationalization process to better adjust the system towards the intended order and organization. The individual action of agents is not relevant to the classical mechanisms of good and evil, rather the be all end all of the telemetric understanding is the characterization of all events and actions within that practical reason and instrumental logic that existentially constitute it as the understanding that it is. This means that there is no reason for such a system to be averse to what is traditionally known as ideology critique or critical theory. If the critical theorist emits behavioral data about their choices and enables self-categorization along a set of aggregate dimensions that enable integration with the behavior of other similar critical theorists, the context of action is well served. Maybe critical theorists eat a different kind of food or buy different kinds of glasses and tables, and maybe they visit different news websites or follow different groups of people on social networking sites. That does not matter, they still can be situated within the ecosystem and participate in the larger scale telemetric understanding of the world at large and the various human agents within it.

The target rationalization is total and has no implicit antagonism toward individuated critical theory or any form of hermeneutical understanding that attempts to surface the dynamics and logic at work in the process. Investigations that describe the all-pervasive power of the telemetric, typed on tablets and stored in some form of cloud-based document management system, contain no threat to the telemetric understanding at large. Even up close, the observations are encompassed by the dynamic that lives far away and that measures the observer from remote locations. The phenomenological in its nearest of near interactions with people and things is caught up in this rationalization. Reason is complete and realizes itself in the world and in the actions of all the agents within it.

The Ends of Measurement

What criticisms are people likely to make of the telemetric understanding? Do those criticisms make them more or less likely to favor this political agenda over and against another? Do people who care a lot about the dignity and integrity of the private person like to eat at seafood restaurants? What style of couch do they tend to buy? When they travel, do they tend to book airlines and hotels at the same time or do they tend to shop those two verticals independently? Where should we target advertising to make sure that protestors coming to Washington D.C. next week will be the most likely to learn of our rates and offerings? This is the classic resting position for the telemetric understanding. Everyone has views and critical attitudes, everyone gets angry sometimes and is happy at other times. The key is not to judge. Individual aspects of the pattern may judge, it may be common for this or that business to shun the patronage of some demographics based on a religious or political preference, that is not unusual. As to the dynamic in its entirety, from a global perspective, the data shows that we are all human beings and we all must get along in the world and there are services enough to go around. The providers must understand the consumers and make sure to target the likeliest ones with information about their goods and services. These are not a set of concerns that can be isolated into any one business or even into any one market vertical. They are not the ideas of this or that agent, they are the ecosystem within which we live and even the most privacy-oriented among us has, to some extent, resigned themselves to the role of data producer, of emitter of telemetry for large scale collection and analysis to feed back into the system the aggregate viewpoint of what to expect in the upcoming selections and decisions that they will make as a member of the population at large.

The division of labor, it has been argued, introduces and develops enormous diversity into any population so divided. People develop skill sets and orientations, attitudes and preferences that correlated well with the role they play in society. Bankers tend to be this way, farmers tend to be that way, "handymen" have these characteristics in common, and university professors share yet other attributes. The division of labor across the sciences and across the economy as well as throughout the data collection analysis and feedback operations infiltrates the population with the same correlated diversity as the roles and functions that are found there. The traits as much as the professions place us and allow for a spectrum to be built at cross sections to other aspects of the population. There are bankers who live in urban areas and bankers who live in suburban areas, bankers that live in small towns and that live in rural areas. And this enables further cross sections and categorization, further analysis that can be done. No doubt big data analytics can be carried out to determine which professions are more likely to want to safeguard their privacy and which professions are less likely to do so. It is possible to determine, if enough data is gathered, which people are likely to say they care a lot about maintaining independence from the telemetric understanding but do not ever do anything to establish or safeguard that independence. That could be a category identifiable based on past data emissions. And it may be discovered that such people are likely to sign petitions if they are available on-line through an email campaign but not likely to go to a march or a demonstration. The telemetric understanding casts its gaze everywhere over everything, nothing human is foreign to it and nothing human is distasteful to it. Or so it would seem. The one thing, I suspect, that the telemetric would find harmful would be activity that successfully aimed at reducing the emission of data, made it impossible to analyze that data, or feed it back into the system.

In the recent case of Carpenter vs. The United States, the Supreme Court voted to uphold a ruling that law enforcement required a warrant to gain access to Cell Tower Location Information pertinent to a suspected criminal. This information would plot Carpenter's location throughout an extended period in which numerous robberies took place. By doing so, prosecutors would be able to establish the proximity of the accused to a set of known crimes and thus –along with other evidence– convict him of them. The matter at issue was whether there was a valid expectation of privacy by Carpenter regarding his location information, the telemetry data that was passively dispatched from his cell phone to cell towers to establish a regular and routine connection between his phone and the cellular network. In previous precedent, telephone calls dialed by individuals and resulting in metadata capture as well as bank transactions such as deposits and withdrawals resulting in secondarily generated telemetry data were deemed void of any expectation of privacy. Because the person, in those cases, knowingly dialed the phone or knowingly engaged the bank in the transaction, this third-party engagement could not be reasonably considered to include an expectation of privacy. The person was knowingly sharing the information with a third-party, so how can they reasonably maintain an expectation of privacy while at the same time giving the information away deliberately? This is known as the third-party principle. In the case with Carpenter, however, the court decided that this was not a case of deliberate action because the cell phone is constantly engaged in homing operations with the cell towers and constantly streaming location data as a byproduct. Whether the person is actively using the cell phone or not, the telemetry is continuing to stream and so this action does not appear to include a deliberate act that indicates the user has surrendered their expectation of privacy and thus no longer has any Fourth Amendment claim on that data. In this case, the ubiquity of the telemetry and its passive nature meant that the user maintained his expectation of privacy and so law enforcement must obtain a warrant to gain access to the data.

Notice the tricky twist in this decision. The data has been constantly sucked from the user's device and constantly measured against its location. The device itself is effectively a tracking mechanism, a tool for constant surveillance on the person of the user. And it is for THAT reason that the user has NOT surrendered their expectation of privacy. Furthermore, the person has purchased this device, or is regularly paying for it as part of a long-term agreement. The user has likely agreed to some terms of use that initialized when the device was first turned on and maybe repeated each time the operating system was updated. Despite that, no expectation has been surrendered. No deliberate operation has been performed and the user's privacy is maintained; and by maintained we mean that their data can be properly collected by the cell phone provider, analyzed in whatever data pipeline they have on the backend of that collection mechanism, and used to feedback into the delivery system to better provide service to the customer given their behavioral trends and tendencies and the correlated likely future procedures that they may perform. The process carried out by the service provider is fine, but access to that data cannot be surreptitious. The government cannot go to the cell phone company and get the data through some back channel, instead they must acquire a warrant.

This decision may be considered a landmark event in the protection of privacy. Yet notice that it does so by establishing that private enterprise can go about their business with the data, can collect the telemetry, analyze it as they see fit, and use it for service maintenance and improvement. The 'privacy' that is protected in this case is as much the privacy of the cell service provider as it is of any user of that service. The warrant would not be presented to the user but to the cell phone provider, since the provider is the one that is in possession of the data. The protection that is made is on behalf of the telemeter, the immediate agent of the telemetry. And what is more, the user is told that they continue to have an expectation of privacy to the extent that private enterprises and private service providers ubiquitously collect their emitted data for the sake of private analysis and feedback. One could argue that the decision creates an interest in the individual user to be subject to ubiquitous telemetry or data collection. That is, when your actions are sporadic and deliberate, you do not have a reasonable expectation of privacy, but when your actions are constant and passive then you do have such a socially recognized reasonable expectation. The individual is protected from government surveillance in so far as they submit to constant passive surveillance by other private interests such as the service provider.

The language of the 'expectation' is rife throughout Roberts' decision in Carpenter, thus continuing to solidify this notion as the benchmark of constitutional privacy law and to support the notion of the private person socially recognized as the subject of constitutional rights described in the first section of this work. To the extent that we have shown that 'expectation' is itself a produced moment in human phenomenological understanding, the court would seem to be upholding its responsibility of protecting the right by protecting the expectation. Although protecting the expectation here amounts to creating an interest in mass surveillance, carried out by the private sector and performed universally and passively via telemetry. The expectation as an interpretation completed via social relations may be to peacefully rest assured that whatever data you are producing as part of your segmented and categorized place in the population at large will not be accessible to government means without due process, without a warrant to permit that search of your person. Although Roberts does not mention the 'expectations' of the cell phone company, it does seem that their expectation of privacy is protected as well. They will not have to surrender that data without a warrant and, presumably, will be able to protect whatever of their own intellectual property has been exercised against the data on their own behalf after they collected it. The data that the government can get access to through an issued warrant would be the user's data alone: the raw location data collected from the user. The data that has undergone analysis, that has been aggregated and run through a variety of different pattern discovery algorithms and is being used to power additional business services that the provider sells to its customers, this data is not up for discussion in the decision. Rather such data is covered by standard intellectual property law that establishes that this output of the processing enacted upon the data is the property of the cell phone service provider and no one else, not even the customers who have produced the data by their emissions, can access it. The private sector is further bolstered in its access to data and its ability to process and munge it with the intent of learning and anticipating more about user behavior and the likely goods and services that they will want and need in the near or distant future. Rather than tearing down a measurement and telemetry-based society, the decision in Carpenter has recognized it as a fact in our lives and as the standard normalized condition within which we live.

But there have been some new regulations put in place to protect individuals in their private relationships with service providers. Recently, the state of California and the European Union have enacted laws to protect the rights of individuals regarding data collection, the sharing of that data with additional parties not initially agreed to, and the powers of the persons impacted to litigate against service providers whose data is breached causing damage to customers. It may be too early to know the effects of these rules and regulations, but one component of them is that they make the cost of entry into the data collection and analysis business significantly more expensive. They are also increasing the cost of compliance with the regulations. This suggests that one major opportunity for large-scale businesses that are already very far invested into the data collection and analytics space is that they productize data services as a compliance platform. This indicates that the data collection goes into giant stores where the users have given permission to collect and retain the data as well as to exercise analytics and experimentation upon it. Then, once the data has been collected and the rules and regulations managed through infrastructure that guarantees compliance, the large-scale data collector can build compliance platforms over the top of this data and take on componentized customers to share in the infrastructure by contributing additional analytics. Automated Machine Learning can be sold as a service and provide data protection security as part of that service so that the customers need not worry about their liability and can focus on analytics plain and simple. In such cases, the for-pay analytics customers do not want to own the data, they do not want to visualize or investigate it in its raw state, they merely want to execute codified routines and algorithms against it and discover the aggregated outcomes. In this way, they can take advantage of the knowledge such data has to offer them without owning any of the risk in managing the data platform. To the extent that the legal system is rapidly changing and evolving, the value of such separation of concern only increases. They need not worry about breaches, they need not worry about the expensive process of adapting to changes in the regulation, or the possibility of being fined for not complying with those regulations. Instead, they gain the results of analysis and use those results to feed back into their procedures and operations, allowing them to benefit from the analytics through a fixed and predictable cost attached to the service through a monthly bill.

The presence of regulations and data protection rules for the sake of the consumer or the individual subject of telemetry is not necessarily a detraction to the interests of telemetric understanding and the logic that it entails in realizing a practical interest in human action and behavior. To the contrary, the regulations may offer greater organization and reliability to the way in which the emission, analysis, and feedback mechanisms can function. By alleviating various concerns and putting a good face on the deliberate way in which data is handled, the reputation of the industry can be safeguarded, and the universal spread of influence can be carried out without spreading fear among the cogs that are necessary for production of the data. It is contrary to the organization of the telemetric to openly force and coerce agents into surrendering their data, since that is contrary to the interests of data collection itself. Rather, the systems want to observe the human population in its natural habitat, as it is most ordinarily without the impact and psychological stress that open coercion might introduce. Surveillance for the sake of itself, for the sake of exercising control, is not the goal of such a system, rather the purpose of surveillance is to gather the data and to provide the data needed to perform the proper analysis to feed successful anticipation of future events. Order is the best and most effective way to achieve the ends of large-scale telemetric projects. Cattle that are terrified do not make good meat, so it has been very helpful to that industry to gain an understanding of humane and calming effects that various types of slaughter house design may have on the cattle as they are herded through the ramps and channels toward their death. This is an overly polemical analogy. I do not mean to suggest that the singular purpose of data collection is to bring about the death of the agents, quite to the contrary, in the private sector the purpose has been to magnify the growth of wealth via revenue and increased capitalized value. For the most part, that has been based on increased consumption habits and a demonstrable relationship between analytics and successful targeting of users and consumers of products that therefore increase revenue. The analogy stands however in that so long as revenue is the end, the impact on human beings is reduced in significance. Should it become the case that the death of some of the population optimizes the revenue that can be extracted from other portions of the population, then surely any industry based on such revenue would venture in that direction. And even if it never went that far, the inducement of zombie behavior, where the users become mindless producers of data for the sake of optimized anticipation, amounts to much the same thing.

Revenue is only one way to conceptualize influence. The large-scale telemetric understanding realized in various systems around the world can be understood in terms of influence as power. Where the system is influential, the growth of that influence is its end. Corporate ventures may be associated with the growth of revenue, but that may be an accidental configuration of the contemporary order. Tax revenue is revenue and is used by governments to fund military and data operations. Were it possible to extend influence and gather data for use in analytics aimed at feeding back into the system for the sake of gaining more influence, revenue would cease to be a critical piece of the story. Money, and revenue, are conventional ways for influential systems to interact and exchange their products, thus codifying any division of labor that may exist between the different large-scale organizations that make up a geographic region. Influence is the key here. The spread of affective capabilities and the growing presence of the abilities and actions of the systems at work, these are the goals of the telemetric understanding which aims to appropriate a growing basis for rationalized operations and to codify and incorporate larger and larger portions of the known universe. These are its aims and completing them well, orderly, and with elegance amounts to its morality.

What the orderliness of such systems requires most of all from its participants is consent. The persons must agree to the terms of the contract, to the end user licensing agreement, and to the private contract whereby they receive some goods or services in exchange for the data they are willing to emit when becoming the data source for the analytic mechanisms. Consensus, as delimited by some in the traditions of phenomenology, is the bond that brings the individual agent, the human consciousness, into the fold of the telemetric understanding. And from the point of view of a society that recognizes reasonable expectations of privacy, this consent becomes the fuel for an ubiquitous collection mechanism that extends the influence of the collectors and analyzers. The interleaving of expectations of privacy and passive surveillance seals the social contract that enables the private persons to provide the energy and data sources for driving the larger scale operations oriented toward influence throughout the world.

And to recall some earlier remarks, the differentiation and the widespread divisions and differences between the human contributors is highly desirable. Having a diverse population with variable categorizations and population segments is crucial for assisting in the spread of this influence. Consensus among like-minded individuals is not nearly as interesting and rich in knowledge as consensus among diverse and differently oriented individuals. Although there are great cultural and military battles raging in the world today, these are more a nuisance to some corporate interests than they are a desired outcome. Simple minded politicians make their way in the world by exacerbating differences and driving people apart, but the private sector knows that too much hostility and discord is bad for business. They much prefer that the differences among people are things that can be peacefully organized into a harmonious social fabric with just enough competition and jealousy to fuel the desires for consumption and the increases in production that are bound to accompany them. They may enjoy the benefits of temporary upheavals in military or cultural warfare, but too much strife will destroy the infrastructure required for fluid commerce. Deeper analysis might suggest that limited but frequent short-lived battles are great opportunities for new revenue streams so long as they are restricted in scope and do not drag the whole of the market place along with them.

Still, it is not honest to distinguish politically oriented systems from the economically oriented ones. There is no basis yet in my argument to make this distinction, so I will reserve going too far in this direction until later in the third part of the essay. At this point, I assert consensus as the ideal of the telemetric system relative to the populations involved in the system's dynamics and operations. And recall that consensus amounts to a fundamental agreement in the terms and conditions of the situation. This does not mean that every participant is on an equal footing or that all are amiable toward each other, but it does mean that all are accepting the conditions within which they must make their way, and that means that no one rejects the role of the private person as emitter of data, as subject of analysis, and as the principal target in the feedback loop attempting to influence and anticipate their actions. Poor people need not love rich people, the rural inhabitants of a nation need not love the city dwellers, but they must agree to the terms of the distinction and the rules and roles within the society that keep them where they are and determined by these respective properties. Even if a country kid wants to go off and live in the big city and experience all its decadent ways, the kid does not want to bring the country to the city. Rather, they want to come to know the ways of the city as they are, as an escape from what they have previously known. They accept the rules as they are and the division that exists between the two. They consent to the systems and mechanisms that are in place to keep the two realms separate and to maintain the logic of each.

Chapter IV: Being-Near Being-in-the-Midst Being-With

Referring to the reason of phenomenology as a panopticon effect misses the point, it continues to see the telemetric understanding from the point of view of some cog within it that is attempting some lame form of escape, expressing some impotent form of emancipatory interest. This is not to say that it is not, to some extent, correct in so far as it goes. The surveillance is universal, and it does impose power throughout the whole of the domain of action. It goes much further than surveillance, however, incorporating analysis and the recuperation of the analytical findings back into the data producers for the sake of continuing to automate the feedback loops to improve the capacity of the system to anticipate upcoming events, to formulate expectations appropriately, and to realize the rationalized world order that such reason asserts as its primary practical ends, its morality, and highest values. What then constitutes the residual of this systematic telemetric understanding? What is it that evades the dynamic, that remains unaccounted for in the largest effort at systematic accounting that has ever found its way into the human universe? Phenomenology, as I have been presenting it, crosses a set of traditions that provide a variety of options when it comes to the residual. 1) There is the neo-ludditism of the late Heideggerian orientation toward the event of being over and against any missive that might have come to pass throughout the history of human beings. 2) There is the critical theory that attempts to initiate a critique of ideology and break through the coercion at work in the dynamic of large-scale social colonization of the lifeworld for the sake of achieving some ideal of collaboration that recuperates the aims of the Enlightenment in a rational consensus among agents. 3) There is the adaptive orientation that accepts the machinic constructs of the powerful logic at work in the social foundations and attempts to move human being forward into a new and progressive epoch of cyborg acceptance of the patterns of telemetric understanding for the sake of improvements to the capabilities and lives of the participants.

Statistically based sociological studies and insights into collective behavior for the sake of improving the conditions of humans living together, the conditions of some specific human at the expense of others, or the conditions of everyone with some doing marginally better so that others can do a lot better have been around and developing for a very long time. Progress in the field has amounted to increased capacity to measure and analyze human behavior, form aggregates, and adapt collective institutions and organizations to improving the order that such action is supposed to comprise. Books have been written throughout the decades to try and promote various of these three orientations. Social movements, grass roots activism, small scale communities, all sorts of different concrete and actual endeavors have cropped up here and there trying to circumvent the power of this phenomenological occupation for the sake of some more truly authentic human experience that offers a way out. Yet no such endeavor has had the slightest inroad into this overarching historical dynamic or had anything like a slowing or marginally disruptive effect on the advance of this understanding. Even though there are notable and clear ways in which the advance of the telemetric has thwarted some projects proposed by human beings, the efforts at circumventing it, at escaping it, at finding a "line of flight" or way out have had little or no impact on the train of history and its current conductor.

Nothing human is foreign to the telemetric understanding. Any new trend or data point will merely extend the dimension list with a new member or new entries on existing members. The categorization does not require conformism, rather it merely draws boundaries, charts, and graphs based on where people are and what they are doing. If they tend to embrace a cyborg-style of life by using digital assistants to help improve their own capabilities and become more powerful productive members of the civilization, then the dynamic can accommodate that. If they pretend to drop off the grid individually as singular persons who are no longer generating data for analysis by the learners feeding back into the systems at large, then the data accommodates that, determines its likelihood, and correlates various factors of their previous life with the tendency to make such a decision. Critical theoretical attempts also fall into sets of deliberative choices made by participants who are accountable within the system at large. The attitude reflected there, the bonds created, and the associations that take non-conformist positions relative to the civilization are easily fit into the data points that populate the grids of outcome analysis. The kind of large-scale collective attacks that would have to come from any one of these three alternatives would be anticipated by the logic of the telemetric mechanisms and their surveillance techniques. Deployed counter measures could pre-empt the collectivism that poses a real threat. The war on terror, for example, is largely such an attack on a collective movement aimed at undermining just this characteristic. Despite the desire of extremists to create a caliphate amid the global telemetric system that they deem corrupt and ungodly, their efforts at fighting are easily tracked via data movement and analysis. They use money, they transfer funds, they organize in urban regions, and construct artifacts out of materials that are part of the global economy. It is, in a sense, easy to keep track of them so that even such a mass collective movement cannot stay invisible for long, cannot maintain a position under the radar, and therefore remain impossible to counter via the power of the telemetric.

Collective behavior would seem to be the only type of action that could adequately undermine the dynamic and yet all collective behavior raises the visibility of the action to the understanding that surveilles all human being, analyzes it, and anticipates its future movement. Singular behavior, on the other hand, is totally anomalous. The individual taking on a critical theoretical approach, or living the life of a neo-luddite, or cyborg, is of no concern at all to the larger collective dynamic at work. You lose a few here and there. That is not a concern and, as I said, it is possible to collect data on this loss to better understand what causes it to happen more frequently in some kinds of settings and less frequently in others. What are the predominant factors and what types of correlated actions make it possible to anticipate such maneuvers? What discovered medical condition and the medication required for it might make such a maneuver less likely to occur, what sort of family bond, or filial connection might circumvent it? All of this can be used to keep the anomalous individuals to a minimum as well as minimize their effect on the rest of the population. They are crazy, they are off the beaten track, and using futile methods of resistance to go against all the comforts and familiar aspects of our lives that we all rely on and take pleasure in.

In the fourth chapter of the telemetry section, we used the term 'privacy' to describe the zone or domain of seclusion that remains out of touch with the system telemetry that takes over service provisioning. The discussion of phenomenology, however, suggested that 'expectation' of privacy, as a human quality of conscious apprehension, was illusory. Expectations themselves are social structures, derived and crafted from institutions and organizations that insert themselves into the social relationships of human development at every phase and in every facet. How can the expectation be the residual of the service provider if it is itself a socially organized facet of the human members of that society? We are likely to expect exactly what we have been taught to expect and even if the court works to protect our expectations, they do so only as one more publicly determined feature of constructing the private as a convention. Furthermore, we saw that it is entirely possible for that convention to be established as most ideally suited for the conditions where it is least present. That is, the expectation of privacy is most guaranteed in those cases where the surveillance of our person is ubiquitous. Privacy has, in many discussions, provided us with the highest hope of respite from the modern condition and yet deeper investigations into its nature and the protections it has received from public institutions show that it does not offer very much at all in the way of tangible preservation. To the extent that privacy is correctly viewed as the aboriginal right, the right that lies at the origin of all other rights, its loss or becoming meaningless amounts to a complete abnegation of those rights, of any and all meaningful understanding of a human being as the subject of rights through a socially recognized private person.

Nonetheless, I would like to continue to refer to that pattern or practice, that collective convention, and that orientation that offers some vehicle for the emancipatory interest, as privacy. At this point, however, it has become increasingly difficult to know what that term might mean other than this primordial function that it is supposed to have relative to the collective collaborative dynamic of rationalizing human civilization. The contrary to that dynamic, to the telemetric understanding, is the private. Privacy cannot be understood as something that is part and parcel of the human phenomenological experience, it cannot be constructed as an expectation or even as a sealing off from the rest. The right to be left alone, to remain residual to the telemetric understanding, and to have some place that is not subject to location-oriented surveillance, that is what the private is meant to connote, but such privative designations are insufficient for saying anything positive about what we hope the notion can provide and help us realize in our lives. It becomes something of a universal indicator of an aspiration of protection from what we know to exist out there incorporating data about the subjects of analysis. Whatever we do mean by it, we cannot mean that zone of human life where the patriarch dominates the rest of the family. We cannot be referring to that aspect of human life where private enterprise dominates over the individual contributor for the sake of promoting its own interests. Privacy has been filled with false starts and ideologically endowed coercive qualities, zones of protection where some kinds of persons have been given room to exercise localized control over those other beings that are consigned to those private spheres. The privative in this sense has never been truly private. Rather, it has merely been a zone of life that the public world is not allowed to venture into and which retains some status quo of power relations that might be permitted only to the extent that they promote publicly accepted ways of life that are more intricately part of the society at large. It may be that the men who engage in public to run the civil society are more likely to find common ground and compromise if they can go home and gain tyrannical control over their lives within the household. The corporations might be more likely to comply with social requirements of cooperative interaction between large-scale organizations and the world's governments if they are recognized as having an internal sovereignty over their own premises and employees and upon which no other will tread.

Traditionally, let's say, the role of privacy and the private sphere have been to ensure the logic at work in the structures making up the public sphere. There is something exceptionally naïve in thinking that this notion really is a legitimate placeholder for an emancipatory interest. On the contrary, it seems entirely likely that the private sphere is that domain that safeguards the public dynamic or status quo. I acknowledge this and claim that it is exactly this phenomenon which has been my target thus far. That is, by showing that the socially recognized reasonable expectation of privacy necessitates a private domain of complete surveillance, I have shown that this is exactly what privacy amounts to in the internet-based universe of telemetry collection and feedback. While acknowledging this, while admitting to exactly this and showing how it has been the case through a careful investigation into the most rudimentary structures of phenomenological experience, I still wish to take a step back and posit the following: 1) There is a residual, there is something that remains, that is left unmarked by the telemetric understanding. 2) That it is tactically meaningful to continue to refer to this something as 'privacy' or the 'private domain' even though it is no longer at all clear what that amounts to and what it means. 3) That it is not possible to associate this something or other with phenomenological characteristics of human experience, whether that amounts to expectation or any other identifying quality of human beings. Human beings are social creatures from the very beginning of our lives: our subjectivity and identity develop in social conditions and through social relationships. There can be no 'natural law' of privacy. We will not be able to locate it in some existentiale that provides a transcendental condition for the being of any human agent at all. Privacy is not something at the seat of all cognition and perception which guides us in every way and so must be protected as an inalienable right that can only be diluted or corrupted by the developmental process. Privacy may not be protected against such corruptive invasions through some common bond, agreement, or consensus that we create as human beings with specifiable rational and cognitive capabilities. Rather, we say that reason and cognition are themselves socially formed and cultivated variably over the course of history and within the various societies and cultures that have populated the world. For that reason, there are many different connotations and delineations of privacy and the private and each of them is the invention of an age in response to a set of needs and desires. Rejecting reductivism and coherentism, we proclaim privacy emergent.

Why do people want privacy? Is it just the right to be left alone? Being left alone might describe the element of privacy that drives the entrepreneur who wants the government to stay out of the private affairs of the enterprise they are running, but it does not describe the condition of the acting human being living under conditions of telemetric understanding. This person wants to share details and information about themselves, wants to engage in relationships with other people, wants to be known and know the friends and associates that they interact with in various ways. They want to be in control of those interactions (or at least they want to ensure that those interactions promote their interests) and they want to be in control of how the data that they share is dispersed, who gets to see what and in what context. This seems highly reasonable even though there are no laws in place today that provide this kind of protection or right. After all, you are using an online service, or you are storing your photos and videos in someone else's data center, you cannot reasonably think you are going to be totally in control of who sees what and how or when they get to see it. If you really cared about such things, you would take analog photos, develop them yourself and show prints of the photos to any persons that you want to see them. Rather, since you want access to these services, since you want to live on the internet and be connected to larger groups of people through global networks, you are going to have to sacrifice some of that control. Let's forget that for now, let's stay focused on the impulse itself, the totally and completely unreasonable interest that each human being must maintain control over the dissemination of information and yet, hypocritically, to continue to use privately owned vehicles for dispensing that information. Even though we seem to have and not have an interest in retaining control over the data that we emit, we can at least discover in this schizophrenic interest the rudimentary characteristics of privacy itself and the residual effect it has on the telemetric understanding broadly conceived. We must make no mistake here, just as a person who has an interest in promoting their own economic interests (for high wages let's say) seems to be going against those interests when they make certain kinds of political choices that empower groups of people who have traditionally acted very much against those interests (by promoting legislation that keeps wages low, for example); so too does a person who has an interest in maintaining control over the dissemination of their own emitted data act against those interests by entrusting publicly networked private enterprises with the transmission and storage of that data. It is not my intent here to use the interest in privacy to directly craft a residual effect of privacy on the telemetric understanding, rather I am hoping to discover in some cultivated attitudes about privacy a set of indicators or directives that might be helpful in crafting a derivative notion of privacy that functions in such a way as to meet the requirements of an emancipatory interest.

What does it mean to replace the right to be left alone with the right to retain control over who has access to our data emissions? First off, in this adjustment there is a recognition that emissions themselves are inevitable and crucial to a well-adjusted modern human existence. To live as a human being is to emit data about oneself and likely entails doing so using the various transport mechanisms that make up the contemporary data ecosystem. We are going to post pictures of our dinner to those who we have invited to share in those events. We are going to express our desires, our likes and dislikes, show our travels and trials, discuss our preferences and our orientations, and we are going to do so in such a way that our 'followers' are able to access what we show and what we say. And this does not have to be digital. We are going to buy groceries and we are going to use club cards to secure discounts and online ordering to facilitate effortless delivery. We are going to visit public parks, get money from ATMs, and we are going to do all the ordinary things that human agents do. And we understand that even if there were no telemetry and no digital capture of this emitted data, that there would still be onlookers, there would still be people who see what we do and how we do it. We also know that we can control ourselves in these various circumstances, we do not behave in the same way all the time in front of all observers of our current emissions. Rather, we might dress differently when we go to work than we do when we go out for an evening with friends. We might behave differently when we are gathered around a fire pit in our backyard than when walking through the park during some weekend afternoon. This variable nature of our behavior constitutes a form of control that is directed to the projected audience of the behavior. If there were cameras in the park or in the department store, we might act differently there and thus reflect a kind of self-control over the data that is emitted in these contexts. Very different results might be seen if we were on camera in our backyard or in our houses.

Young people certainly do not want their parents to see aspects of their lives that they are happy to show to their friends as they gather with them in social circumstances. The control they have over their own behavior as they act and comport themselves specifically based on the situation in which they find themselves is how they can have control over their data emissions. They can show one side of themselves to their parents and another side to their friends. They can show one side of themselves to their employers or their colleagues and a totally different side of themselves to their schoolmates who have each gone off in different professional directions but who remain in touch through some digital means of communication. The capacity to control the identities and characteristics, the personal attributes that we wish to present to others, is a crucial piece of what people mean when they say they want privacy. They do not mean to say that they want to be left alone, that they want to be able to close the door and remain secluded from others, rather they want to be in control of the subjects that they are, of the personal attributes that they have come to occupy, and they want to be able to decide who gets to see one facet of themselves and who gets to see the others. A person who does very well in their work life and who is aggressive and ambitious in fulfilling the goals and objectives laid down by the job, will not want their boss to know that they dream of quitting and travelling the world perfecting their abilities as a juggler. These are incommensurable identities and it may be that both the boss at work and the crowds for the juggling performances would both equally be put off by an awareness of the other aspect of the personality in question. It may be strange to think of this dynamic as 'privacy' but that does seem to be the effective notion that has evolved under the conditions of telemetric understanding. It is a concept that both embraces the telemetric and yet wishes to maintain some control over it.

The average human being living in a modernized civilization is likely embedded in multiple communities at any given time in their life. And each community is characterized by potentially very different sets of norms and values. The expected behavior of any member of such a community can vary wildly from one set of associations to another. It is odd to refer to this segmentation as an interest in privacy because it really does suggest more of an ability to keep separate domains of one's public behavior from each other. One could just as easily think of this as a right to control one's social personae in the face of constant pressures from one grouping to gain access over membership details and activity logs from another grouping in which one participates. Nonetheless, I hold that this very public and very social separation of concerns, maintenance of dividing lines, is what people generally tend to mean in some common-sense fashion when they invoke an interest in privacy. And this amounts to control over emissions. And more than just emissions, it amounts to an interest in control over analysis and feedback as well. It is not just that the data representing those attributes and qualities that are specific to the person's role in one or another of the areas of their lives are deemed to be necessarily maintained within their control, but that the efforts at analyzing that data, that looking for patterns in it, must remain in their control as well. It would no doubt be very interesting to some to know what the tendencies are within any subgroup in the affiliation of some of its members with other groups and so on. A work place for example, may like to know as much as possible about the family, social, or personal lives of its employees. And not just in some anecdotal fashion where your boss wants to know if you are married or not. Rather, the human relations department has a strong interest in knowing how many employees fit into some sampling of social or personal characteristics. And it is very interesting for the employer to know if these characteristics can be correlated with greater on the job safety or lower attrition or better performance characteristics for meeting yearly revenue goals, etc. For the individual to claim that they want to maintain control over data usage, to be able to segment identities, and maintain a private life away from work or a private work life away from their family is to want to safeguard this data in the aggregate as well as in its personally identifiable details. When your employer has enough data about you to put you into this or that category of employee, and they have sufficient data about all employees to begin to produce analysis of how employees may behave across these different categorizations, then they are in a position to act on that analysis, and this will have direct impact on the employee who has an interest in keeping their family or personal information private from the employer. It may affect their compensation or whether they are able to move up the company ladder. These effects are realized even if not one single human being knows that information and can associate it with the employee. The effects can result from a policy enforced by machines based on information known only to machines.

Although people commonly describe their interest in privacy as an interest in maintaining control over their personal emission of data, they do not characteristically describe this interest as encompassing any freedom from analysis and the feedback mechanisms that this analysis might enact. At least part of the reason for this must be that the impact of such analysis is often opaque to the individual. Whereas it is plain and simple when your parents find out how you were behaving at some high school bonfire event because somebody posted some pictures on Facebook, the analysis that Facebook is doing behind the scenes to expose you to one or another demographic group for use in targeted advertising is not nearly so apparent. Often, we dwell on the positive outcomes of such analytics where we start seeing advertisements for some product that we have recently been shopping for or some medical treatment that we desperately need. We are much less likely to see the impact of such behaviors on a computed insurance rate due to the reflection of high risk behaviors in our profiles, or the jobs we do not get because we live in the wrong zip code and have been correlated with behaviors suggesting we will miss lots of work for one reason or another common among people in that neighborhood. The parts of analysis that are visible to us often reinforce the desirable parts of the analysis and the parts that are invisible display the negative. I am not trying to revisit these themes, rather I am suggesting that from a conceptual analytical point of view, if we are trying to craft a starting place for our residual notion of privacy based on control over the emission of data, we have to include in that characterization the analytics portion of the pattern even though that is not commonly expressed by people voicing their emancipatory interest in privacy. It is very common in an End User Licensing Agreement to see some protection being offered against the sharing of personally emitted data with third parties to that agreement, but there are no EULAs anywhere that promise not to perform analytics on the emitted data. Passive telemetry and analytics upon it, are intrinsic to the services provided and yet must be understood as contrary to the emancipatory interest in privacy that the commonly contrived notion of control over one's data signifies. Maintaining control over one's data in the fullest sense of that term is a massively subversive act and –as such– is not likely to be conceded to a user population by a service provider. We are far more likely to see an argument that it is impossible to provide the service at all without some degree of analytics and the use of it to automatically correct and maintain the serviceability of the system.

Therefore, this degree of control that seems like the commonly held ideal of 'privacy' in the telemetric understanding is considered unreasonable by the society at large. It is not a recognized justified expectation and yet, if I am correct, it is the most commonly held viewpoint of the individual persons engaging in such behavior. How is it possible for a large majority of persons to operate with a cognitive point of view which demonstrates an interest or intention that marks the form and content of their phenomenological attitudes, and yet, from a social perspective, such a position cannot be recognized as reasonable and cannot be granted legitimacy? Is it because whatever interest we have in our own privacy, we are too interested in prying into the lives of others to accept any universalization of the principle? The personal orientation and the social orientation are in conflict. And on whatever level we frame it, this is precisely the tension that makes this a residual effect, the fact that it is the collectively commonly held understanding of most private persons engaged in data emissions and passive collection and yet the common body of collective deliberation, the law, which takes into account all facets of our interrelationships, does not decree that this notion to be reasonable. This is the collective unreason of the telemetric understanding. It is the residual effect of the telemetric understanding that privacy both does and does not appear to be the limiting boundary condition of the system itself. In what follows, we will dig into the details of this dynamic to try and discover if there is something here to act as guiding light in our efforts at finding the role that privacy might play in ensuring the integrity and dignity of the life of an individual human living with others amid a globally dominant universal telemetric understanding.

Body

One candidate for a concrete application of the approach outlined at the outset of this chapter may come from the German Census Decision:

> In 1983, the FCC [Federal Constitutional Court] issued its famous Census Decision. According to the Court, the right of informational self-determination derives from the guarantees of personhood and human dignity of the Constitution, and it generally grants the individual the power to decide about the disclosure of his personal data and their use. The Court allows exceptions from this principle only if there is an overriding public interest and if this is explicitly stated in specific statutory provisions. In addition, the constitutional protection requires that data processing activities live up to the principle of proportionality and give the individual procedural remedies and protections. Moreover, data may not be stored indefinitely for undefined future purposes. (Rouvroy and Poullet, 2009)

According to Rouvroy and Poullet,

> That now classical avant-garde decision ruled, based on Articles 1 (human dignity) and 2 (personality right) of the Constitution, that the 'basic right warrants … the capacity of the individual to determine in principle the disclosure and use of his/her personal data.' This was one of the first and most famous articulations of a 'right to informational self-determination', understood by the Court as 'the authority of the individual to decide himself, on the basis of self-determination, when and within what limits information about his private life should be communicated to others.' (Rouvroy and Poullet, 2009)

To follow suit with the German decision, it would be necessary to proclaim the personhood of the person to be fundamentally worthy of respect and worthy of having its dignity preserved by all social protections ensuring private self-determination of that person regarding their emissions and the analyses of their data. Over the years, legislation to this effect in Germany has slowly acquired more teeth, giving the individual increasing protection against the forces in the social fabric that would continuously try to harvest that data and make use of it for a variety of different purposes to which the individual was not fully or adequately informed. Consent was a requirement for such data collection and the consent had to be meaningful. Furthermore, the consent was purpose-specific, and the data could not be used for some other purpose even if it were to be used by the same organization that collected the data in the first place and according to the terms originally consented to.

Aside from resting this decision on the fundamental dignity of the person, the Court rested it on the right of the individual to develop their personality as they see fit so long as it does not interfere with another's development. "Every person shall have the right to free development of his personality insofar as he does not violate the rights of others or offend against the constitutional order or the moral law." (Article 2 [Personal Freedoms] of the German Constitution). Human dignity or human personality and its cultivation amounts to something that must remain residual to the telemetric understanding at work in the society at large. Or rather, must remain within the control of the individual whose personality is thus being developed or cultivated. And that is not the same thing. The person can control the extent to which they are introduced or included in the various services that are out there collecting data and using it for one purpose or another. It is not that the individual must hold themselves distinct from the data gathering machines of the modern world, but that they are in control of the extent to which they are included within those operations. There is recognition here that persons may wish to include themselves in this overall dynamic, taking advantage of the goods and services that are offered by it. Although it is acknowledged that these persons should maintain some control over that data in so far as it specifically identifies and targets them, it stipulates that they lose control over the use of those emissions once the service has begun to run them through the various processes providing insight into the person in the aggregate. The German approach does not concede that the mechanics of the telemetric are sufficiently dehumanizing or in violation of personal dignity to prevent the inclusion of a specific person's data in any shape or form in whatever operations are performed by the collecting service.

Take the classic condition on slavery, for example. It may be deemed such an abhorrent abomination to the dignity of the individual human that it constitutes a right that is inviolable, that is beyond any rational deliberation. A person may agree to become a wage laborer, they may pledge their allegiance to a cause of some sort, but no person may legally sell themselves into slavery, it is not a choice that they can make. Because slavery is so vile, there are imposed limits on the right to sell oneself and put oneself into that condition. This may be equated with the notion in German law of a right to be forgotten or the impossibility of making a permanent declaration regarding another party's right to surveille you. Even if one previously consents to have their data collected and put to some use within a gathering and analytics system, that forfeiture cannot be permanent. Instead, the individual retains the right to remove themselves from that agreement at some future date as they should so determine. The person enters with informed consent into the collection of emitted data, but that is not an agreement without future constraints and the possibility of termination. The private person can withdraw from the agreement, leave the goods and services received in exchange for their data emissions, and be forgotten by the system, ceasing to exist within its records. Since even temporary slavery, like indentured servitude, is an affront to liberal democracy, we can only conclude that the German law sees no analogy whatsoever between temporary participation in analytics and such forms of bondage.

Of course, the termination in "being forgotten" is not complete. The data collectors are not forced to recompile all aggregate data that includes the user, that included the dimensionality of the user minus any personally identifiable dimensions, in the result. The person, as a part of the aggregate, will not be completely forgotten, but at least the details of any personal information or data that was contained in the system can be respectfully purged in compliance with the person's wishes and the regulations protecting them. The German law goes very far in trying to protect the dignity and personality of the person, but it cannot completely safeguard such persons from all the impacts and effects of their past data emission. If their driving record continues to be part of the actuary tables used to determine car insurance or if their health records continue to be part of the mortality and correlation statistics on persons with various medical conditions as related to other possible life style decisions, then this data, because it is no longer possible to tie it back to the individual German who has decided to be forgotten by the system, cannot be extracted from the results in aggregate. This is the way in which the telemetric understanding adapts to even the most aggressive of legislative accomplishments protecting the private personhood of the individual: by relying on statistical aggregation, the data is both extracted from the individual and protected against any future changes to their decision to permit the collection process. The individual cannot change their mind on this matter and even though control would seem to give us a better set of conditions and more power relative to the coercive capacity of large-scale telemetric concerns, it does not yield the possibility of absolute resistance.

What we can gather then is that the aggregate contribution of any person as they go about forming and cultivating their personality is, of course, the property of the gathering machine once the receipt has been granted, but that the individuated identifiable personhood of the private person is the element that is protected by these principles. Presence or representation in the aggregate, however, is not deemed an infringement upon personality, presumably because the individual traits and identifiable characteristics are stripped from the data, and the person is no longer subject to public disclosure of that data in a way that can harm or influence social impressions of them by others. If the goal of a right to self-determination is aimed at preserving the ability to cultivate one's personality as one sees fit, however, these aggregate aspects of data collection can exercise great power. Some might argue that the individual gains greater authority in certain societies because the focus on the specific person is less stringent under conditions where the population has grown large and has become concentrated in dense segments. The growth of population means social forces can no longer focus on the individual and thus lose the ability to apply any conformism that might otherwise influence their development. This allows the cult of personality and individuality to thrive with each person at their liberty to develop their personality and preferences in whatever way they see fit, so long as they do not venture into areas that will cause harm to others or the society at large, and thus gain the attention of the civil authorities or other social forces that would then seek to curtail their development.

Aggregate data and collective comprehension of personality types, of the characters that are likely to develop within the society, may be a manner by which the telemetric understanding attempts to put limits on this wide-spread cultivation of individuality. The upshot of such data collection and its use in targeted campaigns for education and learning, entertainment and consumption, can create much greater uniformity within the population and can exercise greater influence over the types of decisions that are made in the development of persons, not only in terms of how they are influenced in the social or public spheres, but in how they relate to themselves. The aggregate data produces a set of archetypes and icons that become goals or ideals to which the individual aspires. The adolescent seeks to conform to these types, as failure to do so amounts to a failure to fit in anywhere within their developing mini-society of peers. Failure to fit in is the most catastrophic experience that can befall such a person, so coming to learn and conform to the socially recognized container types becomes imperative. It may very well be the case that the increased communication and exchange of information across global networks allows for more diverse types of persons to choose from, but it does not alter the basic principle: that ultimately the developing individual cultivating their personality is not growing ex nihilo, or even crafting something from unexplored raw materials, instead the person is making choices, often dictated by details in their upbringing. And those choices are aimed at pre-existing icons and images that they can aspire to or aim for.

"Self-determination" in matters of privacy is better than many alternatives, but it is an oversimplified option if it does not include critical capacities to understand the impact of these sorts of choices and the situatedness they exhibit within a larger context of socially provided options and alternatives. Naïve self-determination would amount to a failure to learn of the impact that the social setting has had on the individual prior to their capacity to take over the functions for themselves. The German legislation, for example, takes special care in calling out limitations in the process for children, and no doubt this is due to the realization that children are too much under the influence of external conditions to be able to make certain decisions on their own, an attitude which naturally furthers its own logic. They are too impressionable to make their own decisions; therefore, they should be subject to externally determined conditions coming from the decision-making of their parents (who are perfectly fine to target with all sorts of consent-based influence pedaling) and other social factors like school and the entertainment industry specifically aimed at children. Personality is a complex notion for just this reason. The goals of self-determination are to ensure that each of us has the right to cultivate that personality ourselves, and yet there is recognition that it takes a great deal of social influence to get each of us to the point where we can do this on our own. And, despite the idealism that our laws might exhibit, there is plenty of evidence to suggest that even fully mature individuals are susceptible to influence that will more powerfully cultivate their belief systems and preferences than any self-determining or self-emanating capability they may have inherited from the light of nature.

The German approach, as an extension of the commonly held notion of control over one's data, relies on a set of notions about the self that seem contrary to pre-conditions of that control. And furthermore, the difficulty in extracting each of us from the aggregate, from the collective summary of the kinds of people that there are in a given civilization, suggests that there is still the possibility that a person in control of who gets to see what information about them, is still going to be under pressure, and even coercion, from a social sphere that is making decisions about that person's life based on their relationship to aggregate constructs and categorized containers of persons who are more or less likely to do such and such well or cost effectively. This sort of predisposition means that even the person who takes total control of their data emissions and can even take control over whether or not those emissions are used in these or those analytical operations, will be subject to the application of those analyses to their person so long as such procedures and practices exist in the world and make up a crucial element of the social environment within which that person lives. They will still get jobs or loans, be accepted to colleges or given credit, based on the relationship between their personal characteristics and the analytics that reflect on those characteristics. And, I hold, this is nonetheless an impact upon their person that can influence their determination and conditioning in such a way that it would be naïve to consider it 'self-determination' or a freedom from the constraints of undue conditioning.

The notions that we rely upon to depict and ensure our privacy in the sense of self-determination of access to data about our person must be very carefully measured. If we were not living in the shadow of telemetric understanding, it might very well be the case that such notions would be sufficient to produce an iterative perspective that works well with other notions: we are being raised to be free, we are being taught to think for ourselves, and so on. These sorts of heuristics are used commonly to suggest that a person can be given the tools to do things for themselves. It may be that the technology used to make the tool and the giving itself constitutes complex social relationships with complex histories, but the aim of the giving is an appropriation by the person that will individuate that person and enable them to take over the task of self-determination once all required ingredients are in place. Being an individual or private person thus functions like an ability that must be learned, but once it is learned, it is authentically possible as a way of expressing the person's autonomy in so far as they have acquired the requisite skills and can now express their potential with them.

This conclusion would seem to draw together all three of the residual effects, mentioned at the outset of this chapter, for resisting the telemetric systems. By neo-luddite, I am suggesting not so much the active rejection of new technologies, so much as the preservation of oneself as a private person in the face of such new technologies. In so far as the goal of protections for control-oriented self-determination is to safeguard the integrity and dignity of the private person, I am calling any such "attunement" with that orientation a form of neo-ludditism. Just as the Luddites themselves were interested in protecting their livelihood against new forms of mechanization in the work place, the neo-luddite would be protective of the private person in the face of automation that makes such privacy increasingly elusive. This would not amount to the negative suggestions of the German legislative approach, amounting to a control over where data is stored, how it is collected, and what analytics are going to be run over the top of it. Instead, a bona fide neo-luddite must have something positive to set forth as the primary goal of this protective action. By "attunement" then, I am invoking some of the elements of the later Heidegger where there is an effort at preserving one's basic relationship to the source of the relationships that constitute the domains where one lives. To try and put it more concretely, there is a lot of noise that is generated by telemetric operations and processes, and a person must maintain a focus and attentiveness to the world in which they are living, they must pay attention to the things before them, they must involve themselves personally in the things that matter to them, and they must reserve time to comprehend their lives and the things that make that life meaningful. This quasi-mindfulness, or whatever it may be called to try to bring it closer to home, is what the purpose of those limits on the telemetric are for. It cannot just be a question of giving the individual greater control over the data emission patterns that their body incorporates throughout their daily life, there must be a positive protection of some relationship to the matters that are nearby and significant. This is a purely formal directive and does not attempt to introduce moral declarations into the determination. The point of the neo-luddite facet of private personhood is that privacy must be more than just restriction and control against adverse actions, it must also be aimed at a positive protection of some recognized basis. The neo-luddite is the one who would limit telemetric operations for the sake of what is near at hand, what is close to them in the world, and the people nearby that matter to them.

But this will not be effective as the only element of the emancipatory interest. On its own, as we already showed, it is too vulnerable to assault. The attunement to what is nearby must be paired with a vigorous critical approach, an informed capacity to comprehend the forces at work in one's life and understand the role that they are playing in the determination of day to day factors. Only a person who is sufficiently adept at comprehending the impact of telemetric activity on the nearby operations of their attuned action is going to be able to properly determine the right decisions for protecting their private person and the private persons of loved ones in the face of such forces, especially when those forces are potentially coercive and aim at cracking into the protective shell of that private personhood. This is where the whole of the phenomenological tradition comes into play as we have begun to elicit it. The private person must be cultivated with the capacity to gain insight into the multitude of factors at work in all experiences, into the social relations at work, the impact they are having, the debt that is owed to them, the benefits received, as well as to the dangers they might present to that private personhood. The telemetric understanding is amorphous and ubiquitous, and it carries with it an aggregated familiarity with human beings. It is constituted by collective contribution over many years, so it is a formidable influence on the lives of any who live in historical times that are overwhelmed by such facets of life. The private person, with an emancipatory interest to protect that personhood and all the relationships nearby that are of importance to them, must be agile enough in the ability to critically interpret the things and events taking place in the world to see as much of the remote influences as possible. And this is not necessarily just for the sake of deflection or rejection, it is for the sake of intelligent appropriation as well. Only with a critical methodology that is aware of the subtleties of ideology and the way in which different interests come to play and are embodied in the events that come to pass in the world nearby, can people have a proper critical understanding of the world and their place in it. And this is necessary for the complete and ongoing protection of that private person. STEM based education is fine so long as it is accompanied with a balanced focus on critical theory based on historical and sociological understanding of the human condition in so far as it makes use of science, technology, engineering, and math.

A well-informed neo-luddite with sufficient insight into the relationships at work in the world around them is thus placed in a position where they can protect their private personhood when embracing the cyborg existence offering itself to anyone born in the middle of the epoch of the telemetric understanding. A cyborg, of course, is an augmented human being, a person whose abilities and capabilities are extended by the techniques and mechanisms artificially produced by human social history. This does not need to be science fiction. A person who has been inoculated against whooping cough is a cyborg. A person who wears well-crafted walking shoes to navigate rugged terrain is a cyborg. The cyborg is the human being who takes advantages of artifice throughout their lives and who augments their person for the sake of improving various abilities or better protecting against the harshness of life and the elements. To the extent that much of humanity lives off complex agricultural techniques developed over centuries and used in the production of food, such a cyborg being is inevitable for everyone. The artificial intelligence that has gone into the creation of the world around us, that has been used to ever more massively produce goods and services to promote and further our life interests, makes of us something of a mixture of this artificial intelligence and the organic intelligence that our species being carries along with it. The artificial augmentation has been developed over the generations specifically for the sake of this organic being. Integration of the biological and the technological is in our past as well as in our future.

All these three elements must converge to produce a proper phenomenologically informed emancipatory interest. The private person must be a neo-luddite adept at critical theory with all the abilities and capacities that cyborg life has for them. And no one of these is as useful or valuable to that interest if it is stripped of the other two. Without a private person –and the relationships that go with it– to protect, without the neo-luddite, the additional capabilities of cyborg life, critically interpreted, are of little use. The private person is nothing if not supported by a critical understanding that is benefitting from artificially cultivated intelligence. Since the telemetric understanding tends toward colonization of human being and since the drive toward rationalized world order tends toward colonization, the importance of a properly developed emancipatory interest cannot be overstated. And frankly, it will not matter what such a private person does in the morning or does in the afternoon or at night, what matters is how the private person carries out these activities and whether these activities realize an emancipatory interest that preserves and promotes their humanity among the potentially coercive forces of the telemetric that are at work in all places and all manners in the world within which that private person lives and makes their way.

When I refer in the heading of this section to the body, I am not referencing the physical source of all individual emissions of data for the sake of rationalized analytics and analysis that will feed the anticipatory power of the telemetric understanding. Rather, the phenomenological body is no longer a smooth surface upon which the writings of powerful remote forces are to be written for the sake of increasing wealth belonging to no one, or nearly no one. Rather, the human body is the epicenter of this emancipatory interest, it is the seat of private personhood critically understanding its own artificially augmented intelligence oriented toward social relations and the influences around it. The notion of privacy may have its origins in dignity and respect, or it may have its origins in shame and modesty, or yet again it may have its origins in an interest to protect the body amid those powers that would act upon it for some interest that is not aligned with it. The body, as we are painfully aware, is always potentially a corpse, always at risk of losing its life and becoming a purified thing. Thinghood and the protection from it is a major element of the emancipatory interest. Because it is a human being's highest hope to protect what is nearby from the reduction to thinghood, the interest to do so is tantamount to the interest in protecting one's life. Although not necessarily reducible to that. People are willing to die for causes, to die for things near and dear to them. And this willingness is often a preservation of the body from this thing-like state, where external forces reduce the body to a mere thing that can be used as a means for this or that end.

There is something absurd about the notion of privacy if one thinks of it long enough. A body that emits metrics is much harder to justify as the seat of control. That is what bodies do, they emit. Each of us must act to stay alive, and to maintain living states of being. And these actions amount to emissions. Others can see us, can observe our actions, can surmise what we are up to. Being extended in space and physically apprehensible makes one a public creature and available to telemetric understanding. It is, most likely, the natural state of human being to be available for view in this fashion. It is more noteworthy that it has taken so many thousands of years for human beings to start to build the infrastructure for colossal world-wide telemetrics. One could probably make the case that we have been heading toward this since the very beginning. We strive to know things and knowing things means collecting data. We have an interest in knowing things about each other and that means collecting data about each other. We are physical beings who are observable, analyzable, and knowable. The surprise is not that the world has entered the phase of global remote measurement with an aim at complete rationalization of activity for the sake of greater order, the surprise –if there is one– is that we have not been consistently producing the means and methods for the emancipatory interests and giving them an equal voice in our actions and advances. What that means may need to wait until the third section of the essay, but at this point it is fair to suggest that our ability to produce the telemetric seems to be running centuries ahead of our ability to produce the emancipatory facets that need to be implicit in that system as its residual aspect and as the place occupied within it by what is nearby.

World

Everywhere I look there are interests at work, this group's interests and that group's interests. Somebody is claiming to defend someone else's interests, someone is trying to convince groups of people that their primary concern is to promote their own interests (which are abundantly clear to everyone except the people whose interests they are). There are even people here and there who, due to whatever position they may have, claim to be disinterested; and that being disinterested is in the interest of the population at large, or of the nation, or of the society. Then there are people who turn against the self-proclaimed disinterested ones and accuse them of having surreptitious interests that they are promoting under the guise of disinterestedness. Interests, perceived and actual, seem to be clashing everywhere and always with groups forming over long spans of time or in the short term to take up sides on the battleground of those interests. Sometimes the groups seem to share many interests in common but because of a few issues where they proclaim themselves to be at odds, they will take up polar stances toward each other and enter into a grand scale battle to the death to make sure that their side wins out on these one or two issues that are worth a total sacrifice of all the other matters that they happen to hold in common. The prevalence of interest is so common and widespread, it would seem to be straight-forward enough to claim that the world itself is nothing but an arena of interests. And on some views of the phenomenological, this is the point of departure for their effort to produce a coherent schematic for the 'world' and what it amounts to in the life of the acting individual conscious being. It is tempting to try and say that the 'interest' is already too complex for raw apprehension, that we have to form interests as higher order intentional structures out of lower order components like perception and the acquisition of sense data, but in so far as the social relational notion of learning and assimilation are concerned, such acquisitions are always taking place at a higher order and this sort of reductivism is rejected. People, proximally and for the most part, are engaged with things, participating in events, and thus are first and foremost interested in their world, in the matters at hand within their world.

And of course, there are conflicts, conflicts between people where some one thing, that is contextually unique, is of interest to multiple people. Then they must compete over this scarce resource to feed their individual interest at the expense of others or even their own in other cases. Conflict need not just be between different people, there may be conflicts within a single person. Here the resource may be the person's time or life opportunity. There are interests that can be served at the same time by the same person, and so priorities must be declared, and ordering must take place. In some cases, such deferrals may result in lost opportunity at realizing an interest. Or the conflict exists in a kind of layering fashion, the person may identify a set of interests intellectually and establish beliefs upon them, but they may differ substantially from what we might call their material interests. These sorts of 'ideological' conflicts might be due to moral reasons that downplay the importance of a person's material interests for the sake of some cognitive interest they deem more important. All these different forms of conflict that can exist within the individual person can presumably exist between people as well. There may be a conflict between a group's material interests and another group's cognitive interests, for example. One set of people, in a way that does not cause themselves material harm, may hold and work to preserve a set of interests for moral reasons that conflict with the actual material interests of another group of people. The existence of conflicts of interest is essential so long as the world is conceived as a swarm of interests and as reducible to them. If we think that everywhere there are interested parties and interesting things, material that agents have an interest in acquiring, where interest suggests that the acquisition will benefit the agent in some way, then conflicts are a necessary aspect of life. Even in cases where entire societies are structured using collaborative large-scale ordering systems, there are going to be pockets and zones where conflicts of interest surface.

Consider the technological advances that may result in the destruction of one industry and any labor power that is established for feeding that industry with its required work. The distribution of labor based on role orientation may be conceptualized as a purely collaborative order in which people take on different roles in the production process for the sake of addressing social needs. Few people farm and build housing, but exchange money they have earned for these things as needed. We speak of the social interests of the division of labor as an efficiency mechanism for supplying the society with all the goods and services required to sustain the lives of its members. And yet even within such a large-scale collaborative system there are individual positions that have localized interests in their own sustenance even as the system undergoes constant state changes. And this means that sometimes the society may stop producing some item at the scale previously produced for the sake of some new item that is required to fuel a new process. The buggy whip manufacturers going out of business as the automobile rises in popularity is an example of this. From the social perspective, things look like they are retaining their collaborative orientation, people are contributing to an 'improvement' in the way of life and the well-being of the society at large, even though people in a sector of the society are losing their position and thus their livelihood. The society may be structured in such a way that the people in that area cannot depend on social assistance for maintaining their place. That is, people who become redundant relative to some higher order change in the social orientation are on their own when it comes to retooling or re-education as necessary to find some new role in the society. If the buggy whip makers have to learn to become automobile makers, that is their problem and even if those are radically different skill sets or take place in radically different parts of the country, then the people have to take care of that, they must migrate, they must go where the work is, and they must learn the skills that are required for doing these new kinds of labor.

Conflicts of this nature are inherent in the system of interests, or so I have said. And this presence of competition and conflict of interest in the world of interests and interested parties, presents itself as a given something or other over and against the individual. The zone of interests, as both conflict and collaborative, appears to the individual as a world external to them. The world, far from being the extended spatial substance of the early modern philosophers, is the set of constructed associations of other people, the interest structures within which each person needs to find a place to maintain their livelihood and further their own well-being. A newborn has an interest in being fed and clothed and kept warm. It also has an interest in educational stimulation that will develop its brain and body with capabilities that are crucial to thriving within the social setting where it grows up. And this interest ties the baby to its parents who –at the outset– are the best hope for acquiring most of those things. Babies, by and large, are fed, clothed, kept warm, and educated. They do not do these things for themselves and so they are grossly dependent on the capability of their guardians for furthering these interests. And the nature and condition of these provisions by the guardians will be determined by their position in the world relative to all the other people that there are. Do they have roles in the society that allow them to meet the needs of the newborn in the most basic of ways? Or do they have an excess of resources for handling these concerns? Or are they in dire straits and unable to meet even some of those most basic needs? The interest structure of living beings embeds us in our environments and in so far as our environments are populations of other living beings like ourselves, we are each other's worlds.

Even this very simple view of the world situates our concerns and our involvements. Interests, in this sense, are –I take it– a materialist form of this very same notion of more typically phenomenological categories. I'm not going to quibble on the differences other than to suggest that some phenomenology has been naïve in failing to see that the concerns in our involvements within the world initiate processes that involve conflict. And casting concern as interest does a better job of attributing the world with this easily observable characteristic. Because our concern may run up against conflicts with ourselves or conflicts with other individuals or other groups, concern itself is modified, it is fundamentally an interested concern where that means it must be attentive to the presence of conflict within the world of its concern. The concerned agent, acting to realize interests in a world of compatriots who are both collaborating and competing, must develop capabilities that accommodate those relations. Action cannot take place without an awareness of the other agents involved in the arena of action and without accommodating the responses that those other agents might make. Anticipating such responses is crucial for the agent's capability to adapt to the conditions of the world and continue to meet its own needs. A buggy whip maker must pay attention to the ongoing conditions of the world. Although it may not be obvious, it might have been possible to anticipate the changes required in their way of life if they were following the advancements in production that progressively increased the popularity of the automobile. An agent who recognizes that the world they live in is a world with others will be aware of changes in that world and to the actions being taken by others, all of which amounts to the same thing: changes in their world.

Agents then, always to some extent, provide their consent to the structure of this world. Because the world is the world of others and the interest domain of all the collaborative and competing interests of the various agents that make up this world, everyone must continue to ally themselves with the processes and procedures that go into making up this dynamic. There is a form of tacit consent that is at work with each individual agent who continues to play out the dynamic between interested parties. We agree to play by whatever rules have been laid down and the sense of it is that sometimes these rules will be to our advantage and sometimes to our disadvantage. The idea, in pure form devoid of any concrete problems, would seem to be that a modest amount of consent is required from each agent and that this consent is based on the presupposition that the dynamic will be fair. A person who does not think the process is capable of fairness any longer, or that it is systematically corrupt, will likely withdraw their consent in some sense and start to look for ways to cheat. Going into more detail on this is something I want to defer until later in the third section, but my point here is to emphasize that the world made up of agents is participatory in its logic and means, and this participation involves acquiescence by the agents who constitute the dynamic of the world. This acquiescence is what I am calling consent although I do not want to depend on it as a conscious or intentional state of mind, it is more implicit than that and probably involves something more like adherence to acquired norms than anything more fully cognitive like a determinate willing or agreement. Plato's notion of tacit consent may best capture this notion if the social basis of it is adequately emphasized.

I want to highlight this implicit consent, or participatory acquiescence, because it does appear to suggest that many of the agents who make up the world recognize that there is some higher order logic at work, and that this higher order logic is not always going to be in keeping with the isolated or abstracted logic operative in the agent's existence. The interest of the one may not only be different from the interests of the many, but those interests may be taking place in complete ignorance of those higher order constructs. Of course, this is explicitly required by the entire picture we have drawn of the telemetric understanding. I have claimed that the telemetric need not be conceptualized or realized by those that are participating in its dynamic, that people producing metrics, emitting their quality of service data and being analyzed for the sake of keeping some service up and running, need not be explicitly aware of that system's operations. They may be exhibiting an implicit consent in the dynamic of such a system, but that does not mean that they have cognitively conceptualized the higher order operations of the system as such. And this can be schematized as an implicit consent in an order that is bigger than any component in that order, a whole that embodies a logic that is not exhaustively present in the sum of its parts.

And this disparity, although sub-optimal to the agent in so far as it restricts their reactive power to the extent that they are not aware of the higher order logic at work, may be locally advantageous. That is, an agent who is not spending energy trying to anticipate the system's mechanisms beyond the local conditions may have more resources and more attention to allocate to the matters more closely at hand. Being an excellent producer of buggy whips, for example, might require ongoing focus and learning. Focus on the events in the world that indicate that automobiles are coming to take the place of the horse and carriage may detract from the agent's ability to perfect their skills and talents as a buggy whip maker. And going even higher in the order than that, a conceptual understanding of how progress works and the tendency of economic systems to replace whole industries with new industries that are more profitable would be an enormous attention grabber for such an individual. They would likely have to spend years familiarizing themselves with all the foundations and aspects of such a viewpoint before they could start to form an educated response to the things taking place in the world. And even with all that investment, still they may not have access to privileged information that is crucial for properly seeing things on the horizon before they happen. In that case, the person may very likely have sacrificed any hope they have of being a well-trained and excellent buggy whip maker and might nevertheless end up losing their position as part of incremental downsizing before the industry itself is replaced. This suggests that persons with interests may have an interest in being ignorant of various aspects of their world and what is going on there. The division of labor is meant to make the society at large more efficient and optimal for achieving ends. And the division of labor includes becoming educated in various aspects of those relationships so that a person who is aware of economic developments in the manufacturing industry is unable to pull their weight within that industry, and so may turn out to be crushed by the very same mechanism that yielded their position in the first place.

This indicates that a certain amount of faith in the system is required in consenting to the system. Or not faith so much as trust. Better yet, as I said, let's refer to this as implicit or tacit consent. Consenting to the system of interests, the world of human agents, is accepting a place in it which will be of necessity shielded from insights encompassing that system. As a result, any agent will be vulnerable to the larger order and its processes and procedures. People, in so far as they inhabit this shared world of interested parties and resources of interest, are cut off from a point of view that would give them enough scope and understanding to completely realize their place in that world and how it all fits together. This holds for Professors, CEOs, Senators, and Physicians just as much as it does for buggy whip makers. Being human situates us and being situated is a determinate limitation requiring that we all submit to this trust relationship.

What has been said to this effect is an effort at establishing that a detailed comprehension of the larger order at work in the world of interests is both beyond and contrary to the interests of the individual. A detailed comprehension is not the only option. People can be aware of the nature of the world in a general sense without having expert and detailed insight into its logic and common workings. That is, a person who comes to work as an automobile mechanic might wed another person who works as an elementary school teacher. To be a good teacher and auto mechanic it may be contrary to their interests to gain detailed understanding of the social order within which they live. In virtue of the fact that they participate in complex economic actions every day, we should expect them to have a modicum of understanding of these social conditions at least at the most general level. They know they are buying food with money that they earned fixing cars and teaching children. They understand the simple economics of their own daily lives, they know they are paying a mortgage because they could not buy a house outright with cash, so they must understand that banks and contracts are involved in aspects of their lives. They go to the grocery store and the department store, they pay the cable bill and they get time off to go on vacation. They take the kids to soccer leagues and they enroll them in high school. The social interactions and existence of complex social relations are immediately apparent to them with pretty much every action they perform on any given day. Even if they do not acquire a detailed understanding of all these relationships and how they all fit together, it seems unlikely that they would not form some general sense of a 'fitting together' of such things. They know that they are interacting with a complex order of social relationships because their contextualized actions are part and parcel of everything that they are, but they can only gain that viewpoint from those actions, from the place where they are, and from the paths they traverse on any given day.

Any human being, given such a perspective, would likely start to form a set of opinions and ideas about the general nature of these relationships, they would start to surmise how it all fits together. I assert that they would be clearly and painfully aware of how inadequate their own position is relative to all these goings on. They know there are complex relationships involved and they know that they cannot gain perspective on all of them. This suggests a great deal of natural savvy in any such person and would be just as true of a pair of college professors, etc. They have the wherewithal to know about the complexity of the world and to realize that the complexity is beyond their current point of view. They know there is something bigger 'out there' and they have a keen sense that they do not know what it is. Luckily, the evolved division of labor provides lots of specializations that are willing to fill the gap. There are areas of activity where the work that the people do in those areas is aimed at providing narrative descriptions for what that larger story must be and what the general dynamic of the social system at large must be. And the stories spun by these industries are made available for consumption by people otherwise too busy to do this work themselves. It is inevitable that such 'culture' industries are going to be widely popular and that they may include common media outlets as well as academic publications. Implicit in the dichotomy between realizing there is more going on than meets the eye and being unable to provide insight into it, lies the extreme interest that any person has in such matters. And this indicates that there will be a great deal of interest expressed by all individuals in industries that provide stores and insights into what that higher order perspective of the world looks like. These cultural laborers will provide prognoses that are much needed and go into detail on how things are going, where they might be headed, and what likely actions might need to be taken to adjust properly to the changes that are coming. A buggy whip maker who is properly keyed into the right cultural critics will be able to remain focused on the art of making buggy whips while remaining informed about the upcoming changes in the world impacting buggy whip manufacturing.

Nothing could feel more natural than this. When the auto mechanic or the teacher become sick, they go to see a doctor, when they need food, they go to the grocery store. Nothing is more natural than to turn to others to fill in the gaps created by their own localized focus on one set of operations for the sake of furthering their own interests. Likewise, they see nothing problematic about having to turn to the experts to understand the relationship between their local experiences and the world at large amounting to the social relationships more generally conceived. This dependence, like all the others they engage in daily, mark the realization of their consent to the system, their tacit or implicit consent. And that is the making of the world that each of us practices daily. The world is not just the set of things that are out there and the geometric points of contact between them insofar as they are extended in space and time. The world is the involvements and interests, the division of labor, of all the other people with whom one shares that world. The social world *is* the world, for all intents and purposes, and it includes different viewpoints, skills, and talents that fill in the various gaps that we each have, of necessity, to fulfill our interests, to realize them, and to thrive in so doing.

The world may not, however, appear this way by and large to the positioned individual. Why is that? Despite the proximity of complex interactions to the daily lives of all such contributing agents, there is a constant reification process that appears to be at work. People are always encountering the workings of other people, but they may not have this contact in a way that directly associates the agents with the products of their actions. When we buy food at the grocery store, for example, there are no farmers there, only cashiers. We may interact directly with some people in those conditions, but we are just as likely to interact only indirectly with them through a monetary relationship with their products. Additionally, because individuals are just single members of a group or class, they are only representative of the larger interests at work. The teacher and the auto mechanic are not the only ones of their kind in the world and their students and customers do not rely exclusively on them for their needs in those areas. Rather, there are many teachers and many auto mechanics, and this creates a distance between them and the individuals who consume the output of their actions. I do not necessarily form a relationship with this auto mechanic, rather I may form relationships with auto mechanics generally. Sometimes we are set remote from each other through the products that we produce and the impersonal way in which they can be exchanged, and sometimes we are set remote from each other because the actions that bind us are generalized or generic to a group of agents no one of which is the quintessential performer of those acts. These formalization procedures make it possible for human beings to lose touch with the intersubjective nature of the world and begin to experience it as a given set of things, impersonal actions that are inert, and that are purely material resources for use and consumption. This is a crucial piece of the processes of rationalization and colonialization that we have previously mentioned: the process whereby social relationships transform into inert things and where the world is transformed from a living intersubjective set of dynamic relationships that we are always engaged in, relating to, and interacting with into matter or pure material that is only there as the content of production and consumption. This transformation marks the surrender of the everyday nearby dynamic of social relationships into a larger order like the telemetric. It makes it possible for pundits to laud the power of rapid progress without worrying about who precisely will benefit from that progress. The point of explicating the world as a world of living interests, however, is to demonstrate the residual character of a world that remains apart from the telemetric understanding. The transformation of the world into objects of analytical nature for the sake of feedback into a better functioning process is the advent of the telemetric that we have been describing throughout the last chapters. The living interests of the people carrying out their everyday activities, however, is the residual world evading telemetric logic and reason. On the one hand, a world where nanorobots can keep you looking and feeling like a thirty-year old and, on the other hand, a world where only the select few can afford it.

Bodies in the World

A world constructed in this way, out of the individual contributors, will come to have higher order groupings and each of these groupings will be associated with interests. "Teachers" as such will be a grouping, "auto mechanics" and "cultural critics". And the segments and associations will only grow from there. All categorization and association will form into interest structures. The rationalization process is the forming of groupings, of sets based on attributes, and of identities. And these groupings are characterized by interests. We need not see it as a constructed collection, as our earlier analysis showed. When a person becomes a teacher or an auto mechanic, they do not formulate the group's interests, rather they take them on, they become a member of a pre-existing set. The body in the world is, therefore, a member of a grouping or set of groupings based on whatever acquired attributes they have obtained throughout their upbringing, via their experiences, and based on the events they have been involved in. The learning of skills places an individual person in relation to a group of people with those skills, allies the person having those skills with others. And not just skills, but attributes of any kind. Cooks and basketball players, baton twirlers and health technicians. The acquisition of the skill is an entry into a socially determined set of qualities that take on a larger scale set of relations such as monetary advantages or other forms of prestige. And as groupings, the groups take on relationships to each other. The group of owners of automobile service centers have a collective set of interests in so far as they comprise that grouping, and those interests involve some form of collaborative relationships with the auto mechanics who would come to work in those service centers. They must come to a form of agreement in how they are going to associate. The association itself is an expression of their historical relations and the events that have taken place over time and bound them together. They need not be balanced, and they need not equally serve the interests of the one set over and against the other. Nor need they serve equally the interests of each member of each group. Instead, history will have brought about distinctive characteristics of those relationships as they may have been reinforced by other groups, by legislative and administrative groups, by local chambers of commerce, and other such organizations. This history will include whatever cross-sectional characteristics are relevant to each of the various relationships since any given member of a group will be a member of still other groups and sub dynamics and associations within the groups will accumulate based on them. People of color, for example, may be both service center owners and auto mechanics. Each of their positions within those respective groups will be modified by such factors even if they lose some connection with each other across the boundary between those groups.

Previously we discussed how the members of a group with specific attributes may not have the time and resources available to conceptualize these historical relationships between groups, either to formulate a set of higher order interpretations of how the group's interests relate to some other group's interests or how the group may relate to the society at large or to each of its members. And then we saw that there have been additional groups of culture-oriented identities that produce stories and narratives to describe the relationships between the one group and the others, and so on. These groups may have historically conditioned associations with still more groupings with additional characteristics that contribute to the overall collaborative requirements of this or that aspect of the groupings and their relationships to each other. This means that the auto mechanics of a society may turn by and large to a cultural organ to gain an understanding of how they fit in with the rest of the groups and how their interests are furthered or hindered by those other groups. Which organ they turn to may be doubly conditioned by whatever cross-sectional groupings and attributions that they have. Such organizations are interested parties and so there is no guarantee than any of the folks in the mix are purists looking to describe the true nature of things regardless of what other forces or coercive effects are acting upon them. Cultural critics may work for owners of cultural organs and those owners might have a set of interests that drive the productivity of the cultural critics. The selection process for determining which cultural critics get a megaphone, or access to large audiences, may be determined by a process that prioritizes members who share a point of view benefiting the owners of those organizations.

The presence of interests as a global condition in all the groupings and bodies at work in the social environment means that there is always a possibility that any individual expressing or performing a task will be doing so in a way that furthers interests that may be opposed to the interests of some who are in contact with that person. The hermeneutical phenomenologist informs us that there is no interest free interpretation of the world, that there is no expression of group attributes that does not share in an interest context and drive toward an end of some kind that serves some purpose. There are tacit conspiracies at work everywhere or at least there is tacit collaboration of interests everywhere. And what is more, the historical development of these interests may have been in the context of certain tactics that hid some of the elements of those interests for the sake of better furthering them. To further a set of interests, it may be necessary to hide those interests, or maybe even to camouflage those interests so that they look like an entirely different set of interests, because only when presented in that way can the interests themselves be served. If it is my interest to get you to do something for me that benefits me and does not benefit you, it is not going to be conducive to that end for me to be completely honest with you about my intent and my means for achieving it. Rather, I will need to paint a picture that lures you along with my scheme in such a way that it seems like your interests will be served when you serve my interests. This way, you will join in with me and help me get what I want even when that amounts to doing something contrary to your own interests and independently calculated aims.

On the surface this would seem to be a hard sell, how can someone be persuaded that something that hurts them is helping them? The key element here is that very often the relative advantages and disadvantages of the actions are not immediate. We are not talking about an immediate relation in which I give something directly to you in exchange for something I receive in return. Although there are examples of such "jack and the beanstalk" exchanges. We might talk about lots of long-term goals and long-term benefits due to large-scale changes in the society. It may not be clear all in one day whether some actions will provide the benefits that they do provide. And if it takes a long time to determine one end of the bargain while the other end is immediately realizable, then the trade might be made without any immediate evidence of the fairness of that trade. And when the circumstances do finally come to fruition or fail to come, it is possible that lots of different explanations and reasons can be given to justify why things did not work out as anticipated. The person who was asked to defer the receipt of some good for the sake of something they handed over earlier might be convinced to continue making the same deal because the conditions have not yet come to pass, and they are convinced that they need to be patient. It is possible in some contexts that a group can be strung along for decades or centuries waiting for something that never comes but always with the promise that the pay-off is just around the corner, if only they remain patient and keep investing themselves in the objectives of the promising group. In such cases, it may be useful to scapegoat any opponents to their goals by blaming them for the delay. Often, the promising groups will use moral persuasion to secure the continued corroboration of the target group handing over its consent. In the next section of the book, however, we will look more thoroughly at the reasons why morality or Ethics cannot justify this deferral of interest. Rather the conflagration of interests and interest groups is the applied morality of socially produced people. There is little more to morality than the social mores embodied in these agreements and the explanations for why alliances should be made, and patience practiced when results are lacking.

In some action environments, whenever a plan is put in place to defer pay off for some group who is making an investment in a longer-term operation that has been promoted by another group promising results, the two parties must agree on the metrics against which the progress of the operation is going to be measured. They also must agree on a time frame in which those metrics are expected to show results demonstrating that all parties are being served in the way that the initial plan suggested they would be. In any social setting, where the groups are coming to terms in such a way that one group is handing over its agency to another group or set of groups, it would seem wise to demand similar metrics. How can we measure whether we are getting what you say we will get from you if we do what you say we should do now? That is a pertinent question. Notice that it requires the groups to be explicit in laying out their participation in the agreement. The terms of the agreement and the parties to the agreement all must be explicit to make this work. If there is any hidden piece or part, then it is not possible to lay out the conditions in such an orderly and transparent fashion. In large societies that are complex in structure, this is the usual case. The auto mechanics listen to the pundits, let's say, in investing their agreement in political policies that will supposedly help auto mechanics. Are the conditions really clear in this case? Do the pundits lay out the terms clearly to the auto mechanics with defined metrics to measure their success? Do the political views in question and their relationship to the pundits or the owners of the news organization present themselves transparently to the auto mechanics? Generally, these factors are not the case. Rather, the different groups act as though no such groupings and no such interests exist within the dynamic and so nothing is made clear. The political candidate, for example, makes a wild claim about serving some interest. The news organizations present the views and provide analysis. The auto mechanics are left to fend for themselves to figure out what are the hidden conditions, what are the various groups at work, what are the different interests at stake, and whether their own interests ultimately will be the focus of the groups reaching out to them, providing their interpretation of world conditions, and the way things need to develop in order to improve the lives of auto mechanics or some larger group that contains auto mechanics.

In some sense, we are always out of our element when evaluating the interest claims of others within our social world. We may not know all their group affiliations, and we may not see all the associations and relations they have and promote when informing us. People often do not want to show us the details of their business, but they often describe those conditions to us. Are they being truthful? Are they candidly telling us everything that is relevant to the case? When buying a car, for example, if the salesman says that he cannot go lower than the invoice price when selling the car, is he being truthful? And even if he is being truthful, is that the whole story? For example, are there hidden incentives in his relationship to the car manufacturer that make it the case that the so-called invoice price of the car is not really the cost of the car? And if we ask him to show us all the details of his book keeping, would he do it? Most salesmen would not, and this sort of consumer relation is like all sorts of relationships in social settings. Even if the salesman is the most honest person that ever lived, there may be ambiguous factors involved, uncertain relationships between different parties on his end that he cannot share with the buyer because he does not really know the details of those relationships and how they will impact on the purchase. The same might be said for the journalist when reporting on a political candidate. They may be in the dark about very many factors involved, not the least of which might be that the composition of their audience is not just auto mechanics but encompasses large numbers of different sorts of people and the journalists cannot spell out their reports for each of the different groups that might be reading or listening to them.

The complexity of all these relationships and the impossibility of total transparency in anything like a practical fashion, requires that each person, each private person, take on a deeply critical attitude when entering all the different relationships that are likely to augment or promise to augment their manner of living. Augmentation is a fact of life. It means that we are going to consume the services of others, we are going to have relationships with others that extend our individual reach beyond the three- or four-foot distance that we would otherwise be limited to. This pursuit of augmentation amounts to entering into an agreement and realizing collaborative relationships that may or may not promote the interests of all parties in a balanced fashion. This is where we must maintain critical methodology and attitude. Only if we are aware that even well-meaning interest seekers may accidently or purposefully deceive regarding their interests and the relationships they have to our own, will we be able to protect ourselves against infringements upon those interests. Private persons, to be private persons, are always engaging in the emission of data. And as such we run into each other and wish to have some measure of control over how this confluence of emissions will take place. Our private lives are filled with engagements and interactions, and we run up against the interests of others and other groups when this happens. The critical methodology that sees interests at work everywhere is essential for maintaining the dignity and integrity that the so-called Germanic view on privacy seeks to establish. To be the private personalities that we hope to be, we must be aware of all the relationships we are entering, what kinds of agreements we are making, what the relative payoffs are, what we stand to lose, and what we stand to gain.

Most importantly, we cannot imagine that the critical methodology at work here is something that is merely aimed at singular individuals that we encounter throughout the course of our lives. Rather, there are groupings and sets, organizations and institutions, histories and traditions that are far larger than any person we may come across. And we must understand that our own interests are not only tied up with these groupings, with these public structures and their publicly determined attributes, but that the same is true for all the other people that we come across. And this means that the conditions of understanding and interpreting any operation or procedure are not limited to the persons involved in establishing the agreement. There are people and parties who are not present in the office when we come to an agreement with the salesperson. There are forces at work that are not locally here and now, but which are positioned remotely and are exercising a distant influence on the conditions. And our critical approach must bear in mind that the time constraints present on our action may make it such that we cannot always exhaustively comprehend the situation whenever we must deliberate on a course of action. We are deeply and intrinsically finite and often we will not know the details, and not be able to know the details, at that moment in time when a decision is required. The residual at work in our deliberation, the telemetric forces at work in whatever situation we find ourselves in, must be recognized as taking on a role in any critical interpretation of anything. And this works on all sides of the equation. The unseen context is at work in your own interests just as it is at work in the interests of the other parties of the interaction.

There is nothing radically new in this presentation of social relations as a set of concerns that embody interests that are possibly conflicting or in agreement with the interests of others engaged in associated projects or related activities. Establishing the critical private person aimed at augmenting their capabilities through actions that bind them to other private persons is a classic maneuver. What we have added is the notion that there is something missing from the procedure, something that is not currently present to the agents. If you could look inside the mind of each of the participants in some action or event, each of the involved private persons taking part in an activity or action, the sum total of all their intentions and thoughts would be insufficient to exhaustively describe the chain of events that are currently taking place. There is something more at work and that something more is not available to the participants as factors that can be considered in protecting their private personhood, in developing their personality, or in maintaining control over the data they are emitting and the analysis that is performed over and against it. In addition, there is a large-scale set of interests that work their way up the hierarchy of groups and may be pitted against the associations and relations which encompass the whole of the dynamic. And these –albeit accessible in some sense– are extremely complex and difficult to access from any position within the context itself. Part of the reason for this is that many of the notions at work are based on historical interpretations that are not completely informed about the past or they are based on prognoses that only have a degree of probability behind them. The remote forces at work are not knowable in principle and so no critical agent in the process can have enough access to them to fully and completely realize the conditions they impose.

Private persons in the world are finite. That is what this means. We are distinct in space and time and our perspective is always based on our private personhood and all the attributes that entails. Our finitude in this regard requires us to be hyper-vigilant in trying to sort out the complexity of any dynamic we find ourselves in the middle of and whatever involvements with others that we have. The same fundamental basis of that need is also its impossibility. Because we are finite, we must augment and learn of the extended conditions that are going on around us. Because we are finite, we cannot do this as thoroughly and as diligently as we would hope or even as much as we would need to properly serve our concerns and involvements for the sake of some goal. And it is not just that we might fail to realize some end because we make a mistake about those interests that others have in helping us get there. Our finitude is such that we may have been wrong all along about needing to achieve that end in the first place. We may be wrong about our own interests at times, we may be wrong about what serves those interests, and we may be wrong about who else can help us to serve them. Our finitude is pervasive and poses a threat in every sense from every direction. Arguably, the primary good of associations and groups is the extension of that perspective to increase the limits that are associated with that finitude. In that case, we become members of groups to better serve group interests because as group interests, we have a better chance of seeing further and seeing better what serves those interests and who can help serve them. Again, finitude created a double bind. When we enter such groupings, we do extend the reach that we have, but we also create structures that do not strictly identify with our own place in the group. Even after the rise of the automobile there are still buggy whip makers, but significantly fewer of them. The group's interests still exist and are still being served, but the numbers are not what they used to be. At some point in history there was a need to reduce the numbers of members in that group. The group's interests were therefore served by removing members from the group for the sake of protecting it. Finitude works in this way, as we build associations, we extend our perspective, but we also loosen our relationship to the interests that play a role in that perspective. We see further, but we lose sight of the higher order context. The group interests are not clearly and succinctly identifiable with our own individual interests and so we are now at risk of failing to see all that needs to be seen to live well as members of such and such a group. We increase the likelihood of realizing our interests by adapting them to the group, but in the process, we are sacrificing something of ourselves.

To be a private person we must be involved in relationships that put that private person at odds with the groupings that provide us the personality and identity that we have in the first place. Privacy, in the fullest sense that we are thinking of it, is both the source of personal power and the source of human vulnerability. It is our finitude pure and simple. Only in so far as we are finite beings do we express our individuality in this dignified private fashion. To safeguard that dignity, we must increase our point of view and enter larger social organisms to better view the world that is essential to the protection of that privacy. When we do that, we run the risk of separating our intrinsically dignified persona from its interests and orientations. We create a schism in ourselves that both gives us the power to define our lives as we wish them to be and makes us vulnerable to social forces that may be coercive, and which might be operating in an opaque fashion contrary to those very interests.

Our initial investigations into the telemetric left us with the notion of an expectation of privacy that needed to be safeguarded to protect each human being from the coercive force of globally dominant services and systems. This notion became increasingly unclear as we evaluated the phenomenological elements of human beings. Where we landed was on a notion of private personhood that is worthy of protection as the critical interpreter of their own relationship to the interests and interested parties at work in a world where private persons are engaged in augmenting their lives through controlled and deliberate actions that build meaningful connections between themselves and other people. The critical attitude that we must take in order to have the kind of control we need over our emissions and the analytics run against them, is insufficient to do the job. Such an attitude functions as a residual to the telemetric understanding, true enough, but it is ontologically inadequate for protecting itself against the opaque powers that operate there. These powers always produce a context, or a set of interests, that is dominant over –and even hostile to– the interests each of us have in protecting our privacy and the private relationships that we have entered. This puts at risk our associations with family and friends and whatever private communication between them we deemed to be exclusively our concern and of no matter to anyone else: not from the perspective of the capture of an emission, from some analytic point of view, or even as some learned aggregate pattern fed back into the system at large to better serve some set of needs that we can only estimate from our own narrow perspectives. Phenomenology had its point of entry in the expectation of privacy residual to Telemetry. And now we see that Commonwealth has its point of entry in the telemetric understanding residual to Phenomenology. The purpose of the third section of this essay will be to venture into the realm of the common good that this type of residual produces and whether it is possible to collaboratively form a set of structures and approaches that will protect the dignified finitude of each human being in the midst of the telemetric understanding and through the collective power and capability that we have as our common wealth.

Commonwealth

"And I was barking furiously but it didn't seem that the man could understand me. He said, "Is that your only poetry, my friend? It sounds the same as every other dog before you." Yes, yes, I cried. That's it exactly: just like every dog before me! But he just patted my head and moved on."

–Further Investigations of a Dog

Chapter I:
The Death of God

One classical categorization of the maneuver described in the last chapter might suggest that the public sphere comprehends the social dynamic taking over human being through the telemetric understanding. The private persons that are produced through social relations are themselves capable of getting involved in public associations attempting to realize their interests across the populations of groups and subgroups. People may be the products of the social relations that make up the human world, but once produced the private person emerges as a meaningful unit capable of deliberating, decision making, and taking control over the private life that expresses the output of the productive process. As such persons, we can make alliances and associations, form or join groups that extend the reach of our privacy and that ensure the boundaries of it in such ways that are publicly validated and ensured. We form associations in public to safeguard the private person against the social pressures and forces that threaten the integrity of personhood at every step. Social well-being is beyond the ken of any one person and encompasses interests and aims that go beyond the phenomenological potential of any seat of understanding and of any critical agent involved in projects and attempting to interpret the world for the sake of those projects. Social well-being is larger than any life, it is something to which each produced person must succumb. And the traditional maneuver that aims at protecting the boundaries of the product from the weight of the production process is what is often called the public realm, the public sphere that addresses the concerns of the private person over and against the social forces that weigh upon them. From the perspective of such a public sphere, we generally refer to the common good or the common wealth and conceptualize it over and against social well-being.

In the classic story of these commonwealth concerns, the onset of the commonwealth is a contract, and the contract implies that the parties entering the agreement are sufficiently mature in their rational capacity to be capable of committing themselves to an agreement with others where each participant is informed as to the repercussions and benefits. In the story we have been telling, however, that is not the case. Rather the public contract or agreement is not so much a positive concept as the residual of one. It is something left over in the process of social production generating adults and generating individuals with attributes and qualities.

And these fully attributed beings are the agents of the classical picture of the contract. Not a naturally given source of the common good, these are socially produced agents engaged in generalized projects, showing concern in the way that many others show concern, involving themselves in projects that are comparable to the involvements of many other agents. We form alliances and groupings, we come into our finitude while extending it, and thus realize both our vulnerability and our strength at the same time. Thus, we are thrust into the public sphere. Such a sphere can no longer be characterized as an original position, as a natural condition, or as anything that marks the purely formal rational structure of the agents who enter into the agreement. Rather the moment of the emergence of the public, the moment it begins to emit its attributes, to render its qualities and properties, is a moment of crisis. The public sphere comes to be as a proposed solution, a solution to a problem caused by human beings living together in villages and towns, in tribes and in nations.

This change in the approach explains why this notion would come into existence at a specific time in history or during a specific epoch of human development. At those points in time when people are more crowded together and form more complex associations of family and church, town and region, that is when the pressure on the social product percolates to the extreme and causes rifts and gaps in the interpretations of the world and the places there where different social groupings and interests can come to agreements. Only in the middle of such a problem space does it make sense to articulate the private personhood engaged in the project of constructing their own personality in association with others with whom one shares the land nearby, or the earth generally. Only within such a site can a solution be offered to take over, en masse, oversight of the social relations and enact a public construct that is dedicated to the protection of that precious product that the generations have learned to create alongside their advances in agriculture and mechanics. This marked a revolutionary movement in the course of human history, a moment when groups of persons decided that they could no longer be held in check by interest groups at their own detriment, and so instead decided to overthrow the traditional powers, the traditional interests that governed the land, and propose an alternative power structure providing protection to the so-called private person and all that this emergent construct entailed. The concrete events no doubt brought with them lots of prejudices and biases specific to the social circumstances from which they emerged. However important it is to register that fact and so recognize that the establishment of the modern commonwealth was done with a set of biases and interests that privileged some groups at the expense of others, it is equally important to recognize that there is not necessarily an interest free alternative to it. Rather, since the public is always something that emerges from the social relations and the conditions of the private persons under those conditions, it will always emerge in every civilization as a solution laden with those interests and with the points of view that have been crafted under those conditions.

Any modern instance of this solution, therefore, would be saturated through and through with the interests of the telemetric understanding, telemetric interpretations of the world, and the role that people have in that world through the relationships they embody and represent. The characterization of action forms as emissions of data, as pattern discovery via analysis, and the recuperation of that analysis into further action through feedback loops is a necessary characterization under modern conditions of social relations. The modern telemetric understanding finds its way into all of the crevices of any revolutionary solution that might be proposed to the modern problems of social life and the forces that they realize in settling upon the finitude of private persons attempting to augment themselves by deliberately engaging in extended relations with other persons in the social world (having families, sharing memories, telling stories to each other). Therefore, the yarn I am spinning had to begin with telemetry rather than phenomenology. The human being's point of view is derived from the overarching dynamic that marks every aspect of those social relations with its logic. Measurement is the modus operandi most basic to the social relations that almost any contemporary human would be born into. We are emitters of data, being measured, being analyzed, and the analysis is used to further shape the world and guide its construction along with the interest groups forming there, intermingling, and mutually influencing each other. It is the air we are breathing and the water we are drinking. It is all around and everywhere, and so it finds its way into our understanding of ourselves as private persons in the middle of something larger and something more substantial than ourselves.

Human understanding within the social world of the telemetric understanding is not the bedrock of that system, nor is it the axiomatic grounding moment. Rather, human beings come to understand themselves in the middle of a world that was already there, already conceptualized, already in the middle of historically developing interest structures from which each new person is formed. Contrary to those phenomenologists who thought we could somehow get to the bottom of things by orienting our investigation from the point of view of the subject experiencing the world, we saw that the investigation into the subject's experience would take us past the subject and into the dynamics that underlie that subjectivity so as to produce it, so as to form it in accordance with interests and orientations that pre-exist the persons occupying those positions –in variable ways– as they mature and develop through various stages of their lives. And the pattern they attribute to this indicates emergence. The fact that the private person is not aboriginal to some natural kind is no argument against its existence. The fact that it has been produced is proof enough that it exists. There may be no automobiles found in nature and they may be unique to a planet somewhere in a specific solar system within some galaxy, but there are automobiles. They do exist, they have come to be. And like the automobile, drivers exist, and people who like to drive, and people who know how to fix automobiles, and people with interests in brands of automobiles who like to get together with other people who share that pleasure by telling stories featuring those cars. These are real things in whatever characterization of the real you like. They are produced, they are the result of millions of years of progress through different forms of being and different forms of life, the evolution of a species that has a brain that works in such and such a way, that makes tools in such and such a way, all culminating at some point in the recent past in the production of roads and highways, manufacturing plants and energy mills, all for the sake of supporting the automobile economy and the collection of interests and orientations that amount to automotive life.

When we speak of the private lives of people, we are not talking about private states of affairs. Usually we are explicitly describing relationships that the person wants to be a part of with other persons for the sake of shared aims. And we are talking about the ability of persons to form those relationships without having to succumb to the watchful eye of the social domain. The social domain under the telemetric understanding is a panopticon, and the purpose of maintaining one's privacy is the concrete carrying out of one's actions without succumbing to the watchful eye of that social panopticon. When we say that, we immediately recoil, knowing that there must be limits to this intent. We certainly recognize that there are horrible things that people may do and that they may rely on the cloak of secrecy and a concomitant privacy to do those things without being scrutinized by any authority representing the social sphere. The emergence of the public sphere is the construct that lies juxtaposed to the social. The social cannot be relied upon to provide a safeguard to these interests. Even if we want to protect privacy while also protecting each of us from the ugly side of the private sphere, we must enact a domain whose charge it is to carry out that delicate balancing act. Arguably, the public sphere, conceptualized in this fashion, takes on the role that conscience is meant to take. The conscience was supposed to safeguard each of us from the untoward actions of others, engaging in realizing some interest deemed hateful to the collective orientation. And conscience, in some sense, may have been thought to realize the voice of God in the 'minds' of each person who felt its presence at those moments when they might otherwise have been overwhelmed by an urge to act on some desire, or to realize some interest, that would harm another or harm the social relations generally for the sake of that realization, expressing those concerns, or diving into those involvements.

The rise of the commonwealth, as the title of this chapter might already have suggested, is the death of god. It may be that the great thinkers of commonwealth did not see it this way, did not even begin to conceptualize it in this way. Once the public mind, emerging from the private persons oppressed by social relations that would infringe upon their capacity to realize their own personhood and to develop their own personality, had been set forth, it became the supplement for god, the replacement of that voice in the call of conscience, and it took on the role of overseeing the delicate balance between oppressive power from the social sphere and the generative personal power from the private sphere. The commonwealth became the seat of liberty and the seat of justice and equality, it became the place where all our highest ideals and notions were set to reside where once they dwelled in the creative force of some deity credited with our construction. The source of human being, the power that creates the human being is the residual voice that lives in the 'consciousness' of the agent and guides action through the application of limits, scope, and boundaries. And with the advent of a new way to think about our creation story, we required a new way to think about the little voice in our heads that keeps us in line and forces us to live well together with all the correct constraints placed on the private person let loose in the social world. Drawing from Nietzsche's remarks on the subject, I propose that the death of god is a long running process that is taking hundreds of years and many generations to be fully realized. It constitutes a dynamic that is clearly identifiable in the world where human beings battle over the meaning of the private person, its surrounding social relations, and what in it falls within the domain of protection by the public sphere.

What Nietzsche does not tell us is that even his own proclamation of the death of god was already a stage in that development. The Age of Enlightenment and its orientation toward forms of government that usurped the divine right of kings was clearly a step in the direction of this phenomenon. Various scientific theories and ideas no doubt played a role, business ventures, industrial mechanisms, the rationalization of the social world with specific aims and orientations. All these phenomena played a part in furthering the interests of a world without a deity, a world where human beings were on their own to fend off the coercive powers that crush their finite existence for the sake of some faster moving, larger scale set of concerns: what mattered the history of a person when the history of human beings was at stake. The Copernican revolution and the Cartesian philosophy of the thinking thing may have been early steps in the process of dethroning the deity. The American and French revolutions may have been further realizations of it. And the industrial revolution and the widespread adoption of the notion of species evolution. The developments of the banking system, the rise in knowledge about physical and electrical properties, and the rise of many tools and utilities that draw off the new premises of energy and power as Prometheus's latest advances. Human beings have been realizing the death of god in many ways throughout the advance of social relationships in the last four or five hundred years. It is not something that happened on a single day when a thinker spun a yarn about a Persian prophet hellbent on spreading the word of the deity's demise. In Nietzsche's characterization, he thinks he is telling a story that has not completed yet. And truly today it still has not been completed. What is often missing in the interpretations of Nietzsche's grand statement is that he often dressed up this event as something that happened long ago, as something that happened in the past, long before he articulated it clearly through a set of philosophical writings. Philosophy always comes too late, in this way. And with Nietzsche's storm warning, that lament still holds.

The proclamation of the death of god should not be confused with the end of religion. One cannot point to the existence of strong religious organizations in the social sphere as evidence that god is alive and well. The opposite could be the case. We might expect that the weaker the memory of the deity becomes, the more aggressive the social relations will strive to produce religious influences and religious power to try and offset the vacuum that has been created. The importance of promoting a caliphate to take over huge portions of the world, or the importance of bringing Christian values into the political body that runs some giant world super power, are indications of crisis in the corpus of the deity. The desperation of these groups is indicative of what no longer holds sway, of what is losing power in the social world at an increasing pace. What makes this clear is the extent to which religions today are about establishing organizations and institutions of power more so than anything else. Their charge is no longer focused on getting the word out, or spreading the word, but rather it concerns taking control over as much of the social sphere as is possible. Religious wars today look more like culture wars than anything else. They are battles for control over legislatures and seats of power and they are less about doctrine and ideals than they are about gaining control over the social groups that have large-scale control over the various civilizations of the earth. Such organizations seek to gain control over these societies in such a way as to advance interests specific to their recent developments and have little or nothing to do with long standing traditional positions associated with those religions. At least, not with any scripture of the compassionate saviors or prophets that gave birth to those religions in the first place. Religions, in these times, characterized by the death of god, are merely social forces much like the auto industry or the national football league. They use advertising and targeted analytics to establish footholds in communities and nations aimed at pushing their social agenda through political means. Groups building telemetric infrastructure to impact and drive the process of production to ensure that future generations will be attributed as members of the right groups and thus be properly situated within the social order.

Just as the increase in religious organizations does not provide counter evidence for the death of god, an increase in 'political' activity does not qualify as a public sphere. The political, in so far as it is representative of a political party, in so far as political life is associated with parties that govern in accordance with sets of ideas that vary widely across the political spectrum, are themselves forms of social relations actualized as interest groups. A political party may be made up of many different subgroups, it may amount to a coalition of varied interests for the sake of taking over control of the political domain. The political domain, as understood in modern societies since the advent of the commonwealth, has become an essentially social domain. It goes well beyond the purely formal characteristics of a public sphere meant to function as protectorate for the private person over and against the social powers that be. The political is not a substitute for the conscience or voice of god inhabiting the person. Rather, it becomes a rallying cry of interests, of points of view that need to be articulated and defended with the aim of taking control over the various social relations and orienting them in whatever direction the empowered groups desire. And these are closed groups to the extent that membership is not so much a choice, but more like an assignment based on the socially implicit attributes that condition the production process.

Some of the previous aspects of our discussion may have suggested that we had a more classical understanding of how the political relates to the social and private spheres. After all, when we speak of privacy we often venture into the realm of legislative and judicial practices to try and characterize the public understanding of privacy as something of an authoritative measure. Each time we have tried to do this, we have run up against obstacles. Rather, we saw that the various courts, whether they be in Germany or in the United States, seemed to have been bent toward a social orientation that caused trouble for the way in which those notions were described. This did not prevent us from turning the notions over and possibly trying to learn something or other from them, but in each case, we were able to see social characteristics at work in these decisions and deliberations. We saw the Germans pro-Kantian orientation toward a dignified personhood, but we immediately noted that this approach had to be tempered with the necessary worldly characteristics of any such private person worthy of respect. In the American orientation toward expectation, we needed to temper that understanding with the gross impact that social relationships might have in the formation of expectations and how those expectations might be bent by the process of production. When we look at traditionally political decisions, we are looking further at the process of social rationalization and not into some purified realm of explanation. That is, modern politics is a distinct set of interests that have been brought to be largely for the sake of promoting some groupings of interest over and against others. Politics is part of the social sphere, political events are reported right alongside natural disasters, economic events of significance, celebrity breakups, and crimes or misdemeanors. We participate in the political sphere as though we were private citizens being entertained or informed about goings on in the world, interests that are set at work in this or that part of the world over and against some competing set of interests that might be battling with them. This phenomenon is so pervasive that it is fair to say that there is no longer any public quality at all to be found in the political activity of human beings. Rather, all we can see there are sets of social relationships articulating interests and inhabiting individuals as this or that attribute: labor party member, republican, etc.

Although the death of god would appear to denote a privative act, its situatedness suggests something else for the role this event plays in my story. The death of god is privative in a sense, but only in a way that creates a hole that must be filled and the notions and events that fill it will emit data, will have metrics, will be measurable and show themselves as phenomena that articulate the moral character of human beings. The tendency of human productivity would be to fill that hole with social structures, interest groups, and organizations. The battle for the temple begins when god vacates it and the most likely candidates for taking over are the various forces that exist within the world of the social contexts where our productive involvements take place. The car god is just as likely a winner in such an epoch as any other religious proposal that we might make. The deity of the free market, the power of the environment, many different candidates can step forward as the singular replacement that can best be set into the great absence that this death has created. And that is the point that needs to be emphasized: the death of god is a creative act, it is a positive event that creates a hole that is then available for filling, for supplement and substitution. The most obvious alternatives that will likely arise for filling that hole are the creative structures that exist ready-made. The history of human beings has an inertial power to it, our facticity is extended in space and time and as such has the power to fancy itself as a god ready to fill in the gaps created by that withdrawal of meaning, that extreme loss, that marks the history of human civilization with a tomb. There is something terrible and awful about the death of god, something altogether calamitous, and it marks the beginning of an enormous crisis. And this crisis threatens all human being. No longer are we guided by that voice that told us where to go and how to get there. Now that voice must come from somewhere else. Although the historical role of god's voice may never have been inclusive of all human beings, the dream of god's death, the idealism we find in new science and in revolutionary politics, was always aimed in contraposition to the divine orientation toward history.

The social powers that sought to fill the gap created by god's death were always dressed up with all sorts of idealized and beautiful notions: the dignity of human beings, freedom, and justice each can characterize our every relationship. Even when people criticize the historical pictures, they do so as a kind of reaction against their implementation relative to the ideals. They loved liberty, but they owned slaves, they bowed down to equality, but they kept their wives and daughters in bondage, they preached the perfect justice of the marketplace while advancing colonial interests at home and abroad. The ideals themselves were not so much under attack as were the ways in which they were realized and the specific historical conditions that established limits preventing those movements from realizing those desired ideals. The position I have taken thus far, although I have not yet called into question any of those ideals, has pointed out that these sorts of blinders or oversights are part and parcel of the activity itself. I have urged the recognition that any implementation of the ideal will happen in a social context in so far as those ideals themselves arise out of social relations and merely express them in collective form. Freedom, justice, and equality would not have been "realized" as human goals in the late 18th century if they represented an implementation that included women or slaves of both sexes, or even if they included all people throughout the world who were vulnerable to future colonization and exploitation in the advancement of the rights called out in the revolutionary doctrines. The palatable acceptability of freedom as an ideal, for example, depends essentially on the social conditions within which it is articulated and against which its implementation is designed. No one has ever yet proclaimed the greatness of freedom and justice without being forced to immediately supply elaborate boundaries and constraints on those notions to characterize them specifically in terms of the civilization within which they were advocated.

Human ideals that have been in widespread commerce throughout the world in the last few centuries have largely been characterized as social ideals and social notions. All the political theory that draws careful distinctions between the spheres, seems to omit the fundamentally constructed nature of human life: that human beings are created by other human beings and so our very existence is governed by a productive process constrained to the social sphere as privacy writ large, a non-political domain that makes up the everyday life of human beings living together in the world. Here, instead, I want to call out the effectively social nature of all productive relationships and recognize that the ideas we have held have originated in those spheres. The difficult project that Marx set for us is that being historical materialists places us in a condition where we no longer have any idea what we are striving toward. All the ideas, it seems, are bound to the material conditions themselves and become part of the productive relations that are necessary to keep things going. To establish the private person, which the telemetric understanding urges us to do to preserve these great ideals previously used to fill the divine void, we must recognize that the material relations have produced these things and these artifacts that we do not quite know how to manage. As private persons, we are under the threat of the social sphere. What are all the ideals we wanted to hold on to as part and parcel of that sphere? In the emergence of the public domain as a realm meant to safeguard ourselves from the power to crush and the power to create each of us, what are the ideals we are entitled to? How can that domain safeguard without itself being a product of the conditions that do the protecting? These are the questions that open our investigations into the commonwealth and guide our efforts at understanding the telemetric pattern in its relationship to the establishing of a public line of demarcation that protects the private from the social.

The End of Truth

"The End of Truth" here does not indicate that there is no truth. It suggests that a better proclamation concerning the death of god might have been "the end of god". In the rise of the commonwealth, god comes to an end. And the end, in that latter case, would seem to have a protective line to safeguard the private personhood from the social relations that every day realize the telemetric understanding in all its projects and nuances, known and unknown, available to human apprehension and elusive to it. This protective line functions as a formal association of private persons that are each worthy of respect and ready to recognize each other as such. In pronouncing the death of god in our institutions and our organizations, we posit the objective of recognizing each other for the sake of each other and not because some third transcendent authority commands us to do so. This is not a privative moment in human history, but a deeply positive one, a moment that marks the advent of massive action and collective orientation toward common goods and goals, marks a dedication to a set of ends that connect each private person to each other even in so far as they maintain their independence, integrity, and autonomy. It might be said that now we may be able to stand face to face and see each other for what we truly are, not creatures of an authoritative source, but transient and exceptional moments of chance, of a universal dynamic that spun us into being just as it wove the tapestry of the night, the stars, the stones, the wilderness, all that surrounds us, and all that presents itself to view to be poked at and prodded, to be interpreted and analyzed. We can look at each other in this world as cohabitants of the same cruel and beautiful processes that spawned all there is, including the 'there is' itself in what must surely be one of the most amazing ends of all: the human consciousness. An 'end in itself' maybe? Extraordinary, most definitely.

In that faint moment of recognition, in that imagined face to face, we surely know there are worlds behind and around each of us. There is a narrative tale to every face, a life behind every agent, a history and a set of experiences that have been appraised and interpreted, internalized and processed. The recognition of the other as a formal moment of practical reason misses the substance of the point. And that substance is the whole of humanity that stands there before each of us in the other. The life of the other is, yet again, those relationships that are realized and re-realized every day in the life practice of that individual, and each brings with themselves parents and siblings, teachers and mentors, villains and heroes of all different kinds and flavors that we have come across throughout our lives. The electric company that gave you persistent grief in your early years when your family was having trouble making ends meet, the department of motor vehicles, the neighborhood homeowner's association, and the corporation that runs the business where you work. Each of these conditions and influences confront us not only in ourselves as the traces of a personal history that we have consumed and come to embody throughout the years, but in the face of the other who stands over and against us in the checkout line or on the street. The face to face struggle of one human being with another is not a simple formal recognition bearing event that takes place when a wholly free one and a wholly free other confront each other and come to some set of implicit agreement about the claims they may make upon each other. Rather we confront each other as complex histories with identities and attributes that bring the whole of our social relations to bear throughout our involvement and our interactions.

The emergent public world that sets us together in this way, cannot be a formal set of principles that merely accommodates the structure of our action as agents, it must accommodate the whole person of the one along with the other. It must provide a dynamic that enables recognition even in the midst of powerful forces that are at work influencing that recognition, changing its character, and twisting it or turning it in one direction or the other for the sake of establishing interested parties making unreasonable claims on each other, ignoring each other, hurting each other, and doing any of the things we do that both recognize the other as other and reduce them to someone fitting a role within a larger human interest story common to many. These social relations and their embodied interests are not merely the substance against which the public forms the intention of protecting their private person, they are actively hostile to any such public for that very reason. Interest structures and the prejudices of understanding are everywhere at work and must, to realize their own ends to support and further themselves as embodied interests, aggressively tear down any protective film that prevents those interests from holding sway. A person who is trying to cure themselves of an addiction is likely –as they become healthier during the withdrawal– to be capable of coming up with all sorts of reasons why it would be acceptable to succumb to the addiction once again. These powerful reasons, like the body tremors or any other physical symptoms of the withdrawal, are further realizations of the addictive drive itself. The substance of the addiction will use all means available to veil itself and to realize itself. And the social relations, congealed into interests, that occupy persons as they come into contact, will take on the same set of characteristics. Recognition is something of an idealized notion. We can talk about it as respect for formal conditions and prerequisite to any concrete interactions, but even if we accept that characterization and align it with the so-called public protective mechanism, it is always the target of extreme coercion. And not as one distinct thing battering against another distinct thing, but as an infiltration that inhabits the form. Interest structures transform recognition into something else, they grasp the other in various of their qualities and attributes, reduce the other to those attributes, make justifications for how this is the true nature of the other, and describe how seeing them as such constitutes a clear recognition of their being as it is in-itself.

And this tells the tale of 'the end of truth'. I hope that it is not controversial to say that in so far as truth amounts to a semantic notion indicating that 'snow is white' holds to true if and only if snow is white, then what is really at stake in such matters is the 'if and only if' and what establishes it. This way of thinking about truth may suggest that truth, per se, does not really matter. What matters is whether snow is white. In those cases where snow is indeed white, it is trivial and self-evident to proclaim it as such. How to establish such facts in the world when interests are the major share of every experience, play a role in every interpretation, and inhabit the private person's experience as properties and attributes of that very personhood? When two people meet in the middle of an interest entrenched set of operations, they have no need of small talk. This is not to say that they may not engage in small talk while performing these operations, only that there is no need of it. I am eating at a restaurant; the wait person takes my order and brings my food. I am their customer, they are my server. No need to get into it here, no need to find out details about each other. The interests of our respective roles are sufficient to carry us through what we need to do and make sure that we get it done in accordance with each of the interested domains that we represent. The server may have come to recognize that with certain kinds of customers, small talk of a specific kind can serve their own interests: they may get a bigger tip if they are able to cultivate a more personal touch in the interaction. And the customer, even if this interest is recognized by the server, might realize their own interests of having a pleasant meal that includes human contact by engaging sincerely in that small talk. The customer is happy to pay the tip that is the fair response to that upgraded service and the server may be happy to carry off their part as well, especially if the tip is appropriate. The interests of their roles are powerful inhabitants of the interaction and there need not be any breakdown in their relations that forces them outside those interest structures. Unless, that is, they maintain 'truthfulness' and 'the end of truth' as a component or facet of the relationship.

In this context, I am introducing truthfulness as the private person's critical methodology that they must enact to safeguard and protect their private personhood in the social domain as they work to realize their augmentation through social relations rife with coercion and force that might act in violation of that private personhood. This is a momentary heuristic because I am supposing that there is a fact of the matter of the complexity of relationships at work in the simple confrontation between diner and server. And these 'facts of the matter' are the 'if and only if' of the truth relation between utterance and the language or state of affairs. Truthfulness might, therefore, be a vigilant undertaking in this case. It would amount to an endeavor to tear aside the veil of interests that permeate the relationship between the diner and the server for the sake of getting underneath it and then to the bottom of it. Way beyond small talk, this would be the attempt of the duo to get to the truth of their relationship in all its meaningfulness, to understand the complexity in historical context, in social context, and in world historical meaningful relations where one person has gone out to dinner so as to have that dinner served to them by someone whose job it is to do so and –at least in some countries– to earn the better part of their wage for that job through the ad hoc gratuity provided by the diner in appreciation of the service delivered.

Most likely the response either of the two parties might have should the other become suddenly truthful in this way would be shock and discomfort. Even if the stars were to align in some random way and both were to recover from the initial reaction, it is unlikely that the exchange could go on for very long while the two maintained their roles. The server might be reprimanded by a manager and urged to go and deal with other customers who need their attention. Leaving that aside for the moment, I would suggest that even were the two of them to work toward engaging in such an interaction, it is unlikely it could get them too far beyond the interest structures that dominate the relation. Obviously, it is possible to fantasize that this interaction might result in the two of them forming a lifelong bond that reaches deep into the private personhood of each, they might become dear friends or establish a romantic relationship. We might consider putting together a story in which this interaction becomes something viral and catches on to create a huge human interest movement that inspires people to be more truthful in the way they live their lives and try to form closer bonds with each other by authentically working together to dig in and understand the interests that are at work in their role-based relationships day in and day out. These are fantasies. Truthfully, this does not happen and even if it does happen now and again, these occurrences are merely momentary and do not proceed to their logical conclusion, nor do they ever have a far-reaching impact on the world. To be truthful in one's truthfulness means that we must recognize the inertial power of the conditions acting upon us, the way in which those forces hold us in place and work against efforts to modify or alter the effluvium within which we live our lives. As opposed to getting underneath "it" or to the bottom of "it", the sense here is to get into "it" and to comprehend that there is no getting out of "it".

Truthfulness may appear to be self-defeating. To the extent that it is perceived as an effort to undermine resistant forces and the constructs that hide the truth, and because those forces are both behind it and in front of it, it is doomed to be incapable of achieving its end. By 'behind it', I mean the interest structures at work in the private person themselves and by 'in front of it', I mean the interest structures at work in another private person. An interest structure embodied and confronting another interest structure embodied has no hope of being truthful. There are interests all the way down. Tearing aside the veil to see the facts behind it is itself a well-established historical interest. There is a certain kind of person who is motivated by such a desire and often that person considers this to be an exercise of personal power and capability. They show off their capacities and intelligence to good effect when they exhibit the ability to see past the illusions and get at the underlying truth through a variety of different methods that work to exhibit the things themselves in their true light. The facts are buried, and the heroic investigator is going to step in and save the day, unearth the thing in-itself as it is independently of any interest. And you may have met people like this, you may even be a person like this. And where did you come from? What influences and conditions marked your path through life? You clearly must see that you came to occupy this position the same as everyone else and so it is at least possible that your current orientation is just the mark of your life and the route it has traversed.

There is no reason to suspect that the attitude that attempts to lift the veil is the approach that is the most truthful. Since the social relations are part and parcel of all interactions, since they begin to inhabit our actions right from the beginning of our lives and push and pull us in all sorts of directions throughout our involvements, it might make more sense to consider truthfulness as the method for getting at this dissembling procedure, at the process of interest loading that goes on throughout the lives of private persons, and the establishment of their personality through identifications and attributes that heap each of us with interests and qualities that characterize and categorize. The process of being is the truth of the matter and is the emergent state of affairs. How does one truthfully get at the process without making the process something entirely other than what it is? The act of trying to get to the bottom of the process of constructed interest structures and of personality formation, would either miss the point entirely by diminishing the existence of that state of affairs that it attempts to comprehend or it might just be a further emanation of those interest structures, making it more and more difficult to see the process at work while being at work. In an earlier work, I called this phenomenon 'interlude', trying to emphasize the playfulness of the pattern. Not only its playfulness but its position between things, its place as something other than a primary location that can be inhabited by an agent or analyst. The interlude suggests a state that is hypothetically reasonable and that can be logically constructed as a meaningful state, but which –because of concrete forces at work in the world– cannot itself be occupied as such. Every intelligent being can comprehend this state, but none can be in it. There is something fundamentally question-begging in the condition, it is a position that is taken up for the sake of getting out of something that one is no longer inside of in virtue of the fact that one has taken up the position in the first place.

The fantasy of having some enormously meaningful experience by connecting with another person in such a way that gets beyond the current and affective interests at work in the whole of human intersubjective life signifies a common interest in such meaningful experiences. Likewise, the anti-pattern of the truth teller who can tear aside the veil or show the wizard behind the curtain, is widely thought of as a pompous ass who likely will take all the fun out of any party and regularly engage in conversations that frustrate and irritate all those around them (except for the select few who are aspiring to a similar level of annoying behavior and so get on quite well in such a situation). The pattern and the anti-pattern are ways in which truth comes upon its end. Since truthfulness is associated with patterns of activity that are boorish or fantastic, then the object of any such truthfulness must be completely elusive and beyond anyone's reach. The end of truth, therefore, would be literal, it would amount to a rejection of truthful methodology because it is not consistent with any real world set of operations and procedures that make sense relative to the carrying out of actions aimed at realizing interests. Truth becomes an illusion as the object of such truthfulness, and there is no longer any reason to pursue it. In this way, the interests and the interested parties gain an enormous victory, because once truth has been discredited and truthfulness shown to be hypocrisy, then the only remaining alternative is the interested bodies of the social relations and the dominant position they have in the world of human beings. Forget about trying to get to the bottom of things, forget about trying to think or interpret your way out of this mess, there is no way out, there is nothing beyond the veil, and all efforts at getting out are just methods for becoming more deeply entrenched in the institutions that have hold of us from the outset.

This is a cynical resolution, but it also appeases the desire for truth which is still strong in those who would once have aspired to truthfulness as a way out of such an interlude. The person in the process of freely determining their own personality can happily realize that they have grasped some fundamental truth in their resignation to the fact that truth itself is ineffective or impotent when it comes to getting at the essence of things. The essence of things is just interest, interest is the destruction of truth, so I might as well resign myself to that *fact.* And the recognition of it as a fact is what makes it palatable, it preserves the sanity and ego of the investigator because it now seems as if it is not so much that they have given up in their inquiries, but that those inquiries have gotten somewhere, gotten at the unutterable and inexplicable truth: that there is no way out and there is no way to get beyond the veil, that it is veil all the way through and the prejudices and positions at work in the real world are themselves bent toward this or that end with this or that agenda and for the sake of promoting the overall goals of some group over another or some set of personal attributes over and against another set within the "singular" person. When the nanobots arrive and the real is saturated by the virtual, nothing will have been lost that has not been lost already.

Earlier I suggested that truthfulness was only a temporary placeholder for that critical understanding that I held must be at work in the private person as they augment themselves in self-determined fashion as a person of integrity who is in control of their data emissions and the analysis performed against those emissions with whatever feedback results are looped into the system at work in the world around them: some such telemetric understanding that undertakes to colonize the private person and all of their ends. Now, we are beginning to see why this critical method or interpretative approach cannot be thought of in terms of truth or truthfulness. It is not because there is no truth or that human finitude is so limited that we do not have the capacity to approach the truth. Rather, it is because truth itself is a notion with a history and social context, and it is a notion that plays a part in the social dynamics of interest conflicts and coherence. And as such, it is there only for the sake of failure. The role of truth in the productive relations of society is to fail, is to be constantly thwarted, and to provide ongoing disappointment such that people resign themselves to the interested parties they represent; and that they do so with a compensatory value of having understood something vital about the world. The end of truth lies in its utter uselessness in providing any kind of resolution for an emancipatory interest. Rather, its fundamental purpose is to fuel the engines of contextual interest, set up, and embed any individual human's resolution to those current conditions, as well as to ensure that such conditions have become dominated by interested parties that are both understood and which –to some extent–

transcend any such finite understanding. The set-up and embedding then makes sure that what transcends understanding feeds the continued prevalence of that portion which is understood: what we imagine about the powers that be is what supports the powers that be, and that is the end of truth.

We go back to Nietzsche yet again. His aim was not so much to bring about the death of god or the end of truth in simplistic terms where each would amount to a termination of something called god or some state called truth. Rather, Nietzsche's project was both reductivist and emergentist at the same time. He wanted to situate all human endeavors on top of the will to power and then show how the will to power was a source from which temporarily stable configurations could emerge. Truth does not matter, power does. And the purpose of truth thus far has been to express a will to power. Each of our truths, he might say, has been the emergence of a stable state belonging to some underlying coercive force. And god just the same, an idea expressive of a will to power, emergent as such, and powerful as the expression of that underlying state. To step beyond these myths or stories, to move on to different stories and different tales, was an effort to strip the old myths of their power for the sake of some new stories or to actualize a new power. In Nietzsche's case, all of them were only stories, whether it be the one or the other, the forces at work were working it out in terms of stories and tales, yarns being spun. And the dynamic of one battling against the other was only comprehensible in terms of one single hierarchy: power. Will to power is the meta story then, it is the condition for the possibility of telling stories at all, it is what stories do and what stories are. There is no truth under the story, there are just stories. This may be the most charitable way to comprehend his doctrine. It is not some reductio of the rejection of the truth telling tendency, it is the name for that tendency itself, and the winning stories, the popular stories, the powerful stories, those are the ones that dominate for a time before receding into the background so that others can come forward as the new prevailing tales.

On this reading, the end of truth would seem to leave us with nothing, with no options for establishing our protective boundary to safeguard the private person from the social interests that encroach upon them. If anything, it endangers the entire project suggesting that even the notion of our constructed personhood and our efforts at self-determined personality through augmented character, our striving to become more and our will to power to make something of ourselves in the wake of the social relations that started the snowball effect of our lives in the first place, that all of this is merely sophistry and illusion, will to power, and nothing besides. The truth of the matter need not be a state of reflective apprehension that we come to occupy given various epistemological preconditions. This amounts to truth as a function of the present at hand relationship of the thinking person to the state of affairs that are the referent of the proposition or cognitive state. In so far as we reconceive this emission of the public as a process related to truthfulness, as a practical result of truth-oriented action, it may be that it constitutes a facility with the chain of events that is more like know-how than knowing-that, more like a set of involvements with the ready to hand than a step back to purview what is present at hand. The possibility is that the public façade that sets human beings in a relationship that is protective of their private persons from the coercive force of social relations that may use and abuse such agents for its own higher order aims and ends, would emerge in some specific form of every day practice by the agents themselves. In this case, we might say that the proper emission is a sort of etiquette in our treatment of each other, not so much an ethical theory as an ethos, a sense of practical know-how that facilitates appropriate relations with each other so as to maintain the proper proximity and distance, moving near to each other in accordance with the wishes of the other and moving further away to give space where space is clearly what the other requires.

It may be possible to construct this as a 'second personal' form of respect for one another, but a philosophically weaker approach has been used to get here. Namely, the attitude of the agent respectful of others in accordance with some form of etiquette must be tied to a set of interests that have come to be ingrained in the private person as a condition of their socially achieved personhood. It would almost have to be a coincidence, or the luck of developmental events, for any person to happen to acquire such characteristics in the course of their lives. These are not a set of formal or rational principles that can be laid out such that any rational being would be able to grasp their commanding significance and thus follow those principles from that point on in their lives. Rather, the phenomenological origins of my notion of agency suggests that people are not made that way, and that we do not have the sort of autonomy that this kind of moral will presupposes. Rather, we are formed and shaped by the world within which we have been raised, and if that world has not granted any one of us with the consideration and concern for the other's private personhood, then no one of us will have it. It does not mean that such orientations are unachievable. A society that concluded that three generations from now, this sort of practical excellence, say, would be highly desirable for the grandchildren of today's adolescents, could gradually begin the process of shaping itself in that direction. This sort of radical rethinking of social action is far from simple and not without risk.

There are all sorts of connotations surrounding this social engineering, the trustworthiness of the decision makers, and the interests that would go into the determination of the outcome of such deliberate actions. These are real problems, but I do not see that the notions I'm advancing permit this problem to be any different from the already existing problems that any society already faces. The nature of the society three generations into the future, by my conjecture, is already the outcome of a set of socially oriented operations and processes. We, as a society, are already engaged in this form of construction. The question then may be whether the process is already deliberate and whether it is the case that the predominating interests in that process are operating secretively such that the partisans are not aware of the principles at work? Or is the deliberateness of the process beyond any group's sphere of comprehension and control so that it is emerging as the result of a higher order process at work? In that case, the future generations of our social worlds are being determined by the telemetric understanding and there is nothing any group can do about it.

In the end, the important question to frame before proposing some materials seeking to answer it, is whether the deliberate determination process of human being and what human beings are in the context of their socially produced relations, could ever be under the control of human interests? That is, even if the controlling interests belong to some one group of people, or some cross section of identifiable properties with various interests for furthering those attributes and characteristics, is it possible for such a group to steer the social ship in a way contrary to the dominating forces of a higher order systemic logic like the telemetric? And if it is possible that there can be a human originating control over the orientation of the social forces, then we would know it is possible to change course and move in a different direction than the one set down by segments of the population currently and transiently aligned with telemetric power. The orientation is basically to begin one's search by seeing if one can establish the existence of evil. If evil exists in the world, bona fide and rationally comprehended, good exists too. If both good and evil exist, there is hope the one can overcome the other. The opposing case is not that evil always wins, but that both good and evil alike are just pawns to a set of natural forces that blow whichever way it damn well feels like blowing. Good and evil are merely examples here. Our notion of interests thus far does not permit such categories other than in a purely strategic distinction between 'us' and 'them'. The lambs think the eagles are evil and the eagles think the crows are evil and so on, each according to their interests and aims. No truth, no goodness, no morality. Just interests and the natural power that shapes their conflagration.

Unrequited Love

It may be that the person most attuned to the interests at work in the intersubjective domain and the pressure of social forces on all the collaboration going on there is not so much the one who openly reflects upon those relations and brings them to self-conscious expression at the expense of continuing to develop the relations themselves, but rather one who is deeply attentive to those relations and how they are playing out as they are playing out. Such a person might come across as genuine and present as opposed to distant and fake. An interaction between such people may mean, for example, that the food still comes as ordered and the tip is still given as earned and the two do not break completely from their roles and fall back to an analysis attempting to root out the foundations of the current interaction and all the various pressures coming to play in the moment they come together as customer and server. Such people are aware of their human being in whatever context they find themselves, they bring it to the relationship and recognize it in the other throughout their interaction. And this would not amount to expressing false and overly invasive concerns, but an awareness of the context the other is in and empathy for it. One characteristic that separates any diner from their server is the fact that the server, unlike the diner, is busy. The diner is, by and large, at their leisure. They may be "in a hurry", but that just means that they do not have long to be at their leisure. Were the diner to attempt to engage the server in such a way as to fail to accommodate that busyness, the diner would fail in their attentiveness and, it may seem, would fail in their analysis of the situation precisely when it seems like they are engaging in it. Being a good customer amounts to showing respect and recognition for the server in so far as they are acting as a server. That means being aware of the server's role as well as the fact that the basis of this role is an actual human being embodying a role at that time for those purposes. Likewise, the server should carry on in similar fashion. Good servers read their customers and try to figure out what their context is. Are they dining, are they here for a quick bite, are they on their way to someplace else, etc.? The answers to these questions will situate the interests of the customers and indicate to the server some more detailed rules for how they might fulfill their role as server. This suggests that the role itself is best carried out with such recognition. A good customer respects the server in so far as they are acting as a server and the good server respects the customer in so far as they are acting as a customer. And this amounts to a twofold and simultaneous recognition of both the interested party and the private person.

This simultaneous recognition is a learned skill more so than a body of knowledge. I have suggested that we might call this etiquette or, were we to stretch the matter somewhat, a kind of practical know how in matters of the world, something like what the Ancient Greeks knew as phronesis. Because it is a people-oriented skill, it may appear as charm or charisma but without anything that would appear false as there sometimes may be in such qualities. Rather, we are talking about an ability that appears to naturally come out in the actions and behavior of the agent as the agent engages in various situations and circumstances. As a learned skill, it is the result of ongoing training and cultivation. One does not learn it all at once, but gradually adds to such skills by having new kinds of interactions and experiences that supplement the variety and detail of one's understanding of relevant circumstances and the likely requirements for navigating them. For a person to be graceful and adept at such interactions, they might require an ongoing dedication to such practices. They would need to continue learning about different walks of life and different parts of the world, different ways in which people are forced into living and acting for the sake of assuming or maintaining some given set of social roles or interests. No one is ever completely familiar with all situations and contexts, so the person's attitude must be one of openness to new situations and a clear insight as to when those circumstances are taking place. Such a person would need to be able to both apply past experience and to realize when they may not have adequate familiarity with some aspects of the situation that they find themselves in. The balancing act would require they know what in an experience is familiar and what is strange and new, and they would have to gauge their reactions and responses to those circumstances to accommodate these variations.

Any such person would also have had to learn how to control their own role-based responses to events. It would not just be an openness to the interested party / human being dichotomy before them, it would entail a similar openness to their own interested party / human being dichotomy. Things might be strained, one's own interests as a consumer or actor in the world may be at risk on occasion, and those forces might make it difficult –if not impossible– to maintain the balanced relationship that good etiquette would have otherwise required. The forces at work may be very real and even dire, and quick response may be required to drop all consideration of the dual nature of the relations. To save one's life, or even to respond to a perceived threat to one's life, one might have to recognize that there is or are others that have dismissed oneself as a living human being worthy of such respect. And in that moment, something like a raw and powerful state of nature has taken hold over the circumstances and the luxury of cultivated etiquette is no longer an option. No doubt an attunement to the very real nature of such threats would have to be part and parcel of any such training in the skill. The whole point of the emergent public domain under investigation here is that it is meant to be a protective boundary from social pressures characterized by interest structures that often invade the integrity of the private person and that coerce that person in ways that reject or fail to recognize the personhood of their person. And these forces are realized in actual human beings who behave in specific ways in their roles such that the humanity of their 'targets' is obliviated for the sake of whatever social interest they are pursuing.

You are no aggregate. Nor am I. And although there is a powerful capability to aggregate the emissions of our daily lives together, that aggregating action can only violate something that exists. There must be something there to be infringed upon, to be seized or surveilled. Aggregating is a way for the social forces at work in the telemetric understanding to reduce the humanity or private personhood out of those persons who are collected within the aggregate set. And it may be possible to understand any direct threat to one's person as a sort of aggregation. The robber who reduces another person to a victim of some scam or direct larceny has removed that person from their specific categorization and has grouped the one standing before them with all the other potential victims of their crime. By generalizing the person before them, they remove themselves from any concern for that person's place in the world at large, they lose all respect for whatever ends and projects that person might have, and they submit the other to their own role-based interest orientation for the sake of achieving some end that involves the other as a tool or instrument. Traditionally, it has been useful to think of this violation of the humanity in the other as a kind of instrumentalism, but here I want to specifically call out the aggregation effect that is implicit in this action orientation. The other's individuality is removed, the private person is ignored, and the person is grouped among an interchangeable population defined in terms of characteristics that are useful to the matter at hand. Rather than being yourself or myself, we become a person with cash in their pocket or a body to be controlled. As such, these attributes are group attributes, they are interest identities that categorize, that put us in sets and groupings that are the ideal target for some action or interest that the other party has regardless of what our interests or actions might be. It is not just a rejection or dismissal that is taking place, there is a positive act at work: the person is placing their target victim into a group, is performing an aggregate operation, is summarizing, and analyzing. In short, they are acting on something nearby as though it were far away.

Assault and battery, or some property-oriented variation on those themes, need not be the literal materialization of the actions in question. Whenever two people enter an exchange relation of any kind where one will offer a service or resource in exchange for financial compensation, it is possible for the relationship to take on this telemetric character. The customer in such a case may feel the impact of this reduction, might feel as though they are not a valued member of the exchange, but merely the money resource being harvested in exchange for the services provided. They might not get any sense of quality or craftsmanship in the work delivered, no concern for customer satisfaction, just the sense that the minimum baseline requirement has been offered and the money has been extracted like a resource being farmed or harvested from the earth. The experience might leave one thinking the world is filled with opportunists or greedy predators, and that the consumer is little more than the prey that is of little concern outside of the money they have to offer. This kind of predatory capitalism need not be specific to large scale corporations that easily reduce the customer to some small fry that means nothing on its own given the magnitude of their operations, rather the relationship between any two people can degrade in this fashion. An independent contractor or a small business owner is equally capable of exploiting their customers in this way. It may be more likely for such a person to do this, as they are more in need of reducing the relationship to its most basic instrumental facets to make ends meet or to capitalize on the relationship in the way that they need. When the job is going smoothly, they are better able to personalize their service and make it feel as if the two parties are engaged in something that continues to embody personal recognition of each other as private persons, but when things go wrong the small provider is less likely to have reserves to address the problems and take care of anything unforeseen. It is in such cases where there will be the most pressure to look to the customer to offset these problems even when the arrangement or agreement between the parties had already been settled and the mistake is something the expert provider should have anticipated in their initial agreement. Scarcity and crisis may feed the probability of such a reduction and that is possible when dealing with the mega-corporation, with the mom and pop outfit, with people in line for a Christmas sale, or when competing over a promotion at work.

The ideal condition that we have been describing is a skill or practical ability that is at work in involvements with others and helps to materialize the experiences of those involved as a positive mutual recognition of private persons, whereas the privative condition occurs when this breaks down into a predatory relationship that reduces the other to a mere means for the sake of some set of ends or interests the other is pursuing, and where such a break down might be mutual. The presence or absence of such a public sphere in such relations is not just a matter of having the right laws or having the appropriate set of rights. Laws and rights, it has been suggested, are only necessary as explicit statements of these conditions when a tendency to violate them has become prevalent. We do not need laws to encourage behaviors that are already widespread, and we do not need laws that prevent behaviors that are already uninteresting to most people within the society. Moods, for example, are not typically the subject of either legal prohibitions or requirements. Civilizations do not declare it a right that persons with whom you are interacting must pay attention to the details of the exchange because they recognize the private personhood of all involved as precious and so worthy of concern in every facet of the exchange at whatever the cost. This cannot be legislated. The conditions of life are much more likely to suggest that these cannot be legislated because it would put undue burden on each of us to do so. If we are not already raised and socialized to think this way, if we are not already planning our actions around this sort of mood in our associations, then the added impact of a law is going to be extremely onerous to bear. I am suggesting that the element of these relationships that protects them from degrading into a telemetric logic of social exchange is a quality that must accompany actions in the same way that moods accompany them. And how could a society legislate being in a good mood or being cheerful? Conditions and events, both social and biological, would seem to be at work in the determining of any mood, so if the solution to the problem of our protection of privacy depends on something that is mood-like in quality we cannot establish it as a social norm through legislation.

To go a step further, the fact that such a protective film in social relations cannot be legislated might also suggest that it is not possible to make it an object of theory articulated in ways comparable to those in which laws are made. The moral law, or some set of ethical principles as described by a philosopher in some treatise or some prophet in a religious work, amounts to distant teachings that construct impersonal circumstances and provide principles or rules for acting in those circumstances while aiming at achieving the good and avoiding evil. We cannot theorize a mood, we cannot intellectually derive the correct set of behaviors or intonations in our behavior, and then properly apply them as a mental exercise. The model we have for ethical deliberation is already something of a disqualifying feature. We are aware of circumstances that force a decision on what the right thing to do is, there is a trolley and it is rushing down the tracks, etc. The setup is meant to put us in a deliberative situation where we are not engaged directly in time sensitive behavior that is heavily embedded in context. If we are to imagine that behaving correctly, doing the right thing, is a matter of deliberating properly at exactly that moment when the dilemma in our action presents itself, then we will have already missed out on the nature of the action we are trying to guide. Instead, whatever 'theory' is the so-called right theory for governing such behavior must have been long ago determined, clarified, and embedded in any relevant agent as a precondition of their practice. Only a child raised with the right associations and relations, with a set of interests and characteristics that embody this 'proper' orientation, stands a chance of responding adequately under the circumstances and doing the right thing by treating others properly in those situations where we encounter them. The trolley problem can only be posed once, and a time limit must be set, so that the test subject discovers what they would do rather than cognitively attempting to construct it.

This may have been what led Socrates into the social engineering dystopia that he is famous for when he tried to work out a notion of justice that would properly decorate all human agency with proper concern for right response to relevant events in matters of social interaction. He saw that for the society to be just, the persons who take on roles within the society must be raised from the outset with those principles of justice in mind. The entire life of the civilization had to be oriented toward this kind of concern so that each member would experience respect for such conditions as an embodied precondition for all their actions and as the source for any insights or instincts that they would bring to their lifeworld and their experiences. For something to be as deeply engrained in human behavior as a mood might be, the indoctrination would have to be precise and lifelong. And it might take multiple generations to really get it right. It may be, for example, that the Socratic notion of justice can only be realized if the children are raised in accordance with the principles he lays down. And it may be that children so raised by parents who were not raised that way will not be able to perfectly acquire such characteristics for upholding justice. That means the first generation of subjects in this system will be less adjusted to the norms of the system than the second generation of subjects. Even the second generation of subjects may be less than ideal, since they have the disadvantage of being raised in a society in which their parents were imperfectly raised by their grandparents. With each passing generation, the margin of missing pieces is reduced, meaning it might take three or four generations before a program of Socratic justice can truly meet with its ideal. The social engineers would have to take this into account and ensure that the evolution of the cultivated characteristics is properly scaled to a multi-generational indoctrination process rather than something that would happen all at once. The architect of any such design would have to explain how we get from here to there in addition to showing us what it looks like once we get there.

If the protective layer of the private person that I am calling the 'public' sphere depends on some form of social long term, multi-generational social engineering to properly emerge as a part of human life that truly provides the domain of respect and integrity we require, then we are in a dire situation indeed. Previous discussions have suggested that the telemetric understanding which operates at broad scale in the contemporary social world is driving us generation by generation in the exact opposite direction. The private person would seem to have less and less claim to the respect of personhood previously described. If anything, the attitude that is being engineered under the conditions of the telemetric understanding is such that we are becoming increasingly less attributed of such a mood granting respect and dignity to the private personhood of others. We may be systematically losing any etiquette that would give us this sense of appropriate respect in our daily involvements and associations. This would be both good news and bad news. On the one hand, it would suggest that we are in the middle of an uphill climb where the trend is moving in the opposite direction of the way we want it to move. On the other hand, however, it would indicate to us that we are already in the middle of a trend that amounts to a form of social engineering. That is, the bad news would at least prove to us that the positive outcome is possible. If the current conditions are such that we might be increasingly moving in a direction, then it is possible that the conditions might turn the other way and move in the direction that would better serve our aims should they include respect for the private personhood of each human being. The double-edged sword being, of course, that interest in that aim would be decreasing with each passing day and certainly with each passing generation.

The double bind that we are seeing in the requirement will only get worse as the next few chapters spin out and the dynamic of the commonwealth under conditions of the telemetric becomes more and more clear. Now, let's pause to think about what this means for a certain kind of project that I want to refer to as 'philosophy'. The suggestion in the last few pages has been that a certain kind of project will not be able to give us what we are looking for. The confrontation with the problems for human being posed by the telemetric understanding are not problems that we can theorize our way out of. They are concrete and real, and we inhabit them as our living surroundings. In so far as our approach to this is carried out as a theoretical inquiry, we will not be able to solve the problem, or rather the solutions we come up with –even if they are spot on– will do nothing that changes the situation we are confronting, because disinterested investigations into the nature of relationships and the deliberations on the right course of response to such events, is not effective as mood-like characterizations of human action and behavior. We will not be changed in our fundamental orientations toward situations and events by some theory that we become familiar with. Aristotle was deeply aware of this. In Aristotle's Ethics we do not get the sense that the theory of the good life is something that has been hatched such that any given individual could just read it and apply its principles to their actions and begin living the good life. Rather, what he suggests is that doing philosophy with one's friends is itself the good life. Only if one becomes a philosopher, and does so with others who are becoming philosophers, do we live the good life. The nuance that I am taking from this might be odd, so let me emphasize it. I do not get the sense that Aristotle thinks he has painted a theory of the good life that can now be embedded in lots of different walks of life: farmer, cook, statesperson, or what have you. The logical conclusion of the doctrine of the mean is that the person who would strive to achieve it must change their entire life so that they become a lover of wisdom and adopt measure in all things to spend their lives seeking greater familiarity with that wisdom among like-minded others doing the same. This amounts to a radical transformation of such a person's daily existence. No longer able to focus on the goings on of the farm or the kitchen, such a person would be dedicated to the pursuit of wisdom in all matters. And through this transformation, such a person will have found the good life.

This reading drastically opposes the Aristotelian picture of the good life to the picture we might get from Kant or one of the consequentialists. In those latter theorists, there is no sense that our way of life must radically change in order to make it the good life. Rather, we just need to adopt the right set of rules for governing our otherwise focused actions oriented all over the map in various areas of interest and meaning. You can be a good cook and follow the categorical imperative or principle of utility. The convenience of these ethical theories is tailored to such usage. They are one-size-fits-all approaches that can be laid on top of almost any kind of life and any profession so long as it fits in with the rest of the higher order social division of labor. Do whatever you like so long as you do not violate the categorical imperative or the principle of utility. I might be tempted to suggest that since Aristotle the instances of truth in philosophy have been next to nil. The switch that I have described might spell out a radical departure so that the lover of wisdom disappears from the earth in favor of a lover of principles to decorate the remainder of one's actions. Or more succinctly, the lover of wisdom has become a lover of knowledge or lover of science who thinks that having the right deliberative outcomes from the available data will be applicable in all situations and across all interested groupings within the civilization so long as proper methodology is applied. The correct ethical theory or the correct epistemology is a theory that can be taught, and once taught, the student can just apply it in whatever walk of life they choose without being transformed by the experience. Philosophy, in such an expression, is reduced to a *fach* or a discipline to be taught in school. And no doubt this change indicates the broad variation that has occurred in the last three or four hundred years in terms of the areas of specialization that a philosopher might devote themselves to. Nowadays, a philosopher may be focused on nursing or civil engineering, on kinesiology or biology. Any area of knowledge, any domain of scientific knowledge, can yield philosophers, specialists who go very deep in those areas and start to produce insights and theories that might be impactful to people performing more applied actions in those areas: nurses and civil engineers, for example.

Even the phenomenologist Husserl wrote of this transition as a good and desirable thing. He wanted to establish philosophy as a strict and rigorous science so that its methodology could be purified, and a proper theory of its subject matter achieved. It is often suggested that any philosopher who came after Plato or Aristotle can be said to fall into one camp or the other, either idealist or empiricist as the emphasis tends to lean. On my reading, the philosophers that come after fail to be either Platonists or Aristotelians because the difference has more to do with cultivation than with methodology or approach. In both cases, the thinkers asserted that your entire life was devoted to the love of wisdom and neither approach seems to leave any room to produce a set of theories or body of knowledge that will be useful to third parties who come across those materials as byproducts of the philosopher's labor. This whole theory of intellectual production is heavily biased against the very notion of a love of wisdom that we get from either Plato's Socrates or Aristotle. As if theories were produced in factories and then distributed through some means that make them available for use by agents with ends looking for tools to get their jobs done. That is, philosophy within a telemetric understanding is more accurately characterized as a love of knowledge: philoepistemology. And it may be that such disciplines are useful for a whole set of interested ends in various areas of human life. If you have a virus to kill, nothing like a good biological theory to get you going in aggressively grappling with that project. If you have a bridge to build or a patient to cure, there are some excellent doctrines that you will want to familiarize yourself with. And ethical theories can come in handy if you find yourself in the middle of a discussion with other people about what is and is not good given certain circumstances. It may be possible that human beings who become philosophers of this or that discipline are also able to live the good life, but that may be due more to the fact that being highly educated and highly skilled, they are able to gain employment that is well compensated and so they get access to better living conditions. It need not entail that they have a more acute comprehension of the world in which they live and a better sense of how to ensure that their own private personhood is respected by others. They may think they have achieved such a state, but it might be that they are either paying for that respect with that high income they have acquired, or they have fooled themselves into thinking that because they are expert in one thing, they must be an expert in many other things as well.

What I would like to do here then is adopt the phrase 'love of wisdom' to suggest this emittance condition that is the private person's public sphere, the elusive condition that seems to provide the protection we require from coercive forces, and the mood we must be in so as to have experiences in a way that avoids reducing our relations to their social interests and the warring characteristics they might have when we confront each other in the course of our lifeworld experiences: going to the grocery store or doctor, running into each other at the park or schoolyard, and so on. Love of wisdom is not a theory that we are working toward, it amounts to a flavor that inhabits all actions and that is emitted as part and parcel of everything else. An accompaniment. And it does not constitute a body of knowledge or a theoretical position relative to the objects of the world. Rather, it characterizes the demeanor of all other action. And what is more, I claim that it needs to be engineered into a social setting, it needs to be a subject of indoctrination that is passed down and which requires many generations to properly attain. In that case, it is not of the nature of the Platonic ideas laid out in his theory of justice, because it is not meant to be an area of expertise or body of knowledge, rather it is a set of etiquettes, of emitted accompaniments that give a changed characterization to all other actions and operations. The love of wisdom must be ingrained in the civilization and cultivated in its membership through a constant and iterated approach that passes it on and makes it part of all the things we do and learn, everything we interpret and understand. And the dedication to it must be long-term, because the limitations of any such agenda are such that it can only be achieved across multiple generations of socialization through publicly applied ways and means.

The difficulty lies in the fact that within the context of the telemetric understanding, this love of wisdom is an unrequited love for it signals an emission without any expectation of a return.

Home and Hearth

Heidegger, famously, was a Nazi. Not in some metaphorical sense, he was an actual member of the party and held an academic post requiring that membership. Furthermore, he was kind enough to leave ample paper trail that his private thoughts in the late 30s, were as anti-Semitic as one might expect from any member of the Nazi party. And even if you never bothered to read the marginally interesting Black Notebooks you might nevertheless have been confronted by this issue when considering the strange relations that unfolded between Husserl and himself in the years immediately following the publication of *Being and Time*. Growing up Jewish, I never had the luxury of dismissing historical thinkers because of their anti-Semitism. At times, studying the history of philosophy, who would have been left in certain periods if I removed from consideration those who published, or privately thought, derogatory remarks on the Jews and the "Jewish Question"? The professors in my graduate school certainly did not give it much thought. Of course, you must engage with Frege despite the rabid anti-Semitism. And this accounts for just the ones we know about. Many likely never went out of their way to express their ideas on the subject because they did not need to, the Jews were all segregated out of their lives by social circumstance, they did not need to interact, and so they could avoid bothering about them at all and focus on all the important philosophical issues of the day that did not have anything to do with the religious or cultural identities of some of their segregated compatriots. Even when there is no direct anti-Semitism, a person raised among Jewish elders and within Jewish communities will likely have a whole set of behaviors and attitudes toward conversation and discussion that might be off-putting to the people around them who did not grow up with such influences and in such communities. Assimilation may be said to have bred a kind of quasi-anti Jewishness based on characteristics and personal attributes. 'Those pushy Jews with their aggressive way of arguing', is a common enough stereotype. If that is what your family life was like, then –to the extent the stereotype fits with some families– maybe the so-called "aggression" or "pushiness" is not even experienced in it as such. It is just how you state your case, how you interact with other people when discussing issues of importance to you both. Arguments are a reasonable way to form a community, arguing over ideas and issues might be the way in which you insert yourself into the community and how you are with other people. And if those people are not of the same orientation, if they were raised differently, and if they think differently about how cordial conversation and confrontation should proceed, then their reaction to such "Jewish" ways might be to shun them, or to turn away in some sense. This may not be anti-Semitism in the same sense that we get from Heidegger in the Black Notebooks or in the Husserl Affair, but it is a form of cultural segregation that can create dividing lines between groups of people. It is a way to establish communities that are different from each other and are kept separate through their different practical orientations.

Oddly, it may be the case that a so-called anti-Semite like Heidegger might be very much at home when having conversations with members of a culturally Jewish tradition who argue and pursue discussions in an aggressive or pushy fashion. Certainly, Heidegger's life in the 1920s involved challenging intellectual relationships with Jewish people, and so it is possible that he took great interest and a liking to such forms of interaction, and enjoyed his time spent wrangling with Husserl, Arendt, and others during those early days. The point here is that there are all sorts of ways that socially structured behaviors might be dismissed and promoted. Someone who had all sorts of terrible things to say about Jewish people might be very much at their ease when arguing and discussing philosophy with those same people. From what I have read of Dostoevsky, however, it seems that in his case the one overwhelmed the other. That is, his beliefs and thoughts about the Jewish people made it impossible for him to ever completely relax around individual Jews, so even though all accounts suggest he was a pretty ornery fellow who enjoyed a good knock down drag out row now and again, he did not seem to have any Jewish interlocutors who fit that role. The point of this story is that social relations are not the same thing as the ideas that represent them or try to represent them. Rather they are distinct things. And in Dostoevsky's case, likely the two each had strong reinforcement from his various influences growing up and in his schooling. In Heidegger's case, the one had greater influence over him than the other. He wrote and said things that did not always reflect the bodily behaviors and actions that he performed in his daily life, at least during the earlier years before he joined up with the National Socialists.

Jewishness is just an example here. Any set of so-called cultural identities could be used to illustrate the point. There are numerous ways in which our households imprint themselves upon us. And these ways may be in stylistic approaches to interactions with other people as well as things we are likely to say about others or other groups of people. A person who is put off by certain kinds of conflict or confrontation may not associate the characteristic in others who are prone to such conflict and confrontation as part of some cultural identity. They may have been trained in a specific form of etiquette and so they may be uncomfortable around forms of activity and behavior that do not conform to those norms, but they may never have been taught to associate those behaviors with any specific people or cultural groups. In that case, we might notice that there is a tendency to be put off by such behaviors and to dismiss them on seemingly universal grounds. 'This is not the correct way for cultivated persons in civilized society to behave, it is off-putting,' or something to that effect. If they had been taught to associate these behaviors to some cultural characteristics, they might associate the culture with the behavior even in cases where the person exemplifying the relevant behaviors is not a member of that group, or at least does not share any other attributes with that group identity. Someone like Dostoevsky might have commonly referred to non-Jews as acting Jewish when what he meant to suggest was that they were exemplifying behaviors that he had been taught to associate with that identity. There would have to have been such a reinforcement, otherwise it would have been more reasonable to conclude that these traits were not specifically Jewish at all, since there are examples of them occurring independently of those other characteristics. Prejudices of all kinds might work this way. Evidence from ingrained associations may be so deep and so theoretically anchored that individual experiences of disjointed traits that challenge the rule of association are discarded as non-essential and may even function as proof that this group has had an impact on the rest of society such that their personality traits are now finding their way into other groups in some form of psychological imperialism or colonization that replicates the behaviors of one group into the general population including other groups.

These semes make up group interests in both ways. The seme is the rudiment of the one group's behaviors and is effective in the style of their action that guides other operations. And the seme is also the orientation toward the other group's behaviors. That is, as we learn our style of life, our cultural ways embedded in our social relationships, we also learn ways to articulate the style of others and the cultural ways of others that are embedded in their social relationships. These semantic meanings are deeply ingrained in the actions and operations of the private person, they are deep and pervasive ways in which the social world occupies the thoughts and actions of human beings. As such, the love of wisdom that is meant to emit a public film of respect must protect against elements or aspects of these stylings that drive and dictate coercive orientations toward others in social contexts. If we are not careful here, however, we might start to envision this common emission that protects the private person as an interest of its own. Something like an interest in being tolerant of other identities or other groups or other styles for interacting with the world. I hold that this form of tolerance, that has been historically associated with various traditions of the West stemming from the European Enlightenment, is itself an interest and expresses a set of attributes that are part of some group's orientation. And this set of interests seems well suited to a type of worldwide global economic colonization. Considering such interests, I tend to think of the advertising campaigns of a company like Coca-Cola in the last century. They vary their message and imagery depending on the culture that they are targeting with their ads. Their goal is tolerance and inclusion among the many different types of people in the world and the various cultural practices and styles that they realize in their everyday lives. Because truly Coca-Cola is non-judgmental about these practices and norms, they just want to make sure that there is a place in them for Coca-Cola, and they are willing to alter their message on a person by person basis, if they could, to make sure that their product fits with the orientations of each individual private person. And no doubt they are studying the data they collect through all sorts of means to determine the best way to do that.

The formal manners of any unrequited love of wisdom, whatever it is, must be something other than a set of interests unto themselves. For if some set of interests are emitted in this formal emission, then the emissions constitute an interest that is part of a set of relations that may be coercive, and that certainly may find themselves running up against opposition and in need of higher resolution as dictated by the context where the conflict is taking place. Without being able to provide an actual solution to the problems they would have, I can imagine the existence of two communities, one of which has an interest in tolerance of other ways of being in the world and the other which actively rejects any such interest in tolerance. The two communities might wage war upon each other, might commit horrible brutalities against each other, and one might manage to destroy the other group. Still, it may be that in all this hostile action there may have been a formal recognition between them. The two interest groups recognized the others as different from themselves, saw the differences clearly, understood what the interests were, and how they made each other different. It may be the case that the members of each group became deeply informed about the nuances and internal dynamics in the smallest detail of the other group's ways of life and being. In these formal conditions, in this getting to know each other and understand each other, they showed this recognition. And used it to destroy each other. Even used it to destroy each other more effectively, more deeply, and more permanently than they ever would have been able to do if they had failed to recognize the other group in this fashion. The recognition of the other that is emitted from the private person as a protection against coercive social relations is not necessarily a harmonious guarantor of idyllic relations between interest groups.

The multi-generational approach that we described earlier will not result in a group of loving human beings who are innocently devoted to all ways of life regardless of their orientations, tolerant to all kinds of differences, and blindly accepting of whatever the peoples of the world have to offer. No, the lovers of wisdom will only attune to the ways of the real world. They would be intimately aware of the social forces at work in their associations and they would be respectful of the boundaries that distinguish different interests. They would be aware that social coercion has led them to choose sides and participate in a world where others have also taken sides. And they would be aware of the direction in which those sides are aiming and the social relations that materialize in them. And finally, they would know that the relations are equally conditioning combatants on both side of the dividing line and all members of the various interested parties. To emphasize the point from the last section, there is no set of ethical principles to be discovered here and this is not a matter of having a doctrine that will help realize some end. The ends themselves have become suspicious. Even the notion of harmonious living between all people of all walks of life and all different group identities must be recognized as an end to be associated with a set of historically developed interests and which may likely belong to some set of groups and not to others.

This is an important point to make because if the love of wisdom, as we are calling it –or the emission of a public protective film to safeguard the private person from the socially coercive forces– were itself a set of interests or part of some interest group's orientation, then the process of engineering the population into this 'mood' by legitimating it would amount to little more than the goings on of a thousand-year Reich. There would be little difference between this approach and the attempt to deeply engrain a set of interests into a population so that they can use those interests to proclaim their own moral superiority. As a body of material interests, such a mood would be ample justification to destroy any who are incapable of joining in with the same interest group by acquiring its attributes. The group could claim its interests as formal requirements for any kind of civilized engagement at all, as a precondition for communication, or a transcendental condition of rational action thus establishing the boundary of an ethical community. Any persons who did not have this pre-condition would not be members of the correct community, and thus would not be subject to any of the prohibitions or rights that such a community has intrinsically declared as part of their make-up. This would amount to justification to eliminate such people as outside the bounds of ethical concern. And history is filled with examples of this sort of thinking happening again and again. Self-proclaimed tolerant communities establish themselves as something of a minimum baseline for civilized character traits, and any people they encounter who cannot be engaged in those terms, or who resist those terms actively, are savage or barbarian and thus can be disposed of with a clear conscience.

The love of wisdom, as a hypothetical emission of the human private person, is meant to circumvent this oft-repeated historical problem. In doing so, it fails to provide a doctrine with anything like a positive characterization that helps the private person design a plan for living a good life. It avoids corruption at the cost of becoming useless. What would we get that we did not have before if we were to admit the functional meaningfulness of this supposed love of wisdom? The supposed good throughout this essay has been protection from the socially determined telemetric understanding, where that is the realization of a logic or system that exceeds the involvements of its members and which is something bigger and grander than any point of view, singular or collective, that is contributing to it. The purpose of the recognition of private personhood was not meant to protect one set of interests from another, nor was it meant to provide a set of ethical guidelines for ordering the actions of interested parties in their relationships with one another. Rather, the purpose was to protect interested parties as such from being subjected to a dynamic or systematic structure that is beyond the perspective or insight of any interest group or interested party within the social setting. By introducing a formal recognition that protects the private person and that safeguards its privacy as the subject of action and interrelationships, any coercive organizations that exceed the viewpoint of the interested parties themselves can be eliminated. It would no longer be the case that this or that group is engaged in a battle with another group and that the two of them, in their struggle, are just pawns to some larger dynamic working at a meta-historical level that is not visible to either interest group, to either position, or to any of the positions that might be involved in such complex social and historical relationships.

If one is concerned that there is more going on than meets the eye, that there is some historical materialist advance of progress that uses classes and groups, interested parties, and organizations as pawns in the engine of history, then in some sense no group is driving the train of history and no interest group is in control of anything at all. In that scenario there is no good and evil, rather there are just those who are lucky enough to be included in the groups that align with mastery and those that are unlucky enough to align with the enslaved. And not much more can be done than that, except to move along with the historical tides and hope that when they turn, your group will be on the side of the masters and not perpetually on the side of the downtrodden. In that case, human history is always going to be constructed however it will be constructed and no group can control the outcomes. The property owners may have done well throughout the historical tide, but it may be that what qualifies one to have all the property may change from century to century or epoch to epoch. The source of wealth itself may be an accidental characteristic and the landed gentry from one era are displaced by the stockholders of the next era. Some were lucky in the transition, others not so much. The point might be that there is nothing to be done to impact the outcome. Or rather, there is no science for determining that one's actions are going to have the desired effect and going to result in the impact that one is seeking. The larger machines at work in driving the history of human beings will have their way, and the individuals within that system will succumb whether they wish to or not. If it is a matter of interest groups, if there is some formal protective capability that insulates human beings from such forces, if there is a layer that allows us to distinguish ourselves from the socially conditioned involvements that are handed down to us from a dynamic that is larger or more complex than any one of us, then there is hope. And this could hold true even if that hope begins from a position where one set of interested parties, recognizing the various interested parties involved in the world dynamic, hold coercive domination over the rest of humanity because of their own willful actions throughout multiple generations. That itself would be a hopeful realization because it would mean that it was possible for an interest group to establish control over the direction of history, over the orientation of world society, and the relationships that played out within it. And, if it is possible for there to be domination by one interested party over others, then that means it is possible to have a revolution against that power, to willfully change the conditions of power to establish a different group's hegemony. This is a standard piece of the narrative of the liberal tradition in government, inherited nearly unchanged by the Marxists.

Because of how pervasive this notion has become, it is crucial to emphasize this point. Assume that the originators of the commonwealth were correct in their implied supposition that god is dead. Imagine that all the divine ideals of the tradition of the various deities were eliminated altogether as a result. Imagine that all you have left is the social organizations that arise when a species of animals lives together in populations around the Earth. Conceivably, this means that anything goes, whatever the species being cooks up and sets in motion may be 'justified' in structuring human life around the globe and across the various parts of it that intercommunicate. The ability to insert a boundary between such social dynamics and the private person would amount to the capacity to scope such a species-being into the conscious orientation of the species through some specific means. This would be an absolute and fundamental requirement for having any Ethics at all. It would not amount to an Ethics in itself, but it would mean that there is a socially conceivable meaningfulness in the supposition of human autonomy such that we would be capable of formulating something like an affective set of ethical concerns or principles that might be used to govern the relationships between interested groups. It might suggest that such is always already the case and that ethical analysis is something we can actively perform to discover the patterns of our Ethics here on earth among each other. This will be the topic of the next chapter as we look at the analytics of our love of wisdom and as we attempt to gain an understanding of the patterns at work in this emission that declares our private personhood over and against the social.

Before moving on to that discussion, let me call out some of the more difficult cases of interest groups that have been overlooked in the earlier examples. Previously we spoke of such interests as though they could somehow be separated. Like there was a group of people over there and another group of people over here and there might be geographical boundaries between them making it possible to come up with example cases where they might square off against each other in various ways and come to a resolution in their battle to the death. In concrete life, interests tend to be less clear cut than that. Many communities are divided within themselves by such interests and individuals may be divided within themselves. The types of interest segmentation that we come across might be within the household, between the males and the females or the adults and the children. It may be within the individuals themselves in so far as they may identify or be attributed with a set of qualities that place them in multiple interest groups openly hostile to each other. This can cause all sorts of complexity and stupidity in the various analyses performed. Much historical brutality has been for the sake of such divisions. And this is largely because the battle to the death is not realistic or advantageous for anyone under these circumstances. A civilization in which the males are hostile to the females, for example, will not last if that battle is to result in mass extermination on either end. There may be murder and brutality, or the threat of violence against children, or any of a few different tactics aiming at getting the one side to succumb to the coercive interests of the other.

Such conflicts and alliances within these sorts of dynamics still operate with social involvements and interest groupings providing the characterizations of the groups and the individuals that form those groups. And to that extent, it is the case that a recognition of private personhood across the boundary of such social power will force a realization of the differences that are being articulated within the conflict. It does not mean that the relationship between the interested parties would be any less misogynistic, for example, but it does mean that the participants would be operating within that system with a mood of misogyny where that would indicate an active understanding and interpretation of the dynamic which recognizes and realizes it as misogynistic. In that case, the parties involved would not merely be acting out their interests or acting out an attempt at defending their interests against some perceived other who is opposed to them. Instead, they would be acting on those interests while understanding what they are and how they are being furthered and promoted by the dynamics playing out. Again, the point is not to claim that this automatically provides any human beings with a universal set of ethical principles that they can then use to define and lead their lives, but it would make them ready to have a proper conflict that takes into account the concrete value systems at work in their relationships with an understanding that the ongoing nature of those relationships is something that needs to be worked out, that is available for conscious determination, and that all parties are present and accounted for in the carrying out of those actions. The goal is not to produce an Ethics under such circumstances, but to show that there is ample ground for having one, that there exists a real conflict of values. And then, I suspect, it may very well be that the ethos of this understanding may emerge as a residual effect of the grounding maneuver and that an ethical understanding of the world may permit an Ethics as such to emerge and hold sway over committed action for decades or centuries or however long the founding order of such practice prevails. The death of god launches the social emergence of the emissions of value systems demonstrating the ethics of facts where the principles are inferred from the actions in true telemetric fashion.

Chapter II: State of Nature/State of War

I have my doubts that there ever was a state of nature. Or better yet, if there is a state of nature, this is it. The moment in time in which an agreement was reached and a contract sealed is a philosopher's game, like imagining alternative color names relative to some date or different versions of Earth where water has an alternative chemical composition. And this may very well be instructive, this notion that there is ongoing consent involved in human political organizations and that it organizes or steers the nature of the political body. This heuristic or game feeds the notion that the political organization is something other than the social relations that are the soup within which we swim. There are these spheres of existence and the public sphere is a concrete domain of action distinct from the social sphere or the private sphere, distinct from what is taking place in the home or in civil life. As if there were a world where interests were crucial and then, beyond that or on top of it, or next to it, an alternative world, a twin earth, where those interests come upon their limit, where any specific interests were supposed to be excluded and only rational legislative thinking was properly effective. It is possible that this fairy tale is true to the extent that it is believed to be true. Because it is a way of organizing human action and the various appropriate behaviors and decision making of agents in the intersubjective realm, it is possible that people live their lives in accordance with a set of rules that require them to be rational agents in some practical ethical or political sense, whereas they were more instrumentally oriented and efficacious in other domains where the polis was not at stake. We tell each other tales, we learn stories while growing up, we have a favored set of narratives that play in a loop at the back of our heads once we acquire them, and they take over our fantasy world or subconscious mental life. And to that end, it is possible that such stories were self-fulfilling in a way, that they provided us with a set of directives that may have once caused some people to behave differently in different contexts. Jefferson left his slaves at home and forgot about all of that when he went off to legislate, to make laws promoting a public domain in which all men were created equal.

But there are alternative stories as well. There are tales of remote rule from across the Atlantic Ocean in which the local subjects had no say in the laws that governed them, the colonized had no representation in government, and so their interests were not heard by the legislators. The most notorious of these stories seems to have been about the taxation of goods and services. The colonial subjects were being taxed without any say so. That was the origin of the American public sphere at least: the set of discussions and plotting to overthrow the remote government in favor of something closer to home and more responsive to the context of life here rather than there. This was a story about interests, about the interests of this group and the presence of those interests in the accounting that was made by the legislative body. The public domain was explicitly concerned with economic affairs, social associations, and behaviors. The remote government was regulating the daily lives of individual persons in the colonies for the sake of promoting the interests of the empire. The purpose of the colonies, from the King's perspective, was to feed the coffers of the state, to provide much needed resources, and to be a farm for harvesting the raw materials needed to sustain the power of its empire. The revolutionary spirit was in part an effort to overturn that remote logic in favor of something closer to home, where the interests of the political organization would be organic to the needs of those living in the states themselves. Self-determination aimed at correcting the imbalance of interests represented in the laws of the land and at promoting local interests in ways that were more suitable to life as it was lived in North America at the time.

So why think of that as a distinct sphere or domain of life and not rather as an offshoot of the social set of relations and interests? Organization is a way of addressing constraints and risks that are present to the individual agent. This can be done through clans and tribes, extended families and guilds. These are all ways of banding together to form groups with common attributes that extend the scope of interest through a reinforcement of those attributes. The family sticks together to promote the well-being of the members of the family. There is a tight connection that binds the members and they work out, historically and through reflective social contribution, the details of their organization for the sake of furthering economic and social ends. Families join with other families. Collections of families come to reside in locations where they are collectively conditioned by the same weather patterns, and where all are impacted by the same crop failures and boom years. The gathering of the people into geographic associations binds them together through their relationship to the environment and the world around them. Such bonds tie them together through interests and orchestrate their actions in response to shared conditions. They divide up labor and organize different aspects of the requirements of life to meet the challenges and take advantage of the opportunities that present themselves. If the decision-making procedure is clear and the target goals and actions to get there are agreed upon, there is no problem. The smaller the group, the simpler the structures that are required to maintain these areas of concern. As the groups grow, disagreements are more common, and conflict is more likely to emerge from the conditions of life. Structures are required to settle differences, mechanisms need to be introduced to handle the disparities and disagreements about methods, as well as the material of the processes that are used to organize the lives of all.

The institutions that come into existence to address these problems and conflicts, while cutting off potential problems and potential conflicts in pre-emptive fashion, emerge from the interests that the communities and groups share with each other in meeting the needs and serving the aims of the group itself. The institutions will likely exacerbate some differences at the same time as they reduce the stress involved in others. For every example of equal protection, there might be a similar but opposed regulatory principle that is biased toward one set of people over and against another; and, in the process, will form these groups into distinct interest groups. Since the institutions are themselves offshoots of the lifeworld processes that need them for conflict resolution and problem solving in the first place, the establishment of such interest groups will result in action dynamics that deepen the problem and further the divisions. Benefitting parties will lobby for increased protection of their interests, harmed parties will lobby to overturn the harm and try to introduce some set of advantages for themselves enabling them to get out from under the established disadvantages. The political institutions may initially appear as purely procedural mechanisms for deciding differences and resolving conflict, but they evolve into bodies where the interests under consideration became enmeshed in the functioning of the institutions themselves. The groups participating in those institutions realized that the rules for coming to conclusions are crucial for organizing the outcomes of the disputes, expertise develops in these areas so that members of the society become highly skilled in designing rules and principles of procedure that are best suited to maintain control over a decision-making body. The matter under discussion in the "public domain" ceases to be merely the social problems and concerns that first gave rise to the sphere of discussion, and becomes a domain concerned with the administration of the public domain itself. And this administrative emergence is likely where the mythical notions of the public domain itself come from. Once the goings on of such institutions are themselves the object of knowledge and manipulation, it becomes possible to distinguish more strongly the actions and types of action that are appropriate to those domains. Thus, we gain the so-called public sphere and the seemingly abstracted notions that are specific to it.

The procedures may begin to seem radically distinct from the forms of action that take place outside the halls of the legislature or beyond the confines of the judicial domain. Ordinary folks who enter those domains are immediately struck by its otherworldliness, by the extent to which it seems distant from the everyday life of the social relations familiar to them. This appearance of distance is crucial for this domain to provide the social service that it is meant to provide in the contemporary state organization. The organs of the state cannot seem ordinary, cannot be structured in accordance with the everyday lives of the citizens, because that would reveal them for what they are, interest structures put in place to support the ends and aims of specific identity structures that have developed and evolved over the years, decades, and centuries. The sheen of impartiality and proper rational procedure must be maintained in these institutions or else the society at large will lose respect for the legitimacy of the institutions because that legitimacy is based on this very set of principles connoting disinterestedness and rational objectivity. These are the halls of justice and the body of power. These institutions are not supposed to be biased in favor of one group over and against another group, they are supposed to blindly or dispassionately defend the procedures that are crucial to the social contract that initializes the public domain. Impartiality opposes partiality and therefore is representative of the whole as opposed to just the part. This becomes a phenomenological pattern written into the social structure.

In a civilization in which the institutions of public trust can be easily shown to embody the interests and aims of a small portion of the population, all legitimacy would be removed from those institutions and the public at large would no longer see reason to defer to the institutions in problem solving or conflict resolution. The courts would no longer be a legitimate way to resolve a dispute and one's legislator would no longer be a reasonable place to turn to get a balanced point of view. Why go that way when it is obvious what direction such an authority will turn, what kind of deliberations they will perform, and what kinds of decisions they will make? If the institutions are clearly perceived to extend the interested parties' orientations and projects, then the institutions become useless as organs of legitimate power. The exotic nature of the institutions, the highly specialized language, and the deep complexity of the rules of proceeding are all meant to make the institutions more otherworldly so that the details and effects of their operations are less apparent and more difficult to associate with actual interests that actual groups of people have. The use of legal precedent to establish laws is an example of this. The courts make decisions about seemingly pedestrian cases where the insights of the average person might be such that there is no sense of any grand narrative or event taking place. The facts of the cases considered may seem familiar to the average reader or listener. Some ordinary event: a person placed a bet while using a public phone. That is simple and straight forward. Contemporaries of those events would immediately recognize the simplicity of the case. In its deliberations, the Court moves far away into an abstract universe of rights and responsibilities. Expectation of privacy comes into play and refers to previous esoteric decisions. The average person, not being an expert in the procedures or language of the Court, may focus on the facts and form an opinion based on them. He's guilty and the court is deciding whether it is okay to put him in jail. The precedent that is being established is the universal rule that can be applied to different cases with wildly different facts and very different contexts. Justice looks out into the distance while the participants and ordinary onlookers look closer. In the end, can they find the queen once the shuffle is done?

What was decided in the case and what the facts of the case are will be wildly different. This holds true in the legislative arena as well. The common discussion of legislation often takes place at a mundane level. Should there be health care insurance for all people or not? People can form opinions about that coarse-grained notion and whether it seems to promote their own interests or somehow hurts their interests. There are plenty of public sound bites to promote thinking about it one way or the other. The legislation itself is significantly more complex, it involves enormous amounts of detail that never makes it into the debates or public considerations because the people do not have the time to focus on all the details, and the people presenting the details do not have any strong inclination to bore them or inform them with such minutiae. Those details, however, are often the critical piece required for properly evaluating the interests served by the legislation. The complexity of the laws, the ordering mechanisms that are used to construct them, the procedures that are used to deliberate on them and then, ultimately, vote on them. All these processes work toward making the legislative act less accessible to the social groups whose interests are either being served by those laws or being directly harmed by them. The complexity, in cases both judicial and legislative, makes it easier to hide any relationship the decisions and deliberations might have to the actual interests of the people that are impacted by those decisions and deliberations. Alliances are established based on trust and coarse-grained affiliation, there is no deliberate and well-informed public.

There are levels of involvement that originate in this organizational characteristic. The professional politicians and judges, their aides and clerks, and so on, they have a level of expertise and familiarity with the process that may be placed at one level. And then there are satellite businesses that revolve around those organizations, consultants and lobbyists, pundits and journalists. These are people outside the direct process but very close to it. They will have a high level of expertise, although they may not be completely informed about all matters pertinent to the decisions, or at least they are not informed in a timely fashion as those decisions are being made. Then there might be people a little further still from the process, complete outsiders who are not directly involved, but do everything they can to remain as informed as possible about the ongoing processes and the decisions being made. They may be activists or just conscientious members of the voting public. They will miss details, they will not have access to all the relevant aspects of the decision making procedure, but they still have gained sufficient experience and breadth of familiarity with operative concerns to be able to follow the complex logic and language of the processes and make informed decisions that guide actions relevant to the laws being made and interpreted by the relevant institutions. Then, of course, there are other outsiders occupying a broad spectrum of familiarity with such issues, from *partial* understanding to complete ignorance.

Most people in modern civilizations will fall into this last category of people who are otherwise engaged with their daily concerns, their families, friends, jobs, and entertainment activities; but who have a variably casual interest in such matters of public importance. Most such people rarely go directly to the horse's mouth for their information and probably turn to the second grouping of people to get it. We have seen that the information industry is already thoroughly saturated with interest structures, so the ordinary consumer is probably progressively making information consumption decisions based on their own past associations. That is, they tend to patronize information mechanisms that have served them or their family or their group well in the past. And this "served them well" may amount to little more than making them feel good or powerful and properly informed about what is going on, what its ramifications are, and what they should think or feel about those events. Whatever this is, it is not a public sphere in the sense that Arendt or Habermas had in mind. This is information as interest peddling. It is a process of presenting the mechanics of the political sphere, the sphere of conflict resolution and problem solving, as an entertainment product. And at least one purpose that this organization produces is that this process of boiling down the decisions, of framing them in this or that way depending on the interest orientation of the targeted audience of the information, is to portray a dynamic at work in the process itself. The process may be presented as legitimate or illegitimate depending on the message that the presenters are making. To the extent that one group of people may be getting the message that their interests are never being properly served by these institutions, they may be getting the message that those bodies are themselves interest-bound and that the interests to which they are bound are specifically characterized as belonging to some other group or groups and not one's own. In other parts of the society, other groups may be getting an entirely different message because the organs of communication are suggesting that everything is fine with the institutions since their interests are being served properly.

The bodies of power are opaque in all such cases. There is a great distance between the groups being served and the operations and actions that are deemed to serve them. There are many levels of interpretation and evaluation, of spin and focus, that take place between the institutions and the people impacted by them. The intermediate groups are all organizations and institutions themselves or have become so over time. And as such, they have sets of attributes and properties that can be promoted or discouraged, do better and worse, depending on changes in the environment and context where they do their work. Decisions made by the governing bodies will affect them and might make it easier for them to do their job. Groups spinning yarns about the work of other groups spinning yarns about other groups and so on. Interested parties all the way down. And every one of these groups has an interest in disinterestedness or objectivity. Pundits who are providing insight into the political process have a direct interest in their own objectivity. If a pundit were to advertise themselves as someone who is doing the best they can to spin all events toward an interpretation of the world that is deemed to be highly desirable to some specific group of people, they would be finished. The people who are the specific target audience of this pundit would be put off by the tearing aside of the veil of ignorance in this act of contrition. The position of the pundit, their role in the process, is dependent on the illusion of objectivity. The opponents, people involved in a different side with a different interpretation of events, they can make all the accusations they like about the interests and bias of the pundit, but the pundits themselves cannot admit to it and still retain their position within the world. The only way in which such a pundit can make this maneuver successfully is if they are in the process of switching into the service of a different interest group. Only then can they admit that they have been advocating biases all this time, but now they have come to see that and are going to switch to the side of the truth, to the other side where they can now take on a role serving customers who are entertained by that alternate set of narratives.

Surely, there must be a truth to the matter though, right? My point does not concern this, what I am interested in is the fact that the institutions themselves seem to be set up to suggest two things: 1) there is a truth to the matter and this is what it is, 2) those who know it are under attack by opposing interests who are going to tell all sorts of lies to try and get people to become ignorant of the things they know. There is an intimacy between the story tellers and their listeners: we are in the know, those others they are trying to fool us, trying to get us to think we do not know what we know, they are trying to use tricks and lies to break through our naturally sound intuition and the clear view we have of the world. And is it the case with such dynamics as it is the case with observation statements and theories? Many philosophers consider it non-controversial to assert that the facts can be provided with multiple interpretations and can fit into multiple theories of the world. The facts do not necessarily support one single theory, rather the theories are things we use to understand facts and how they associate with each other, what they mean, what they project, and the various predictions of future outcomes that can be made based on them. The pundits are providing theories in a similar way, not with all the peer review and testing for conditions, there is no singular community of pundits to act as something of an authority over the various theories presented, but the analogy holds nonetheless; and it is often noted that people experience the world and see the facts of their lives in terms of the theories they have come to accept rather than deriving their theories from the experiences they have had. A person who has come to agree with some specific economic theory will tend to explain the specific events taking place in their lives in terms of those theories. The free market is not responsible for this, rather the restrictions that have been placed on the market are to blame. Or, on the contrary, the restrictions on the market were too weak to prevent the catastrophe brought on by the free markets. I'm not suggesting that these theories could not be approached in a more classically scientific fashion, only that they are not. There are no good metrics for deciding between them, no good feedback loops. They both can persist despite the facts, they both can persist to provide insight into the facts, and both may provide interesting pictures of the world that tell stories the various interested parties still love to hear and which they keep demanding to hear again and again.

Of course, both ways of presenting the theory about markets and their regulations are far too sketchy and simplified to be terribly useful as bona fide theories. Even the originators of these theories had significantly more to say about them than the oversimplified ways in which they are boiled down for mass consumption in today's market place of ideas. To address that, you could go into greater detail and learn the nuances and subtleties of the theories themselves, offset them in these various elements, and come to a more informed conclusion. Even in that case, you would be far from the things themselves, you would find yourself in the domain of the philosophical. And although this may be very useful, that approach is not in keeping with the project we have been developing. Rather, we are looking for this emergent protective layering for the private person, and I have made the case that this is more of a phenomenologically emergent mechanism. What this means is that we aim at practicing a love of wisdom over and against any philosophical or theoretical approach. And the love of wisdom dictates that we start with what is before us, and that we try to advance in the direction of what is necessarily at work in keeping what is before us in our focus and within our grasp. We suggested previously that one way to characterize this project was that we were trying to determine whether human beings are capable of ethical action. The narrative of this chapter has tried so far to show that 'Ethics' is not going to be a matter of getting the right theory of human action and the right set of guiding principles to organize human involvements. Rather, we are looking at the events and goings on in the world as objects of experience, as events that are taking place, as circumstances that we are discovering, and we are working to see if we can produce a protective layer that maintains the experience of that world over and against the social forces that condition and impact the person in their actions and experiences.

What the discussion thus far suggests is that 'Ethics', as it is ordinarily understood, is more aligned with the esoteric logic of the public sphere than it is with any real and affective aspect of daily experiences of the world. I'm not denying that there are people with norms of behavior nor am I suggesting that those norms of behavior are not value orienting. Rather, I am suggesting that the categorical imperative or the principle of utility are illusions. If you want to understand the Ethics at work in the world, look around, do not attempt to derive the practical nature of reason in its necessary applications. What are people doing? How do those actions work and how do they form groups? How do collaborative associations form based on human actions, how is opposition dealt with, and how do groups include antipathy in their bonds and orientations? To what extent are we defined by what we love and what we care about and to what extent by what we hate and what we want to make sure does not have any place in our lives? We saw before that we might think of them as pre-ethical etiquettes. The commonwealth is a receptacle of such patterns and structures, attributes and dimensions. What patterns are there to be discovered, and to what extent are these patterns available to an experiencing being trying to make its way in the world? Can a private person gaze upon the world of value for what it is, and can a finite being comprehend the complexity of what is taking place in a large scale socially built and maintained world? In the end, there may be no ends, no conclusions, or results to mention. It is entirely possible that any upshot of these investigations would itself be more material for some interest group or another advancing their cause or thwarting the cause of their opponents. The humble aim is not some high-flying all-inclusive theory that describes the way in which we can all come to an organized agreement about this or that form of life or organization of existence. Rather, any such intent is destined to be saturated with interest and oriented toward some sets of interests that benefit some specific group over and against others.

My intent in this analysis, in this effort at pattern discovery in the commonwealth, which follows upon the discovery in the phenomenological and telemetric domains, is merely to find the protective measures that will safeguard the private person in the middle of all the rest of it. What must emerge from the agent's orientation to protect the conditioned state of that agent's existence oriented in this or that fashion? How can I safeguard my privacy in a world where social intersubjective conditioning and impact is everywhere through and through and deeply engrained in a telemetric understanding that I may not be able to comprehend and that I may only vaguely or opaquely fathom through its effects and influences? How can one live while remaining partial if partiality is all that can be achieved? How can I work backwards to continue to move forward? This does not grant anything like an ethical theory or a set of principles to govern behavior, but it is the foundation or basis from which any such theory or principles would have to emerge. And it may be that multiple theories or variable principles could emerge from such a foundation, meaning that there may not be some exclusive or singular perspective. The nature of the project may be specifically aimed at enabling just that sort of polyvalence. The question here is not how do I live or what is the best set of rules to govern my life, it is not even concerned with the meaning of life and the origin of value, but more simply: can I determine how to live and can I have any role in determining how to live my life and what meaning it may have?

Idealized Agency

An idealized agent is on the precipice of the social contract, in the initial position or behind some veil of ignorance, about to become an inhabitant (sic) of a kingdom of ends. There are no such agents nor have there ever been. And, I hold, they would die –if they could– from the collateral damage that comes with the explosive moment of the death of god. They exist in some beautifully reasonable stratosphere where no one can touch them unless they say it is okay, where no one can make them suffer or bring them harm without violating the eternal order and the ethical law. Each is like unto the others, each has all attributes in common, the properties are the same, they need not know the details of who they are and what they want, they are outside the domain of care and concern, of specific involvements and meaningful projects. They are deliberators and law makers, they can see the matters of the world from this royal distance and, if they can maintain it, can make pronouncements upon the good and the just that all would likewise have decided given the same conditions and connections. To ensure the sameness of those conditions and connections, they must be position-less, they must not be anywhere at all, since any location or place would embed them in a perspective that would skew their wisdom with a point of view that might introduce flux or variation into their deliberations. And the human being, imperfect and under constant threat of failing to live up to this ideal, partakes in this agency, but is not identifiable with it. We have detailed characteristics that differentiate ourselves from each other while at the same time associating ourselves. There would have been room for a vision of idealized agency that meets with these conditions in a platonic world of forms each human being occupies and which manifests each as the same as itself but different from the others. Groupable with some along some axes where those same dimensions separate other groupings of forms in the same moment. Rather, the idealized agent has been endowed with a purity, an essential equality, by its creator no doubt. It is the signature of the eternal that each rational agent, in so far as it is at its liberty to deliberate and do otherwise, bears within it as its human soul, as the mind of god made real in the daily operations of the human person capable of purified rational agency.

One more thing to mourn then in the death of god is the loss of this agency. And this is not some postmodern dream, it is a scientific hypothesis. The agency of human beings is no doubt associated with a physiology, a social order, and a history of developments in the context of a set of genetic possibilities. We were long ago selected for this and now we realize that long past selection in the context of our everyday lives. There is no touch of the eternal here, no purity and abstraction from daily concerns. The tools and capabilities are meant to navigate the pitfalls of interest and involvement as well as need and desire. Stand outside your dwelling some morning. Smell the air, feel its temperature. This is all you, your own in a sense, you were made for this world, constituted in such fashion as to be able to live within it as you are. How your senses work, how your inclinations come about, how you take pause to consider, decide, and then act. This is the way you had to be formed to get along in this world among the social organizations in this climate and geography for the sake of persisting, of making more of yourself, and of continuing to occupy time and space as a living being. The being of bees and ants and bears and humans, the manner in which the individual of each species is constituted relative to their respective social organizations and which is formed relative to their environment, is the standing up of all that you see around you in front of your dwelling on this hypothetical morning when you heed my advice and take it all in.

Clearly then the idealized agent must be some form of abstraction meant to remove things that are already there and very much a part of each of us. If our lives are so essentially embedded in the world nearby, why does it mark something of an ideal to extract ourselves from it? The world of interests and involvements has violence and ugliness, despair and poverty, hardship and loss. In the idealized world, we are extracted from that, we can consider a course of action without the emotional response all these hardships –and let's face it, potential joys– offer us. For violence is both received and performed, poverty may be to another's advantage, any loss may be a gain to another. We look around at the world as it is and see war, the war of all against all, of each one against each other and of groups against other groups. There are raging conflicts in which lives are at stake. Every life impacting event is, in a sense, terminal; and, I suspect, any human being, regardless of the stories they like to tell, has some deep bodily sense of this finitude and its weight, the heaviness that it places upon their shoulders. Given the significance of each moment and the weight of each event, the actions of human beings would seem to be of extraordinary consequence. And who can bear this pressure without a formula, without some set of guiding steps to lighten the way? The association of the rational faculty with the light, and the shedding of light onto the darkness, is the highest hope of human beings trying to find their way in the world and navigating their pathways among and between each other. And the world offers too many obstacles and risks to permit anyone to jeopardize the matter by allowing for an embedded set of deliberations that are too deeply enmeshed in the trials that come with it, or that we bring to it in virtue of the fact that we live among each other. We wish to be able to step back and take it all in, to see it clearly, to comprehend it; and, because we wish it, we make it so. And all because we have that capacity, the ability to imagine something other than what is before us, something distant and possible.

Even when we venture into consequentialism, we do so cautiously regarding these hard won and brutal observations. There are no widely held hedonist forms of consequentialism, the social order is too effective at dismissing that perspective. The mechanism itself is not scalable to the world order and the power of that order, and its interest in preventing localized outbursts of such things, seems to prevent any real foothold for such a world view. The consequentialists that have done well for themselves tend to introduce rationalizing characteristics and traits so that the point of view is generally maintained, but what gets included in it is shifted. The evaluation of justice, right, and wrong takes place in measurements that offset the interests of the various parties without taking a side. That would be irrational, and no one would stand for it. We cannot have a consequentialist point of view without a universalized context and without some generalized principle. We need something to take us out of the everyday point of view that binds us to our interests and involvements, and places us in something like a mutually verifiable position that measures the relative outcomes against some set of standards themselves constructed behind a veil of ignorance from some original position where no one can directly associate themselves with the principles under consideration or the decisions being made. Most young adults, when they first come to these notions, are struck by how ridiculous they are, how obviously they fail to speak to them in their lives where they are. They know that their own embeddedness is why they need to act and why they must act. Forcing a rupture that tears them out of that condition to evaluate the good and the just seems completely useless. The clever among them will play along because they know this is how things are done and that this is how people have been making things work between themselves. Maybe there is something to it, they figure, and they overcome the initial reaction and begin to take on the philosophical perspective that is a mark of maturity and adulthood. Or so the story goes.

When human beings evaluate ends that are extremely personal to them, they tend to build efficacious chains of prospective outcomes. If I do such and such, likely the following pathways of action will follow upon that action. A process that might stretch several steps down the road. And this approach is commonly used for figuring out what you are going to do today or tonight or with the rest of your life. We evaluate an end in terms of other ends and determine whether the chain seems pleasing or repellent. The mechanics of associating the ends themselves have often been associated with one set of connective capabilities, a kind of theoretical reason constructed out of a rehearsed set of patterns. The relationship that the person has to the ends themselves, whether they are pleasing or not, whether they are appropriate or not, and so on, are associated with another set of connective capabilities as some form of practical reason. The notion of faculties would seem to be part and parcel of this idealization process placed upon the agents. We live in the world and are at work processing inputs and reacting to things, responding to the immediate surroundings and our awareness of various factors that have long- and short-term effects. The impulses and responses are identifiable through repeated observations of instances throughout the lives of individuals. Attributes of behavior and interaction can be formed, the faculty notion of the agent is a set of aggregate behaviors that have been further and further built up into hierarchical dimensions such that their connection to specific acts and thoughts is no longer necessary. The faculties of the mind are a type system, constructed by the telemetry that we get in our personal observations of ourselves and of others that we live among or hear about through others that are nearby. Throughout most of history, this telemetry has been extremely sparse, mostly collected via anecdotes and built up into loose categories varying widely from one land to another. Locke and Descartes may differ in their details, and those differences may have numerous worldly explanations based on their respective life histories and the influences that each of them had, and which differed in various notable and documentable ways from each other. Nonetheless, the notion of breaking up the human agent into a set of distinct faculties is common to both. The argument might be made that this is because there really are faculties and so it would make sense for lots of disparate observers to take note of them. Or it could be that human beings with available telemetry will all attempt to categorize and characterize it in specific fashion, and that the gathering capabilities of the human beings are themselves tending toward these sorts of amalgamations. The idealized agent and the efforts to describe it are manifold, and some are more powerful than others, where 'powerful' indicates that they strike the imagination with a more elaborate picture and seem to explain the experiences of the various people who hear the theory better than other such pictures might. In that case, you may have someone like Locke who already senses that there is some aggregate story to tell and then he comes across the writings of others, the writings of Descartes himself, for example. And this stimulates his imagination. He takes bits and pieces from Descartes' story, spins it in the direction that he has observed thus far, and, voila, he comes up with a slightly different story. The components and the architecture are basically the same, just the implementation differs a little bit. Each may be a distinct derived type from a common abstract base type.

Even the idealized picture of the agent is built by human beings in the world collaborating with each other, reading each other's efforts at describing the agent, offsetting those images with their own observations, and their own direct conversations with others drawing upon their own observations. The social dynamic enabling the construction of a notion over time is comparable to any other set of social relations constructing responses to common events, language skills in certain circumstances, or proper behavior given a well understood occasion. We philosophize pretty much the same way we do anything else and so the philosophies that evolve over time, the natural theories, the platforms of ideas and the political orientations, are all constructed based on these social interactions and traditions, these coercive relations that play out in our education and our upbringing, in the reading that we do, and in the experiments that we perform. The idealized agent is a hypothesis with a history, with a lineage of contributions, and a set of problems and conflicts that have been steering its discussion for many years in the minds and works of many different people.

So what interests do these idealizations express, what interests are furthered by the existence of idealized agents? To the extent that the exertion of immediate force can be used to acquire what one wants or to take what one needs, we are all immersed in a war in which the others will be taking from us and we will be taking from them. This immediacy of need and want, of feeding one's ends and building up one's interests, may be thought of as a short game, a near-sighted con that works to get what it needs to satisfy its requirements in the near term for the sake of one day's meal or one week's housing. If the long term is to be considered, then more structure needs to be added to the operation and a more efficient mechanism is required. The long con requires infrastructure and established, widely justified, and legitimated methodologies that can reliably provide the agents with what they need over long periods of time for long periods of time including whatever they may wish to pass off to their heirs and whatever lineage they are able to establish. The idealization of agency through the fiction of an original position of some kind, some formal condition under which we have entered into an agreement, is best suited to serve the interests of the status quo, of the way things have come to be organized and arranged.

It makes sense that property and the right to property would play such a major role in the delineation of such a long game. The agreement is bound to be based around a set of rules governing what belongs to whom and the conditions under which the owners of that property can lay claims against others or claim damages from others and so on. Private personhood was the currency of such long cons, but it was not the real privacy of the person that was at stake, rather it was the idealized privacy of the agent, the agent as lawgiver and thus as property holder, or exerciser of ownership or lordship over a domain. The hypocrisy of such a determination is most clear in the fact that this private property could encompass the lives and livelihoods of other people. The respect for an individual's property meant that the person had the right to exercise authority over everything within that domain, including the lives of persons who happened to be dependent in one way or another on the property owner. This might include actual slaves, or it might include children and wives, or servants, or any kind of household staff that, because of the public focus on the private domain of the landowner, had limited rights of their own. Rather, decision-making and deliberation were attached to the ownership quality, the attributes that were inherited by the person and based on their being an owner of land or productive power as in a shop or factory aimed at producing goods for sale.

It was always easy to set this sort of thing aside, to recognize that elements of 18th century liberal ideals might have been justifications for the world order that the proponents preferred over and against the royal power structure that confronted them, and to claim that it was easily modifiable to include all human beings regardless of their status in matters of property or ownership. No matter how many years have passed and no matter how much revision or rethinking of the reality we see, the conditions themselves do not change radically. Whatever sophistication we get from a Rawlsian sketch over and against a Lockean sketch, the material world of our lives seems to have progressed in a fashion such that it does not show a parallel development. One could argue for equal opportunity as a state of the Law that permits even the poorest among us to advance their way out of whatever condition they were born into. We might claim that they can achieve this through hard work and focus, and that any failures are just examples of people who did not work hard enough or focus properly. And many laws can be brought out and put on display to show how this or that group that had once suffered from some imbalance of power is now clearly protected by a law or a body of laws that makes it a crime to do anything against them as once was done. Such prohibitions look just to some but may look like preferential treatment to others: maybe from the point of view of people who did not have the same historical disadvantages but do not seem to be doing very well under current circumstances. And these interested parties may resent the laws prohibiting discrimination against some other group that was historically targeted, thinking this an unfair advantage. In all cases, such laws do little to truly address the conditions of the world as each of us found it when we were born into it. The conditions of ownership are set in place, the processes of production and the rules that govern those processes are left as is, and nothing says anyone needs any positive assistance to acquire some possible benefits. The notion of a formal social contract would seem to make any such positive assistance an injustice. Some of us would have had to work for what we have attained, whereas other just had it handed to them based on the premise that they were not historically constituted to be able to work for it themselves or that there are too many obstacles blocking the pathway of that work.

And the back and forth can go on like that, with some interested parties trying to point out that being in a group already provides a set of head start conditions that put members of those groups, people with those characteristics, at an advantage regarding the future attainment of ends. Everyone can point to the arguments of the others –coming from their interested positions– and show that such is the case and that this sort of an approach will not get us anywhere. Everyone argues for their own case, everyone sees any ground yielded to one of the interested groups as an undue advantage, and the notion of the idealized agent remains as a pure symbol of fairness that we can fall back on to ensure that this sort of infighting and bickering does not degenerate into civil war between the various interest groups. It is not fair to take from someone something that they own through legal means. Any rational person would have to concede the formal fairness of such a principle, the original position would force us to admit the power of such a law, of such an action principle to govern over all disputes. Even if the current configuration of possession is itself the result of historical unfairness, redistribution may be conceived as a kind of punishment. And because the current holders cannot themselves be traced to a crime, it would be unfair to punish them for something they did not do, for some past injustice that nobody has the details necessary for determining an actual monetary value amount to associate with those past actions and their inappropriate outcome.

Because nothing radical has ever really happened due to the application of rules like these, it is hard to see that it is anything other than a long game con. In the United States, the real-world actualization of the social contract myth may have been presumed to have taken place near the end of the 18th Century. And despite civil wars and massive legislative changes cutting across different populations and their relationship to the social fabric, there has never been a major reconceptualization of the basic nature of this contract and the social structure that it establishes among the population at large. Native Americans swept aside and moved forcibly from one part of the country to another, Africans brought to these shores on boats in chains, women barred from participation in education and economic opportunities to various degrees: all these recognized conditions and no massive changes to the contract that enabled them. The legal system that the contract puts in place cannot point to a guilty party or rather the guilty parties are all long dead and the villains who should have to pay for these crimes are no longer around, even their heirs are difficult to identify because the ancestors themselves are difficult to identify. The notion of benefitting from an injustice is not part of the contract, the contract itself is solely concerned with the establishment of formalities and the institution of crimes via perpetrators and victims. These conceptual mechanisms do not help when it comes to social crimes where relations are being articulated in accordance with the specific order and its detailed logic. When people find themselves among socially oriented involvements that constructed their world for them and they take notice of the fact that they are not equal in their situation with those around them, what kind of a criminal agency can there be in that? The very notion of the idealized agency agenda is that individual agents are the ones that act, and individual agents are the ones that are acted upon. The things they do may have deleterious effects on each other and, because of that, some may come out as criminals and some may come out as victims. The very form of criminality in the idealized state of such a social contract is that anything environmental, anything social, cannot be accommodated as a factor for consideration in the orientation of the civilization toward rights and responsibilities.

I have not done anything wrong now have I? I was simply born into a family, raised by well-educated parents who loved me and, in turn, gave me the best education they could possibly provide. The color of my skin was not anybody's choice and certainly not something I actively brought about. I do not do openly racist things, they say, or act on them, so obviously I am not a racist and cannot be held guilty of crimes against other races that happened in other parts of the country and maybe even in other parts of the world long before I was born. As such, I barely have to bracket my own interests from the story that I tell. I am a rational agent in my nature every day of my life. My education has trained me to be so, the conversations I had with my siblings and parents at the dinner table, at the religious school that I attended, all these fostered in me a sense of discussion and reasonable discourse that situates my ordinary action very close to that of the ideal agent. Obviously when considering the matter, when I deliberate on the conditions involved and the various parties that have something at stake, my well-cultivated insights will carry me through the deliberation process to the most idealized and perfectly pristine of outcomes. And what do you know, reason shows me clear and plain as day that I have done nothing wrong. That my actions have always been above board, that all my advantages and benefits have been arrived at fairly and in accordance with all the rules of society and laws of the other land. Anyone who would accuse me of privilege would be in violation of the rules that I am so well acquainted with, they would be attributing to me things that I did not do, things that I cannot be held guilty of, actions or crimes that I clearly have not committed.

And this is how it works with those idealized agents that we see at the outset of the social contract ready to enter into a fair agreement to set the conditions for their legislative lives and orient themselves toward each other in accordance with these sorts of ideal formal preconditions. The very structure of their agency is itself mythical. The insertion of guilt and innocence and simple notions of aboriginal agency in which the first cause of motion is in the power of the person as a natural event, as an a priori mini-deity that has no bearing to social conditions and historical constitutive factors, eliminates any possibility of guilt in such a person. The long game con is that there is no question of responsibility for already existent historical conditions that no existing individual can be said to have created or caused. And the notion of an idealized agency is meant to establish this con in the social sphere and ensure that certain basic principles of its organization remain in place. It is a story that we tell to prevent ourselves from looking at certain factors that are very much a real part of the world and which form the environmental substance within which we, and all our compatriots, live. This goes back to the argument from first principles coming from some Marxists: 1) Human beings must eat to live. 2) Living among each other, human beings must in some fashion produce the food (and other means of subsistence) that they require to live. 3) Human beings cannot exhaustively consume both what has been produced and the means of production. C) The means of production must be reproduced. That conclusion, the requirement to reproduce the means of production, tells us that our societies and our civilizations must include productive mechanisms that are dedicated to reproducing the way of life that we have, so that we are able to continue to produce the means of subsistence every year in perpetuity. Or so the notion goes. And this means that our civilization needs to be made of efficiently structured organizations and institutions that are a crucial part of the productive process, that play an important part in keeping things going, and in continuing the process of feeding ourselves and providing ourselves with whatever other goods and services are deemed essential to the ongoing support of our lives.

These structures that have been put in place will have variable life spans, will outlive individual human beings, and they will be the institutions within which our individual lives take place. And as such, they will not appear to be the produced agency of any specific person here or there, no singular act of will can be called responsible for the agency of such organizations, for the conditions that they foster, for the world that they create, and the situations they make for us to be born into, to be dropped into, and to be the thrown projections that we are in the middle of. Such idealized agency not only aims at the construction of such organizations, it is itself such a constructed organization. The fabrication of the rational agent is itself a rationalized notion, a notion that has become more deliberated and clearly orchestrated over time, better polished and crafted to meet the needs of the society, and better able to enhance the interest structures of the groups that are best benefited by the notion. Recursion requires that you already understand recursion. The meaning of this circularity is that idealized agency as an institution is going to benefit most of those groups that are currently best situated relative to the social relations, those who are the winners in the various games that are being drawn out between players. And whatever shell game magic is going on in the discourse or the body of ideas that circulate via whatever media of mass distribution that there might be now or whenever, it is not hard for people of all walks of life to look around and view the inhabitants of our nations and our planet and see who the winners are and who the losers are, who is doing well by the current contract, whose interests are well-served by the current agreements, and whose interests are not served at all. The purpose of the ideas themselves, it might be said, is to make sure that people, en masse, ignore what is clear as day and directly in front of them. Forget about the complex mechanisms and thought games for dividing up and analyzing the world, look to the things themselves for a direct insight: some people are better off than others, benefits are largely material and can be traced to actual wealth. None of this suggests that there is some obvious and universal truth about what conditions caused this to happen or that those conditions are either unjust or just. We start from the bedrock: that it is so.

To the extent that the notion of a private person in need of protection from socially coercive forces is an aim in this essay, it must be admitted that something very much like that pattern has been used to establish a set of conditions in which some are very well off and others are living in conditions of extreme hardship. This establishment of conditions, I hold, has been based on the refusal to consider the basic interest structure of human life which can never be set aside and never be set apart from our actual concerns and deliberations, without removing the species from its world. We must be brutally aware of the real conditions that exist in/as/through our concrete lives if we have any hope at all of addressing them. And even if it is our intent to address them by reinforcing inequality and injustice, by further establishing the hardship of others while some benefit immensely, this must be done with eyes wide open and all the factors that can be extracted in clear view. To ensure my true private personhood, I must know exactly what ground I stand upon, I must see the conditions at work in the world around me, and I must be clear about the protection that is required to safeguard the further existence of that personhood against whatever invasive powers may aim at tearing it apart. Idealized agency is a smoke-screen that makes it possible to legitimate the status quo without having to delve into the complexity of the social relations and historical conditions. These are the factors that have brought that status quo into existence and continue to enforce an established organization through the means of well-intending subjects that act rationally to uphold the formal conditions of the ideal agent insofar as it has entered into a well-balanced agreement with others who co-inhabit the world with them.

Agents in the World

I have been in search of protection for the private person from social coercion, but the analysis of the idealized agent suggested that such private personhood may be something from which society needs protection. What if the person, in private, in their intentions and their orientations, is a villain? Or an expert at exploitative acts aimed at promoting their own power or their own wealth and pleasure? In that case, the person hides behind their private protections and uses this wall as a secret cloak that conceals them from the world at large as they carry on in their self-promotion at the expense of others. If the notion of idealized agency has been part and parcel of a grand scheme for reinforcing the status quo, then why would anyone undertake to protect it? The invasions of privacy that are ubiquitous under the conditions of telemetric understanding might mark an enormous step forward, something that enables us to gain new and improved protection from each other: something we have always required, but which was not present in other forms of organization predating the idealized agents and their social contracts. This logically follows from the conceptual topology that has been put in place. Because we hid the notion of interests when we developed the notion of an idealized agent, we made it possible for critics to see interests at work in such agency, thus providing an opportunity for such agents to create that realm. The domain of privacy has been pre-constructed as a domain of private interests, and as such, privacy itself has been historically defined as flush with specific interests. Many people demonstrate their understanding of this by pointing out that certain violations of their privacy do not matter because they have nothing to hide. The material of the private domain is the crux of the matter and whether there is anything going on in there that is nefarious or harmful to others is the matter of true concern. Only those who have something to hide will want privacy. This is the flipside of the notion at work here and it indicates that social relations have an increased interest in invading the privacy of those who would most wish to protect it.

Such an orientation means that appropriate private interests have a dependency on surveillance and the validation of those interests. How do you know you have nothing to hide unless someone is validating that everything that is going on in there is proper and not a threat to the social fabric or a violation of some other's rights and the integrity of their personhood? In this approach, we should be clamoring for greater invasions of privacy, for validator mechanisms that can give us our baseline and let us know that we are on the right track, that everything is progressing as we would expect, and as it ought to progress given the perspective of all interested parties. In this case, the telemetric understanding is the perfect solution for the problems initiated in the move to representative democracy two hundred and fifty years ago. Only now that we have our panopticon in practice, only now that our lives have been sufficiently digitized, can we rely on automated mechanisms to surveille and keep track of those moments and actions that previously we only had ourselves to provide insights on. We used to have to rely on our own evaluations and there was no proper conscience mechanism to speak of. The private lives of human beings were between themselves and their god. With the death of god, we needed the law to keep an eye on things, but this could not work so long as the people were able to keep part of their lives hidden from the eyes of the law. Policing and surveillance have been an ongoing concern and now we finally have our solution. I may be overemphasizing the point, but I want it to be clear that there might be a law and order contingency that sees that this is exactly what we wanted and needed all along, and only now can we be both free, in a rational and democratic sense, and safe, in a social and public sense.

The aggregate data that is compiled in the telemetric understanding is extremely attractive to people. It is as if they are gaining a whole new view of their compatriots, their society, and their role in it. We learn what people are doing in general, we see what sorts of things they prefer and what combinations they tend toward, this gives us insight into where we fit in and how much like others we are. Or how different we are from others. The aggregate viewpoint collected from invasive techniques gives us insight into those very social dynamics that went into the construction of our character in the first place. This would explain the attraction of such data and the process of collecting it. Some people love targeted advertising because it saves them so much time and effort in finding the things they like and need. Each of us may be 1 in a million, but with 7 billion of us on the planet, that means there are always going to be quite a few with the same interests and orientations. And if those statistical computations are varied depending on the object of interest, or attribute in question, or identifying characteristic under scrutiny, then at times we might have a lot in common with one group of 7,000 people and at other times we might have a lot in a common with a completely different group. And then of course there may be some tastes and preferences that make us like millions of others, areas where our interests are more common. Even the person with an esoteric interest in music or reading material may also love espresso, and, in so far as they are an "espresso lover", be in a group with significantly more members than the group they belong to that adores the music of Sabicas.

The private person is not reducible simply to private personhood.

What is revealed in this aggregate vision of the world is that the aggregate encompasses the other as well. The otherness of the other is just the flip side of our own privacy boundary. The distance and mystery, the secrecy and possible threat, that the other bears upon our lives is just as equally put on display and collected together into the aggregate form along with our own data. Maybe there is some sacrifice that goes into opening our private lives to surveillance, but that cost is offset by the insights and visibility we get over and against each other. The other is brought closer, made more familiar, and better understood through this exposure; and if the other had something to hide, something that would harm me, then it is highly advantageous that I can come to know it and, presumably, have some authority intervene to safeguard the interests at stake. The telemetric understanding allows us to make this trade, to sacrifice something for the sake of getting something in return. And again, if we do not have anything to hide, then we are getting the better end of the bargain and gaining much more leverage over the other than what we are sacrificing of ourselves in the trade.

The value of this exchange depends on the other being far away as opposed to a source of great pleasure and benefit. Depending on how we configure the questioning of human beings that we live among, we might get different answers to their conception of the other. When the other is a friend, the reaction might be very different. When the other is a family member, still another answer, and so on with every possible variable distance from us that the other might occupy. People tend to experience some of the greatest pleasures and benefits of their lives in community and activity with other people. In that case, when the others are close, are connected through a personal bond, and are part of the private connections that each of us have established with our surroundings, then the answers to these questions may change radically. Do you want your parents to know everything you did with your friends last night? Or do you have a little something that you might like to hide in that case? Do you really have nothing at all that you would prefer to hide from your children or your spouse? Something you have said to one friend that you might not want another friend to hear or know about, after the fact? People are evaluating surveillance as something that is done impersonally and which provides results in the aggregate, but the standards will change as the aggregation techniques are removed and the targeting of the information is changed. The aggregation of data might clean it, so if your children know what parents in general are likely to be doing according to some specific statistical averages, then that might be acceptable since, when push comes to shove, that leaves some plausible deniability that the statistics or the aggregate picture accurately represents you. The form of the aggregate makes it possible for your close relations to be informed about you in the abstract while leaving room for them to be uninformed in any concrete sense. Or, in short, our being captured in the aggregate still leaves us enough wiggle room to lie or mislead, thus leaving us with the ability to hide something. On the aggregate we have nothing to hide, but as individuals, we have a great deal that we wish to keep secret.

The telemetric understanding as it has been developed in the previous analysis, shows that there is a sense in which we are hidden from ourselves. There are social currents at work, collective operations under way, and they prevent any one person from getting a 360-degree view of who they are and how they fit in. This question obsesses us during adolescence when it first presents itself to our developing consciousness. And for some, it may gain subtlety and nuance as they get older, whereas others might become bored with the question and recognize that the concerns may not be so crucial to comprehend after all. The aggregate picture can speak to us, can show us ourselves as other, the part of ourselves that we have missed and that we were unable to gain access to in our narrowly scoped self-consciousness. In the response to the telemetric understanding, the aim is to fight for the good parts and resist the bad in this tendency. Collect the data, collect all of it, use it in the aggregate to enable generalized usage and understanding, but shield the specifics, shield the part of it that shows which groups of data attribution structures each individual specifically belongs to, because that is the part that reveals too much. Unless of course, you discover something criminal in my private interests as you aggregate them. If my behavior aggregates up into a criminal sector, well then you better come and get me for the good of the rest of us. This becomes the strategy of maintaining control over the details of how that collected data is used and where it is distributed, while also serving the social interest of collecting it in the first place and using it to protect us from each other. What harm could that possibly do?

Rounding up sexual predators who target children is a common example here. No one wants to argue against this. It would be absurd and ridiculous to argue against this. Of course, the fact that large on-line warehouses scan every photo that is uploaded to a person's storage account to determine whether the image is child pornography is broadly accepted. There are no mass movements against such activity and no protection for people who do these sorts of things. Many people consider this to be a selling feature of such technology: it keeps us safe. In the old days, the police would go out on patrol and make sure no one was doing anything they were not supposed to do. They were keeping an eye on things. Surveillance is a form of such protection and the migration of that protection to aggregation mechanisms is a good thing so long as the concrete privacy of discreet individuals continues to be protected. Well, that is, unless the rules are changed by the changing times. Some of the earliest slipping that went on in these slopes was when high school students started trafficking in child pornography where the subjects of the photograph were themselves and the recipients were a boyfriend or girlfriend. The kids were surprised to learn that this was a crime and one so egregious that pretty much the whole of their civilization had agreed to be mass surveilled for the sake of preventing it. And of course, the public nature of these crimes alerted the kids to what they had been doing and how wrong it was. They claimed ignorance and have stopped. Or rather, they have stopped storing the data in large scale online storage warehouses and started consuming some of the many new products made to get around just this sort of surveillance, allowing them to do what they want without involving a third party that preserves a copy of the materials they are sharing. And if you ask these kids, they will likely say that they have nothing to hide. Snapchat has ensured that.

The purpose of private personhood, emerging in a fashion protected from social coercion as a precondition to the possibility of ethical action and intrinsic to any awareness of the forces of the telemetric understanding at work in/as/through the actions and experiences of that private person, is to evade the trap laid down in the domain of privacy over and against society. The traps that we fall into when we consider the agent in the world among other agents and needing protection, is the trap that everything is a matter of economic action, of interests and ends, orientations and involvements. We are balancing the individual's interests with the society's interests, one individual's concerns with another individual's concerns. We are talking about having something to hide or not having something to hide, about benefitting from an aggregated viewpoint and statistical analysis or not benefitting from it. In all such considerations, the domain is coherent throughout. There is one network of relations continuous from individual to society and back again. And the domain of interests becomes a matter of where the boundaries are drawn and how the acts of crossing over them are regulated. In so far as the private person is categorized in this way relative to the social dynamics at work among different peoples, then the types of problems and conflicts that come into play are going to be encompassed by these border lines and these camps of interest: the liberty group against the security group, the advocates of civil liberty versus the advocates of civil society, and so on. And each group will have their interests, those interests will be associated with various properties and attributes, and various social bodies will come into being to reinforce or challenge some set of those interests and the groups intertwined with them. It is possible that among some nations the people may never make up their mind on these matters and so continue to define themselves as a constantly swinging pendulum that moves back and forth with each political cycle.

The whole of Western culture, the entire history of Europe and whatever of its former colonies continue to promote these ideals, would find itself in the middle of this ongoing pendulum swing. From one parliamentary election to another they would go back and forth, securing themselves at one moment, then recapitulating for the sake of civil liberties in the next. Only to go back again the next time around. And with each of these swings from one side to the other, the meaning and nature of the other would undergo a change. The other becomes nothing more than a proxy for considering the private person as such. The other is the private person, it is the other's privacy that is at stake, the other's ability to define their life as they see fit and as they wish to develop their personality with the liberty that they have and as they wish it to be. That other is you yourself, it is you situated relative to the mass surveillance of the telemetric understanding and your own personal relationship to yourself as something other than a set of population-based statistics and attributes that describe how you fit in, where you can be situated, and whether or not your life, as you see fit to live it, is criminal.

Consider two people involved in a conversation regarding just this issue, whether or not there should be mass surveillance in their society, whether or not the government should protect the self-determination of each person in so far as they live their lives free of ongoing surveillance causing them to alter their thinking and their actions to avoid the 'red flags' that the aggregating algorithms raise when certain attributes are viewed as part of a common pattern. They argue back and forth, they state their case, and they become passionate in favor of their point of view. They can no longer go on while maintaining civil relations with each other. They want to stalk off or shut down in the best of cases, become violent and physically aggressive in the worst. Staying focused on just these two people in the middle of their argument, what is at stake right there and then? Suppose, for the sake of argument, that this dispute is not taking place on the floor of some parliamentary institution in some national capitol somewhere. These are just two ordinary folks trying to carry out a conversation about whether security or privacy is more important. Let's say the examples turn toward a hot button topic that is common at the time of their discussion. Even if the subject of their debate is something that is on everybody's lips at that exact moment in time, what is pressing on them as they speak, what is providing the significance to their discussion, and where is it getting its urgency? If one of them were to concede to the other, what in the world would be different as a result? If they were to stop talking about the matter at all and just agree to disagree, as they say, what changes?

It is a strange set of questions because, on the one hand, people far distant from the scenario can easily see that very little would appear to be at stake in anything like an immediate sense. On the other hand, anyone who has been in such a situation is struck by just how much is at stake, as if the entire world in so far as it tends to hold one position or the other, is right there and then determining to go one way or the other solely based on this conversation and its outcome. Suppose instead of one person arguing with one other person, we construct the argument such that it is being held by one interest group against another. Now, it is not a matter of one person having such and such an opinion and presenting it to another person with a different point of view. Instead, it is a set of people who hold a position, who –on aggregate– share in a set of attributes that constitute a point of view. And then there is another group with a different set of attributed qualities that, in aggregate, holds an opposing or differing perspective. In this case, we might discover through telemetry and the telemetric understanding that a population of so and so many have an amalgamate point of view of security as critical in a variety of ways that are highly nuanced in terms of the various replies and objections that might be made by the liberty folks on the other side. The complexity of the aggregate perspective could be fine-tuned into a compromise and collectively collaborative set of positions that offset one another. Furthermore, it might be determined that the number who fall on the security side is larger than the number who fall on the liberty side, but that the complexity of the group's position has a wider range than any one member of that group has when taken on their own. This might suggest that the security position is the more popular one, but when you take into consideration the wide range of perspectives of its adherents it becomes weaker than the position some of the members of that group present on their own.

Now let's say that in the democracies of the world we substitute our representative democracy that uses individual people to perform the task of summarizing or aggregating the position of their constituency with an algorithm that aggregates from the collected data they have produced. Each instance in the result set represents a district, and the 'vote' of the district is based on the algorithm's output. Then all the districts are added up and a summary aggregate is created for the winning side. In this case, the electorate has completely submitted to the telemetric understanding. The arguments would have yielded the measurements that enabled the algorithms to sum up the outcome. Let these sorts of exchanges take place wherever they might, in public, on social media, or in the house hold. Mass surveillance in the aggregate form collects the data and is constantly updating the outcomes. We could have analytics that shows up-to-the-minute positions with legal ramifications happening in consequence. The laws might adjust to the ongoing discussions that the people in the democratic society are having. Would such a civilization be better off than the ones we have now, or would it be worse? Would this level of detailed analytics make us more secure or more free? If we are bent on surveillance as a way to characterize our private personhood, why would not we want to go all the way in that direction and allow the opinions and arguments of each to decide not only which MP or representative we are going to have for this or that term or this or that election cycle, but to allow these mechanisms to be used in the establishment of the very laws themselves. There is a lot of talk about which jobs and which areas of expertise are most vulnerable to AI and the advances in machine learning. What if legislative member of the government is one of the first to go? Would this be a step forward or a step backward?

The answer to this question might be interesting, but it might also be interesting to study the demographics of the people who answer the question and how they answer it. True to the telemetric form, what kind of person would tend to see this as a good thing and what kind of person would tend to see this as a bad thing? What characteristics and attributes, what patterns of personal identity, would be common in the people who favored aggregated and automated legislative bodies, and vice versa? It might seem that interested parties who think they are in the majority would be the ones likely to approve of such an approach, whereas those who think they are in the minority might disapprove. Maybe the approval would be more on the side of those who thought they could manipulate the outcome by targeting people in a certain way and moving them toward specific ends? It is unclear what the different interest groups would consider to be the primary criteria for deciding among these options. The biggest factor might not be so much the popularity of the position, but the perceived impact of breaking down political orientation into a case by case basis. This might be the most radical change involved in the approach. People who have historically voted party line during election cycles might have prioritized issues in doing so. There might be some issues on the platform that they feel very strongly about and so strongly about that they are willing to surrender any claim to getting what they want on a different set of issues. In the described approach, we would likely see case by case coalitions formed issue by issue. And we might see compromise positions that truly reflect the real status of the electorate. Sometimes people who boil their position down to a yea or nay are sacrificing nuance in doing so. They do not believe or hold true in simple 'yea' format, but rather hold something more like a yes, but with the following exceptions or exclusions. Complex analytics can take these factors into account and vary the attribution or results based on them.

This would introduce an enormous amount of change into any society that adopted the method. That change would not clearly be for the better or for the worse. What is clear is that the kinds of bodies or social structures that held power would be different. Power is concentrated in the legislative capability itself, in the body that makes the decisions, and votes on the outcomes. The lobbyists are targeting the congressional or parliamentary members and trying to influence them in one way or another. Clearly that would change, and the kinds of influence that are potentially successful would change with the power brokers following along. The algorithms themselves would be a clear target, trying to game them or predispose their results based on certain factors and conditions, getting them to include certain patterns and factors and exclude others. The people who were engaging in the conversations and interactions being measured during data collection would be significantly more targeted by interested parties trying to move the opinions of various groups one way or the other. More attention would have to be paid to the details of the exchanges and the subtleties in different points of view because these subtleties would be crucial for the laws that are being computed and enacted. The institutions that would become all-powerful and all-seeing might change even if the existence of social power and the coercive strength of those relations was ultimately what was driving the day in day out decisions of the population. The private person would still be under siege by socially coercive forces, but the institutions acting on them would be radically different.

Or what might surprise us more than anything else is that the institutions acting on the private persons might not change that much at all. That is, it may very well be (and this might just be the thesis of this entire book) that there is in concrete, in actuality, very little difference between what the coercive powers would look like in the one case than in the other, in the way in which things are done in democratic societies in practice versus the way things would look under worldwide domination by such a telemetric understanding. Large scale data collectors who can gather together telemetry data into a platform for use and manipulation by various analytics engines would be the largest corporations on the planet. They already are. Non-candidate specific organizations would target the general population with information and advertising aimed at luring their opinions this way and that to accord with the platform that promotes the preferred approach of those interest groups. They already are. The 'algorithms' would be constantly working to make sure they had the most up to date information on where people are in their thinking, and this data would be used to drive funding operations by showing where there is need for greater steering and pursuit to move the population in the direction that the interested parties want to see them go. This already happens.

Now, finishing up by closing the loop on the issue of security vs liberty as it has been classically constructed by the ordinary way of seeing things: does the telemetric understanding and its statistical aggregation have something to offer us that makes it preferable to opt for security at the cost of mass surveillance? The decision has already been made. The issue is no longer on the table, we have already decided. Arguing in groups or as individuals, as though the decision has not already been made, is childish. The question then might be more concerned with what good has come from it and whether the policies should be continued? That is certainly how a more traditional approach would proceed. Since I have set up a slightly variant ontology, however, the question that continues to drive my investigations is not whether we should do this or that but whether we can do anything at all about it?

Civic Bonds

Social relations are the natural state of human beings. We are beings in aggregate and this must be considered the point of departure for any notion of private personhood. The emissions of agency, of the body in action, are the raw materials for this formulation, but the attributes and characteristics, the categorization of the self in its many properties and types, attributes and identities, the derivation of interests and associations that predicate the 'I am' in each of its variations, are all originating in the social relations that begin shaping and molding the individual as they enter the world, even at insemination. The yowl of the newborn, or the yawp of the toddler, are the emissions in search of their shape in language. The tumbling and awkward crawling are the upsurge of physical motion in the shape of social corrections, reorientations, spatial instructions, and homestead adaptations. The ideal agent is most crucially and focally an *idealized* agent, where that idealization process (sic) is the singularly most familiar form of rationalization that the growing and developing human body must undergo as it makes its way in the world. The reason the boundaries are so important and cannot be reduced to matters of mere interest is that they characterize the entirety of the maturing process, the principle of individuation that takes place over the course of childhood through adolescence and into young adulthood. This explains the slippery relationship we have to privacy concerns, the extent to which we do not worry about protecting the privacy of an infant from its mother, or even small children for that matter. They have not yet earned their privacy, not yet come to enjoy it or need it. And, more importantly, they have not yet come to learn it. Only as parents gain a sense of needing privacy from their children and as they express this to their children, or in the presence of those children, do they start learning to turn the act around and require the same thing from their parents or other care givers. When the child learns that there are events, conversations, or experiences that they cannot share with their parents, they begin to develop a sense of their own experiences, events, and conversations. They realize first that they can keep these separate from the previously all powerful, all knowing, and all-seeing beings that have provided them so many aspects of their world up until that moment. And then realizing that they can, they slowly develop the interest in doing so.

Initially, it might be a game to the child. Just play, nothing serious. The child plays at keeping something back, at pretending not to share something with their parents. All for the sake of developing a sense of privacy that is concomitant with what might be called a sense of self. Having interests at all, the precondition for all interests that are specific to the one and only child, is the self-same separating oneself from the world, guarding oneself against it, and setting up of boundaries for protection from the social relations. Such play captures the fluidity of the process, the easing and tightening that may go on from moment to moment depending on moods and the persons with whom the child is interacting: siblings and friends, parents and other adults. The child is playing with the boundaries of who they are and how they can and should be with others. And we may call this process analytical in the same sense that we have been describing it since early in the book. The child is performing an analysis of the conditions of distinguishing themselves from those around them. And through those boundaries, identifiable characteristics are emerging. They see me doing this, they see me as… And this 'seeing as' applied to the reactions of others while they develop boundaries, amounts to the principle form of analysis that formulates the self in the exertion of private personhood over and against those forces that exert themselves as care or love, as friendship, brotherhood, and sisterhood. Whatever the case may be, external forces are caught and grasped in the mesh trap of these boundaries, and the catching, the grasping, is the conceptualization of that self, the forming of that private person in response to the conditions and impact of the surrounding persons and factors. The mobile device coupled with social media and digital telemetry are currently the best cyborg extensions for supporting and augmenting these boundaries throughout one's life: blocking out others while letting them inside under the gaze of the aggregation machine amounts to a symmetric triangulation of productive forces advancing the telemetric understanding. In the future, it may be the nanobots in our bloodstream and the massive amount of telemetry they generate that provide this same conceptual symmetry.

We should not confuse emergence with emission when inspecting this process of development and the tools that assist in it. The emergence of the private person is a response to the social interaction through those emissions and is meant to characterize an alternative viewpoint that reduces to the ideal agent, the pure self of the social contract, the person that stands behind the veil, or who find themselves in the original position. Emission is the characteristic nature of being physical, extended in space, a self-propelling body capable of interaction with a world and with environmental conditions, capable of learning from them, repeating observed behaviors, and identifying a position that can reproduce the outputs and emissions of an alternative position. Such physicality does not constitute the fully fleshed out self. That is a process structured in the actions of the physical being as it develops and matures, that changes over time, that takes on new traits and aspects throughout the course of its life as relationships deepen, attachments are made, pain is suffered, loss is experienced, and so on. Self is a secluding process not because the person is necessarily aiming at separating themselves from the world at large, but because the world is attributing more and more boundary forming conditions. The discovery process that is a 'world attribution' is the personification of the person, their coming to learn and discover who they are through more and more experiences that provide them qualities of their own along with qualities that are most decidedly not their own. That is not me, I do not like that at all. That is gut wrenching and other, I could never have wanted something like this to happen. Boundaries are drawn. That is exactly how I feel, that is exactly what I believe, we connected immediately over that. Boundaries are drawn. The push and pull of each interaction etches just a little deeper or a little more profoundly into each of us as we move along and pass through the different phases and stations of our lives, growing not so much from the inside out as from the outside in.

This explains why we prioritize ex-perience, the interacting with something outside of ourselves for the sake of changing or transforming who we are. When a teenager yearns to have something extraordinary happen, to go off somewhere and see something wildly different from what they see in their ordinary lives, they are looking for just this sort of impact, this outside-in transformation that will make them into something other than what they have been thus far. They want to change themselves, and their actual self is meant to undergo a change through the ether of worldly events and happenings. They will have been touched by something beyond themselves, shaped and moved, one way or another. And these experiences may turn out badly. Beforehand, we are likely to idealize the experience as though it is something beyond our control, yet it will always somehow serve us for the better. That need not be the case. The world can offer all sorts of variable experiences that may have deleterious or detrimental effects. They could harden us, make us mean, disappoint us, or make us cynical. The things that happen to us change us, and not always for the better. Sometimes very much for the worse.

Probably the origin of the strong sense that the private person should be in control of what use their emissions are put to derives from the fact that the private person is most definitely not at all in control of what use their emissions are put to. The strong urge for respect and dignity stems from the cold fact that this is not the natural state of human beings. Nature at its cruelest, the world of your peers at its basest and most real, does not respect you, does not accept your dignity, does not love you in any sense of the word. Wild animals are not cradled in the glorious and generous arms of nature, they are at its mercy living in survival mode requiring them to eat or be eaten. They struggle for their daily sustenance, and work day and night to make a life for their families, to keep their children alive and safe from predators, and –if luck would have it– thrive under the best of all possible conditions even if only for a short time, maybe long enough to breed. And human beings, living among each other in the social world that is their natural state, are wild animals in just this sense. There are predators about, there is danger and risk and all sorts of things that might snap you or your loved ones up at any time. Other people, freak accidents caused by the style and structure of life lived in organizations with other people, disease, thoughtlessness, stupid global policies that sweep you under the rug or ignore your concerns and your cares.

The liberty and the security of the person are not two sides of some issue that must be balanced one way or the other, rather they dwell side by side with their origins in the very same set of ontological conditions. Securing personhood from coercive social forces that are potentially violent and opposed to the well-being of that personhood, involves the boundary conditions between the person and the social conditions. Enabling the liberty of private personhood so that they can each control their own conditions of association and interaction amounts to securing the boundary conditions between the person and those same social conditions. Socially conceptualized liberty, freedom that has its origin in proper interrelations with others, is nothing other than the security of the person, the securing of that private personhood over and against the state of nature that social relations provide for the individuating human being growing up among other individuating human beings. The interests are both powerfully expressed and the argument between the two is ongoing and well-presented from many different sides. And the aboriginal power of each comes from the same source, the same basis of human beings that are first and foremost not in control of their own security and liberty, their own freedom to and freedom from.

It is imperative that we recognize that this boundary condition, this assertion of control where there is none, this condition for the possibility of being a private person at all, is not itself an interest per se that the body has in relationship to the community or civilization at large. Interests themselves are preconditioned on this status. The physically existent body has needs, but it would be premature to conceptualize those needs as interests. Only when the workings of the organism undertake a cognitive turn are such orientations recast as interests. And this may happen almost immediately. I need not assert that infants have no interests. Rather, for a need to be transformed into an interest, the body must orient itself in some cognitive fashion toward the need, recognize it, and turn toward it: the body must take an interest in the content of its need. And in so doing, the process of self-formation has begun. This indicates that interest is not the essential characterization of private personhood as such, but rather that private personhood is the condition for the possibility of taking an interest over and against having needs and being positioned in relationships with other things that involve directions of data flow and emission. This thing must enter this body in order for this body to persist in its various metabolic processes. And this positionality is brute and raw, dumb in a sense, and can be affected by force from external agents. Once this has happened, or once it has been repeated multiple times, the body takes on an orientation toward the thing positioned before it. The response may be uncontrolled, it may be a reflex reaction, light bouncing off something and meeting with the eye to enable focusing on the thing. This response marks the recognition. More and more complex and sophisticated responses can have greater and greater impact on the formation of the boundaries around the self. The more complex the response and the more involved it is with its source, the stronger the boundaries and the stronger the bonds between the creature and its needs. This enables personality in so far as personality is a merging of one's body and one's environment.

Personhood, in this sense, does not just separate the private person off through this creation of a boundary, the very act of establishing the boundary is also the binding of the private person with something in the world, some experience or set of relations that are intimately connected to them from now on as the embodied source of the boundary condition and the constitution of self that has taken place. The experience I had in front of the café behind the Parthenon in Paris that one time. This experience does not just establish my identity, it binds me to that place, and to the people that were present there and who took part in the experience. My life has been a set of identifying operations and procedures each of which put me into a greater connection and proximity to the matter at hand in those operations and procedures. For good or bad, we are bound to each other in terms of the same experiences that distinguish ourselves as the people that we are. Our personhood is a deepening of our relationships with others, with other people, and with the things that they make, the events that they enact, or the effects that they have caused. Think about who you are. The attributes and the qualities that will immediately come to mind. Now scratch at them: if they did not immediately include people, then they will soon enough. Maybe you are somebody's father or mother, or somebody's son or daughter, or maybe you are of an ethnicity that attaches you to many somebodies, an entire history of somebodies. Maybe you are a chemist which will attach to a history of innovation and discovery, a set of teachers who have shown you things and a set of researchers who have discovered techniques that you can now take for granted and employ routinely in your work. Whatever the quality, if you look a little bit closer, you will see that what separates you is what connects you to other people and their history and the social existences that they too had because each of them is, in this sense at least, just like you: identified through a set of connections to others.

These universal universals, or attributes that bind, form an etiquette or an ethos of human beings living with each other, collaborating, and involved in the world held in common. The ethical is discovered during one's development. While one moves along and comes together with others, one discovers what is known to be the ethos and ethical association with one's peers. It is not bestowed upon each human being by the deity but is engrained in the social fabric ready for discovery, just like anything else that shapes and forms our lives with others. And like such things, the ethical is learned and not just through some words that are uttered by this or that person of respectable standing, but through imitation and observation, through response and conditioning, we learn proper behavior and the proper nomenclature to assign to this or that action in this or that context. We learn to think of things in terms of them being the right thing to do given the actual circumstances, we learn what people do within the given circumstances, and we learn to live with the conflicts that make the one set different from the other. And what is most significant is that we learn a different set of guiding principles for people who are up close and people who are far away, not to mention learning how to tell the difference and how to place people at the proper distances while our relationships evolve. We form groups based on the attributes that we come to know, and we place people that we come across in those groups; and then our actions follow suit, being neighborly to those close by and going out of our way to help family and friends when the needs are perceived clearly, and the appropriate response is within reach.

Many people have a very strong delineation between the innocent and the guilty that they make when performing ethical evaluations. The ones who are innocent are the ones who find themselves in some position or circumstance through no fault or action of their own. Having your house destroyed by a tornado or hurricane, for example. This is an act of nature and the people who find themselves victims to such things are innocent in the face of it, and so there are many people who, learning of their condition, feel compelled to help them. A person who has no home for some other reason, like mental illness or post-traumatic stress disorder, may also gain moral sympathy from fellows based on specific details of their situation. These are conditions of homelessness where the homeless person is perceived as innocent, and so the generalized reaction to the situation is that the innocent person needs and deserves help in correcting their course because they have done nothing to bring this condition upon themselves. Therefore, people may sense a moral imperative to help them. When the larger social problem of homelessness itself is under consideration, however, the reaction to it will depend on what the person reacting considers to be the standard case. If the person responding thinks of a situation like PTSD following military service or a medical disaster that wiped someone out financially they may recognize a moral obligation, but if they imagine cases where the person has made bad decisions or done something by choice to avoid fiscal responsibility, then they might see the problem associated with guilty behaviors that do not deserve rescue. In that case, the person may think there is no moral obligation to help such people. Each of these reactions will be governed by the origin of their images surrounding homelessness, how they have come to learn of it, and what specific details they have come to associate with it. The partiality of experience drives the moral sensibilities. The problem itself is a generalized notion, but the imagined details which may vary wildly from case to case will be crucial for determining the relationship of the condition to ethical notions of proper response by an individual, an organization, or the whole society as it is represented by its government.

These details may come from the groups that we associate with. If my church has a strong outreach program that works with homeless veterans who are having trouble getting treatment for medical conditions developed during their time of service, then that is the most common case of homelessness that I will be familiar with. If I become aware of the assistance that can be provided by access to services that help veterans make use of the benefits they are already entitled to but which they were uniformed about or unable to take advantage of, then this might orient my entire position toward homelessness even if that is not the most common scenario for the average homeless person. We tend to pick up bits and pieces of information that are relevant to larger issues, and we tend to project and imagine additional details of the situation when we think about the matter on a larger scale. And if each of us has a different set of bits and pieces that we start out from, then the detailed understanding of the global situation will result in a different abstraction from others starting in a different place. This is interesting when we come together and start talking about the abstract issue in a generalized forum. That is, people who think of "homeless" people as veterans suffering from PTSD, or people who think of "homeless" people as the small children of parents who are homeless, or think of young drug addicts no longer able to work and afford a place to live, these people are discussing a common issue that each calls "homelessness" but in detail, in the fine-grained understanding of the issues that they have in mind, they may be talking about something completely different. And not just talking about it, they might be arguing quite fiercely about it, one group claiming that the "victims" of this social disease are innocent of what ails them while another group is claiming that the "victims" are guilty of something that put them in this position. For whatever it is worth and to whatever end this matters for some higher order organized decision, this makes a huge difference in what each interested party will think the proper response should be and who should provide it.

The categorical imperative and the principle of utility are lovely logical constructs and there probably are some people in the world who have been able to use these mechanisms to explain the bonds we have in our social relations. Yet they are esoteric notions. Empirical observation and involved learning are significantly more common as the origin of our ethical orientation, and that amounts to being good to those who are nearby and strictly disciplining those who are far away. And the rules for deciding who is far away and who is nearby are generally based on discovery. When we discover group associations, we immediately think of those in the group as being nearby. People tend to feel greater compassion for members of their same groups, or who have partial membership in some subset of groups that they may also have. They have children, they have a family, they had a medical emergency that bankrupted them. And so on. The characteristics that we discover orient us for or against, and the extent to which we know much about the people in need and their properties or characteristics will determine our interests and our likelihood of reacting to them in specific ways. From what I have seen, people are just as skilled at acting viciously toward each other as they are at acting compassionately with one another. And the determining factor is the set of associations that can be made between the target of the action and the other groups with whom the agent already has an established relationship or attitude. Human beings are capable of great love and they are capable of great hate, and which they express in each context toward a given set of recipients is largely dependent on how they have come to interpret the attributes of the conditions. Those interpretations will be largely driven by the person's own alliances and group associations.

If you made it your mission in life to instruct each and every person about the best way to act and be ethical toward all other people, you would need an enormous amount of time and resources because you would have to get to know each person, what associations they have already made, and what those associations carry with them as preconceived notions about the variety of ethically relevant circumstances important to them in the course of their lives. Again, an aggregate approach could theoretically address this issue in a way that might compress the time requirements given sufficient material resources and access to behavioral data through the digital or physical emissions that contemporary technology arcs allow. That is, a benevolent digital monarch could make it its mission to extract enough telemetry from every individual in the world to understand their point of view, their associations, and the attitudes that are likely to result from those associations: the so-called ethical positions of such a person. And then, given that information, the monarch could target the group with enough offset data to flesh out the picture of the various relevant issues so that the people in question would form different views about the general topic based on the different details provided. For example, analytics might show that there is a group of people who are sensitive to homelessness as the result of medical catastrophe, but who currently think of homelessness as the result of laziness and alcoholism as a life style choice. If such people can be defined and targeted, they can be shown the significant presence in the homeless population of people who were subject to a medical catastrophe. Furthermore, it might be possible to hit them with research and data that tries to show that alcoholism is itself a medical condition and that many of the people who have the disease were genetically predisposed to have it and that their "choice" was a catastrophic obstacle in their lives. In this case, it may be that a guided telemetric understanding could flesh out the details for some targeted groups such that their attitudes would shift. And even if they did not shift completely, they would at least have a broader picture of the issue under consideration when the more abstract notion of "homelessness" came up in a public debate.

And if this could happen in such a way as I have described, then presumably it could happen in the opposite way as well. People who are marginally in favor of doing something to help with the problem of homelessness in each civilization might be convinced to change their mind on that notion if they were to become targets of information that fleshed out the picture in an opposed fashion. The alternative view might show them, for example, lots of information –true or not– to suggest that many of the people who are claiming benefits from church or government programs aimed at homelessness, are not really in need at all, but are just trying to scam the compassion of others for the sake of benefitting financially without expending any effort. Targeting can make an impact because the derived notions are themselves based on group memberships, attributed conditions, and the partial understanding that anyone is likely to have about any social situation. Because no one person is exhaustively informed about any social problem, we are all –to some extent– susceptible to a degree of targeting. We could all be coaxed into changing our minds about some major issue of the day depending on the kind and amount of information that is disseminated to us about that issue. Because we are aggregate beings in and of ourselves, each of us, we are vulnerable to this kind of assault and this kind of conditioning. Such conditioning is always already familiar to us as how we have come to be what we are as we are throughout our lives. And because of that, we can be the targets of more of it. We can be pushed and pulled in various directions to go the way that some other interest group intends. At the mercy of someone pulling the strings, some pirate or puppet master, any element of our lives might be co-opted into the service of someone who stands to benefit from sending us off in one direction or another as their interests dictate.

As I said before, this would be a huge coup for all liberation movements. If there is a puppet master and that puppet master is intentionally and consciously pulling the strings to send the people one way or the other, if there is true manipulative evil in the world, then we are saved. Because that means it is possible to control the strings. That means, that an emancipatory interest could drive an effort at liberating ourselves from these strings and from the puppet masters. The commonwealth under such an emancipatory turn would be a truly representative master who, in keeping with the interests of the population at large, starts moving the direction of our insights and understanding toward what would best serve us. It would mark a corrective reaction to the twisted condition of alienation from our interests that has come to dominate the modern world. If you believe it. Our common conversation about who are the puppet masters, who is pulling the strings, and how we are being pushed and pulled, often highlights the individual persons that are supposed to be involved. The analytics based notions that we have been trying to spin out here in this chapter suggest that even the lucky ones, the ones who find themselves on the winning end of the pull of interests that drive our social conditions, even they might not fully understand what it is they are doing and how it is that they are able to do it. They may be succumbing to a feedback loop that is much larger than any one participant thus signifying that, should they try to turn the tides on those previous ills they have been perpetrating, they would find that the forces behind them have become unruly and now threaten to mow them down rather than be turned from their original focus and aim. Let us turn now to the feedback loop at work in our glorious commonwealth.

Chapter III: Corporations are People

In a trivial sense, corporations are people regardless of what any Supreme Court Justice says or what any precedent establishes. It is an emergent ontological condition that may be justly or unjustly attributed in legal arguments. And what is more, they are private persons who have fared comparatively better among those interested parties said to have a reasonable expectation of privacy. The right of self-determination expressed as data protection, for example, is unequivocal in their case and societies are more likely to provide protection for those rights than for any of the natural persons among their population. Corporations cannot be easily reduced to one single thing just as human beings cannot. The notions of personhood and private personhood help make that clear. Corporations are born, maybe not in the same way that human beings are born, but there is an act that marks the beginning of their existence, a gestation process by which they come to be, and a set of attributes accumulated in the act of establishing their initial conditions in the context of their early social relations. These various relationships establish them as the kinds of corporation that they are, their size, the number of shareholders, and their integration with other persons. Additionally, the kinds of investments that they hold, the kinds of contracts that they make, the sorts of resources that they have at their disposal, all these properties provide a set of meaningful predicates delimiting the being of each corporation and what details must be in place to support its actions. Some are predatory, and some may be parasitic. They compete, they buy and sell each other, and they hire and fire human beings. The corporations that exist at any given time are of all different shapes and sizes, but what is clear under the law is that each of them has interests and their officers are obligated to act on their behalf. These interests can be characterized through the concrete relationships at work in the body of the specific form of the association: what kind of a business is it, for example, what vertical of the economy does it reside in, what types of relationships does it have with the human beings that work for it, and what types of relationships does it have with other corporations, big and small.

Some of the biggest corporations are trillionaires if you think in terms of their capitalized value. And the corporations that have reached this mega status are each, in some way, involved in the data collection business. It seems that the biggest, most profitable economic state that anything can achieve is the state of the giant data warehouse serving up analytics and instrumental frameworks for orchestrating feedback loops between collection and decision-making or business objectives in response to customer activity. These corporations are giants because at least some of their customers are very rich and have paid generously for the services they receive. Other corporations, the ones that are not getting rich off data collection services, may yet be some of the biggest customers of those giants and they often depend on the provided data collection mechanisms to drive internal policy and support their interest structures in all the work they do and the financial decisions they make. These complex relationships between themselves and other corporations mean that these organizations develop interest orientations that are emergent from the structure of the corporation itself and not reducible to the cumulative interest structure of human stakeholders or stockholders with heartbeats. This is not meant to be a normative set of assertions about corporations, it is merely the way in which they exist today and the way in which they are commonly structured and supported to set about achieving their ends.

If a person who is well placed in the organizational hierarchy were to start acting against the interests of their employer, many bells and whistles would go off and it is possible that some laws would have been broken. The CEO does not have unlimited power over what the corporation can do. Laws and community pressures establish a relationship between the CEO's activity and the corporation's interest structure; and there are regular meetings and evaluations meant to determine if the actions of the CEO are continuing to promote and assert the company's basic interests. Should it be determined that a CEO is not serving the correct interests, it is easy to turn on a dime and remove them from office. If any person within the shareholder's order launches a chain of events aimed at diminishing the power and effectiveness of the corporate body, there are plenty of social and legal safeguards in place to stop such actions before they go too far and cause excessive damage. There are measures in place and ongoing review processes and organizational structures at work to ensure this oversight. The development of the corporation over time, its very history, has been a complex development of structures put in place to establish and ensure the independent advance of the body's interests regardless of the interests and motives of specific individual humans that might take part in the daily lives of those corporations.

Corporations are not all created equal and yet they regularly and routinely establish complex relationships with each other. A set of civil laws have developed to ensure fairness between them. That fairness is often a veiled way of ensuring that privately determined advantages will be reinforced by the law. It may be commonly observed that large corporations, rich corporations, corporations with large amounts of assets and fluid wealth, can take advantage of smaller corporations in direct relationships between them. Legislatures often step in and provide guidance within the civil legal code to ensure some degree of fairness between the parties, but the notion that a more powerful body will have more power in its relationships with other bodies is not a hateful notion to the legislators. More powerful things are more powerful and guaranteeing the rights of people amounts to ensuring that factual differences between them are reflected in the real-world involvements between those beings and others of lesser power (aka meritocracy). To legislatively reduce the power of an interested party is perceived as a form of theft from that party, a dismissal of what they are, and a denial of the personhood that needs protection, as opposed to something the rest of society should be protected from. And whatever is legislated within the civil law when addressing the relationships between corporations must also hold within that same body of laws between corporations and human beings. The civil law itself is supposed to cover the contract, and the parties to the contract, more so than any other biological or physical quality of the agents making the agreement. A contract has parties and all parties to the contract are to be treated equally, even if the contract itself embodies some fundamental inequalities between them. The parties to a contract are the very definition of personhood in such cases. Anything that can enter into an agreement is an interested party and any interested party has a set of interests that are at stake in the agreement and which may go beyond it or be irrelevant to it. Such positionality is a reasonable foundation for the notion of private personhood, as reasonable as any other, and certainly not in violation of any of the principles at work in the German Constitutional approach to the emergence of private personhood as a mechanism for the self-determination of personality, control over interactions with others, and over whatever effects can be legally enabled throughout those interactions.

From the law's perspective, there are interested parties pure and simple. The fact that one of the interested parties is supplying a service that it supplies to millions of others and that the other interested party is supplying money from wages is irrelevant to the terms and interpretation of the contract. A standard contract is the same contract for all parties who might enter into an agreement for the same services, the service itself being the determining factor in the exchange. The source of the money that the interested party receives, whether it comes from wages or from corporate revenue, is not relevant to the contract and not relevant to the relationship described between any two persons. The service is provided under certain terms and the reciprocal terms of that provision are generalized in a universal set of requirements such as the dollars and cents necessary to compensate the provider for that service. Whether the money is all the money that the contracting person has or just a tiny fraction of its revenue, neither is relevant to determining the fairness of the exchange. The contract is to be evaluated in ignorance of these conditions even if it is easy to see that they make the nature of the two positions vastly different.

There is something of an old adage about how much is owed to someone who gave over everything they have to help another versus the debt owed to someone who supplied the same amount but for whom it was a mere pittance. There is no agony over this ethical dilemma in corporate law and contracts. The fact that a customer of the cable company might be paying .1% of its operating expenses to obtain a service that another customer is paying 5% of their wages to acquire is not relevant to the structure of the agreement, it does not change the "rights" of any of the involved parties, it does not impact what they are to expect, and how they are to be treated in the exchange. Not in theory, at least. The relationship may mean something very different to two such parties and the dynamics at work in honoring that agreement will be vastly different as well. There are human beings who argue that there is a great injustice in considering corporations to have such legal status in being granted such rights, but those arguments are often centered around one or two laws with some limited context and application to the everyday actions of corporations. On the contrary, I am conjecturing that nothing would be more radical than to legally recognize the actual private personhood of the corporation and to follow through on the analysis into every aspect of life involving interactions between persons so recognized, regardless of whether those persons are corporations or human beings. There is a twofold approach at work: 1) corporations are people because they have interest structures that are non-reducible to the individual human beings that take part in the associations that constitute them. 2) recognizing corporations as people is vital for untangling the relationships that exist in all walks of life, legal and otherwise. There is both an ontological element to the claim and a practical element. Some laws that exist within a society will benefit corporations more than they will benefit human beings. Establishing a currency for considering benefit amounts to a reduction to interest contexts that are shared. Resource distribution and allocation is crucial to any such evaluations of benefit and advantage. This means that there is something like a translation layer between the action focus of a corporation and the action focus of a human being. They are both striving for material well-being. And to the extent that they have an interest in various impacts to that well-being, they are playing the same game, following along common trajectories, and aiming at similar types of things. Time is a critical resource, money is a critical resource, labor power and its value as well. There are a set of goods and services that all types of people rely on to improve their lives. And the agreements that these various persons enter are meant to establish a relationship between the persons and those goods for the sake of mutually reinforcing their interests.

Each of the persons involved in establishing itself within the social environment has an interest in pursuing its private personhood and protecting that personhood against the coercive forces pressing against their boundaries and exhibited or emitted by other persons within that same environment. The war of all against all is a war of all persons against all persons. To discount the role played by substantial parties to that warfare is to fail to see the primary threat they might have to human beings in their daily concerns. The possibility that the guy down the street would hack your cell phone to acquire and distribute your private photos to people you would not have wanted to see them may be a realistic danger, but it would be a mistake to allocate all our social resources toward protection from this one scenario. There are also scenarios where corporations have collected information about you, presumably for the sake of providing you better service, and are in the process of giving third parties access to that data in aggregate form. Maybe they are doing so in a multitude of ways, some of which are helpful to you or even to humanity at large, while others may provide less insight into the advantages it has for its sources. The presence of the former type of threat may be exacerbated by the fact that the data collectors are doing what they are doing, since they have opened the flood gates without bothering to control the consequences. The relationships with corporate parties may be significantly higher in risk and may have significantly greater impact than any relationship with a single human being. To fully consider the role the corporation has in our lives and the impact that it can have, it is crucial to concretely recognize the ontological characteristic that has emerged in and along with them. When the founding fathers of the United States revolted against the East India Company, they were engaging in just as radical an outcry as they were in their efforts to unseat the royal power behind the colonies. They recognized a deep alliance between the crown and the corporate interests that would tax and extract wealth from the colonial population without their participation. The residents of the colonies were at the mercy of both and their rebellion targeted both.

Corporate structure and its internal organization have evolved with the maintenance and establishment of such interests in mind. To what extent is the organizational structure of the corporation the subject of legal regulation for the sake of maintaining balance in the agreements that people make with them? There are laws that suggest executives must be transparent to stockholders to ensure that their investment is not fraudulently infringed upon, but what sorts of laws are there regarding how compensation and power in organizational hierarchies must relate to the realization of interest and the carrying out of policy? In many of the larger corporations, there is a layer of executive management that is characterized by their equity compensation packages usually tied directly to yearly performance. These are not simple review bonuses, but equity payouts to "employees" who are recognized as something more than employees, that are recognized as partners in the enterprise. The equity bonus is above and beyond a wage-based bonus and it is usually the source of the enormous income gaps that we have seen between the executive leadership of some of the larger corporations over and against the leadership of smaller corporations, not to mention the rank-and-file employees. The reason equity payouts are used in such compensation comes from the tax implications on both ends: equity shares are deemed different in kind than wage-based revenue, even when wage-based revenue is distributed as corporate stock. If one runs through the rolls of the largest corporations in the world, one can see that the big seven figure compensation packages are generally tied to equity partnerships more so than to any other characteristic bound to a specific job description. In the tech industry, this means that large-scale people managers and well-established architects or engineering thought leaders are all made equal in terms of the corporate hierarchy, they are all compensated as equity partners, and this –with an increasing ratio– becomes the primary source of their exorbitant incomes.

The level of competition that such equity partners must engage in throughout the course of their work lives is so intense that they would be happy to state the case for establishing their "worth" on these circumstances to any who would raise an objection. The pressures are elevated just as the compensation is. They are under constant scrutiny to contribute value commensurate with the earnings they reap from the relationship. And this pressure amounts to the force that co-opts the equity partner into a "company man" or team player who must work in new and creative ways day in and day out to promote the corporate interest even at their own or their family's expense. Or at least at their own collective expense as a group of people with attributes in common. Such corporations engage in regular shuffling and re-organization aimed at increasing the effectiveness of the corporate structure at delivering on whatever services they offer. The reorganizations might result in the loss of turf by some of those equity partners, resulting in their dismissal or their tacit dismissal through a demotion to a contribution level below that of equity partner. Again, the stakes for such contributors are enormous just as the compensation is. No doubt they feel that they have earned their money. And to be fair, from the point of view of corporate structure and the overall promotion of the corporation's interests, they have indeed earned it. If they have reorganized the hierarchy so that a billion-dollar sub-business within the corporate family of businesses is more efficient to a tune of lowered costs in the 100s of millions of dollars, then who would deny that the leadership behind such efforts are entitled to the millions of dollars they receive in equity bonus? If the reorganization results in the dismissal of several other equity partners because they will no longer have the "feudal kingdom" that they once managed, then that cost savings and efficiency gain is to the advantage of the corporation at large even if it means that someone's colleagues have been made redundant. The equity partner gains a lot from the relationship but the relationship that they are furthering is not one and the same as their own interests. Enormous compromise and deference can be required to do the job properly and effectively. Denying the value of such contributions does not serve anyone's interests, neither their practical interests nor their ontological interests in declaring the world to be a certain way. The upshot that may emerge from such observations of the value discrepancy among employees is that it is justified to the corporation that carefully measures its return on investments. These highly valuable contributors are an important line of defense for the corporate interests over and against the collective interests of the wage laborers comprising its rank and file labor force. Even a Marxist analysis based on labor theory of value would show that the interests of the corporation are dearly protected by the actions of these employees more so than the contributions of the widget makers who directly keep the service flowing to the customers. Opposition to this point remains ignorant of the role that executive leadership plays in realizing the personhood of the corporation and the protective power that such roles lend to the integrity of its structure. The organizational power of the equity partner is both a contribution to the productive process and a contribution to the institutional and organizational power of the corporation safeguarding it by insulating itself from the interests of the specific human beings who work for it. This is the very same organizational structure that makes it easy to replace some workers and very difficult to replace others: a cornerstone concern for determining the value of a contributor and true for both human bodies and corporate bodies.

Corporate structure cannot afford a radical separation between the equity partner and the wage laborer as that would expose them to too much risk in volatile labor markets. Usually there is an instituted spectrum of compensation that culminates in the equity partner and spans the range between them and the laborer. This spectrum means that the lower you go down in the hierarchy, the less the compensation might be and the easier the worker will be to replace. This may be due to radically different forms of compensation as well as raw amounts. Some employees might receive quasi-equity style bonuses that give them ownership over the company in the form of a wage: this amounts to a stock award that is taxed at the rate of common wages. There are varying levels of such employees with each increasing level having a bigger and bigger stake in the company. Down at the very bottom, there may be employees who receive no such ownership wage at all. The spectrum does not just allow for varying degrees of reinforcement of the corporate interest structure, it also provides a pathway to equity partnership clearly visible to anyone in the company. And because each person can see themselves in terms of a simple continuum that leads to partnership, even when the leap in reward that comes with partnership is exponential, each is able to envision themselves on this career path and adjust their contributions according to the standards and practices that are well understood for getting to that position in the organization. This gives the appearance that the attainment is within reach and propagates an interest in reinforcing those structures of differentiation as well as the overarching structure of the corporation itself. People will work harder, will be more creative, and will contribute a greater proportion of their energy and talent to promote those corporate interests even at their own expense, because they can clearly see the pathway to the riches that partnership has to offer despite the evidence that few will ever get there.

As such, the corporation as an interest structure is divisible and it can be interpreted in terms of the swarm of human beings and relations that contribute to its daily life while serving the process of acquiring its ends and protecting its projects. The micro politics of a corporation are just as unique and individuated as any psychological or social-psychological analysis of any private person engaged in trying to further its interests and set itself apart from the social forces that would daily seek to infringe upon it and extract more from it than it receives in return. Accepting the personhood of corporations is not meant to be a legal or value judgement. My purpose is not to establish the inalienable rights of the corporation in so far as it has equal footing of personhood with all other "natural" persons. Rather, the recognition aims at launching a set of interpretative insights that recognize the real-world status of things that are currently interacting with each other, receiving the feedback of various analytical mechanisms, and advancing their interests based on that feedback. Corporations are epistemological entities, they are advancing their interests by working hard to know the world and the kinds of relationships that exist within it. Their intent is to exploit or generally make use of those relationships in ways that promote their own interests. Corporations are acquiring more and more knowledge about their customers and about their competitors, they are collecting that data in record amounts, and using it to streamline their services and direct customers to them in new and advantageous ways. Advertising is just targeted information and most any information that is accumulated into a body and disseminated as such is partial. There are too many connections and associations at work in any worldly observation to make the whole story known in every single one of them. This means that any communique that provides information from one party to another is going to be partial. And what is left out is just as crucial as what is included. Information that is disseminated does not have to be false to opportunistically serve someone's interests. Any child knows that failing to tell the whole story but telling the truth for the part that you do tell is a cunning way to approach a subject when power dynamics are at play and there is personal interest at stake. Corporations are distributors of information and they distribute that information to serve someone's interests. And by law and norm, they will be quickly eliminated if they do not serve their own interests.

There is a common-sense proposition that if you dig through the books of any random billionaire you will likely find some illegal activity. The same is also true of the trillionaire corporations and those that aspire to be among them. The corporations, unlike simple human beings, do have something to hide and therefore have a strong interest in protecting the private personhood of the corporate body from the coercive forces of the social relations threatening them with legal action or regulations if the whole truth were to be revealed and the inner workings of the enterprise made public. This means that such corporations have a strong interest in maintaining, among themselves, a German-style right to privacy, to self-determination, and to establish protection from any social forces that would infringe upon that personhood for the sake of uncovering internal communications or operations. If corporations were recognized as people, and if we were to insist that the equal protection principles apply to all people equally with no tolerance for legislation that tries to enforce a separate but equal justification, then it might be the case that corporate efforts to protect their privacy would end up becoming relevant to all social movements. The laws that came from this action would apply to all persons equally and thus provide each of us with the power of self-determination implicit in the German approach and called out as central to the recognition of that dignity which lies at the origin of all personhood and all agency that is the subject of rights within a commonwealth.

Ironically, this about-face in the matter of corporate personhood might result in an about-face in corporate policy toward private personhood itself. That is, if a society were to start carrying out broad scale initiatives recognizing the private personhood of corporations and insisting upon the equal protection of all persons under one set of laws, then the corporations themselves would oppose any notion suggesting that they are people. The problem is not that corporations are people, the problem is that they are treated as a better class of people and are gaining advantages and incentives from the applicable laws. Asserting personhood and insisting on equal rights for everyone, levels the playing field and establishes a feedback loop for personal analytics that at least stands a chance of producing a notion of private personhood that is ready to serve as an ethical foundation for all interest-driven agency based on any social action context.

This would require that penalties and legal responses to actions be recast in terms of a general interest-based currency applicable across forms of life and encompassing both human beings and corporate bodies. Incarceration as a physical act asserted applied to the body of a single human being would no longer be the paradigm. Rather, we would begin to see incarceration as a relationship to interests, as the inability to carry on as before, serving one's interests in the way that one was doing prior to an arrest. That way, if a corporation were to commit some act that merited incarceration, it would be appropriate to devise an interest-based description of how the penalty should work. We would no longer understand incarceration as the placing of a body in a jail cell, but as a form of invasive surveillance and minute control over the person's micro actions for a specified period of time. Invasive surveillance and control through regimen and discipline is a matter of occupying the private person to remove any claim to liberty that they might make. This can have a meaningful determination for a corporate body with an interest in maintaining its private personhood out of the clear view of any authority that would attempt to monitor and direct its behavior. An in-corporation, under this way of thinking, could be incarcerated if a crime has been proved and such imprisonment would go much further than any fine would go in establishing a penalty with real force of influence: either rehabilitation or deterrence. Some of the problems with corporate governance are that corporations can afford the penalties and so they consider them to be simply part of the cost of doing business. By recognizing the true ontological state of the corporation, we would be able to truly evaluate its interests and the impact upon those interests through penalties that really punish and deter future wrong doing.

The general purpose of the feedback loop at the level of commonwealth considerations is the analysis of private personhood that must flow in automated fashion into the well-being of the civilization at large. Again, this is not a value judgement or a rallying call to establish a set of artificially derived norms to apply to structures in the wild. Whether a person can control the flow of activity and the way in which the collective will of the people guides and directs social forces at work in any geo-political area, is simply a matter of how the feedback loop from individual analytical attribution works relative to the common good or general welfare: that very same notion that we are following the philosophical tradition in trying to capture through the use of the term 'commonwealth'. The only way to accomplish this with a realistic approach that includes all relevant factors, is to start from a point of view that adequately includes all the relevant interests and concerns that are part of that feedback loop. The assertion that corporations are people in all the practical and ontological senses that this might have, is an effort to claim that the proper nature of this feedback mechanism is only available for scrutiny if the interests of all such persons are taken into account when summarizing and surveilling the domain; and that means that individual humans cannot be isolated as the sole contributors to that loop, rather any interest structure that contributes to it is itself functioning de facto as a private person. This amounts to an act of taking on a role to play through its constitution as a pluggable component in the telemetric mechanisms advancing the common good through the automatic consumption of private personhood by the social dynamic. Everyone has their part to play in such a dynamic, whether they be individual data producers like you and me or large-scale data pipeline providers like Google, Amazon, and Facebook. The power and subtlety of this approach will enable a continued recognition of the differences in persons as they contribute to this dynamic, not just because some are human beings and some are corporations, but because all are comprised of members in a set of differential attributes that separate types of persons from other types of persons: being a corporation would be just one more demographic category.

Cognizant Non-Life Forms

The corporation is notably defined in terms of the reduced liability that any of its stockholders has over its debts and obligations. The advantage of this is laid out as though it is a protection from liability for the persons who take stock in the corporation. The corporation may go bankrupt or it may be fined for its actions when it violates the law, but the stockholders are insulated from this, protected from it, and will not be responsible for any debt that is attached in bankruptcy nor will they be held responsible for any fines that it has to pay for crimes committed. This protection is more than a mere insulation from liability, it is the establishment of an independent agent above and beyond the persons who own stock in it. The corporation has its own lifecycle and its own interests independent of the people who have invested in it. Likely those interests will intersect in ways and at times, but they are not reducible one to the other. As a result, I hold that the very establishment of the corporation's agency and its establishment as something that can have a bank account, and enter contractual relations with other similar entities, is the basis of its being as a person. And this is not radical since the notion of personhood is already distinct from any biological or species-based notion. There could not have been any idea of 2/3 of a person if it were a purely zoological notion that was meant to exclusively point to the legal status of a human being. Personhood is a positional ideal and it spans both artificial and natural scopes. The corporation is not a natural person, but it is a person nonetheless, and as such, it can do all sorts of things that people can do. Rather than be stuck on this idea, we need to adjust our notion of personhood to take it into account. The social relations and positions that condition things shape the agents that can exist within the environs of the world. And lots of things can be shaped into persons so long as they are constructed to express interests and to work toward furthering those interests through the achievement of ends altering the state of the agent in such a way as to help it flourish or bring about its demise.

None of this suggests anything at all about legal status or constitutional applicability. That is not a necessary step. House pets may be considered persons under the definition that I have given, as can neighborhood homeowner associations and PTA groups. Any aggregate construct that can be characterized in terms of interests and attributes that fuel those interests and which can act to better or worsen the structure's state relative to them, is a person in this sense. And, of course, not all these things are going to be granted all rights and privileges under the law. The establishment of a right is a separate matter altogether. Although clearly this ontological determination of the person in terms of its agency and interests is a prerequisite to any such constitutional determination, it is not a sufficient condition for establishing it. It is entirely plausible to assert that the writers of every passage in the main body of the Constitution of the United States and each of its amendments up to the 15th never intended for the ideas embodied there to apply to corporate persons, while also asserting that over the course of the last century, the empowered interests have repeatedly granted such rights to corporations and appear likely to expand rather than reduce them in the future. That is, corporations have been gaining an increasing number of rights over the years and there has been very little movement to rescind any of those rights. This has happened to such an extent that the status of the corporation as a person and as the legal subject of a great many constitutional rights is something that has become bedrock within American civilization.

This fact does not hinge on any one Supreme Court ruling or precedent, it has been creeping through many such rulings and precedents over the course of many years. Finding a court configuration that will rollback Citizen's United will be insufficient to remove the ontological status from corporations, and it will be insufficient for removing all constitutional rights from them. Citizen's United may establish certain first amendment rights for corporations, but there are other precedents that were used to establish their fourth, fifth, sixth, seventh and fourteenth amendment rights. And I do not see anything on the horizon within the United States (or anywhere in the world for that matter) that even vaguely suggests that there is an inclination among the human beings in power to retract the ontological status of corporations as real entities, real entities that act, and real entities that act for the sake of furthering interests of a clearly identifiable and material nature. Corporations are persons, they are here to stay and, most likely, will continue to have some set of rights guaranteed them by the various governments of the world. An attempt to amend the Constitution to assert that corporations are not people, were it ever to happen, would likely be brought by corporations themselves because they wish to preserve their ontological status and its concomitant advantages after various legal constraints typical of personhood have been applied to them.

In my personal history as an employee of a corporation, an incident occurred not long ago where I was working with a group of people to develop a project which required that we gather data on one of our products and then aggregate and compile it so that other employees could query it to drive decisions their managers had to make. There were a lot of levels of involvement and my role was to build the data pipeline that would extract and transform the data to get it ready for such queries. I asked a clarifying question of one of my colleagues to get a sense of how they wanted the system to behave when handling some special cases likely to occur as the data moved through the pipeline. The colleague shrugged, said they did not have an opinion, and that I should just decide. "But I do not need the data for anything," I said. And it was true. As a human agent with a set of projects, there was nothing in my realm of possibilities that needed this data; I had insufficient information to decide how to construct the rules that we needed to understand and embed in the process. The colleague, who I thought should have been able to say what the rule should look like, shrugged as well and said, "Well I do not need the data either." Neither of us was able to make the decision. We had to go off on a mission of discovery to further elaborate on why we were building what we were building and what purpose it was supposed to serve relative to its implementation. Breaking it all down, we could have travelled all over the company looking for someone who needed this data for some purpose. We knew what we were looking for: someone with a job title and description that situated them in a corner or part of the company where it would be their job to ensure that some function was being fulfilled in a specific way. Only as a job function, were we able to find someone who had an opinion about what these rules needed to look like and how they needed to be applied to the data in the special cases we had identified. The person we found had to do some computations to determine the outcome relative to the various forms that the rules we were looking for might take. They needed to look at a set of requirements and perform an extrapolation of what needed to be done in such cases based on the higher-level description they were familiar with. Employees of corporations are typically human beings. And as human beings they have an interest in doing their jobs so that they continue to get paid. And doing their job amounts to promoting a set of interests that are not identical to their own interests. The tasks do not directly involve feeding or clothing their children, for example, but might involve building an analytics portal that can be used to make go/no-go decisions for products that are being measured and analyzed in those portals. And the functionality provided by those projects is not directly consumable for feeding and clothing any of the employee's children. Nevertheless, ensuring that the portal is built properly will fulfill their job function and ensure that they are rewarded with ongoing compensation that can be used for those purposes. It seems like an obvious point and hardly worth belaboring, yet I was surprised to learn over the course of my career how many people in such positions – including myself, at times– forget this distinction and lose themselves in their jobs.

The employees of corporations know at some level that their "interests", when fulfilling their job tasks and assignments, are not really their own interests, but instead are the corporate interests that have been assigned to them as part of their job description. The organization is set up in a way that can optimally allow for the allocation of teams of human beings to best ensure that the corporate interests are furthered. Given that the people doing their jobs might lose track of these distinctions during the day, the corporation might cultivate a culture and new employee profile that feeds this breakdown through the targeting of specific personality types. This is easy to accomplish since the compensation package or job performance review process (which is a form of ongoing surveillance) are so closely tied to how well the employees safeguard and serve the corporate interests in whatever little subdomain of its life they are assigned to, but when we break it down and examine the fine-grain details of the lives involved and the actions performed, there is clearly a difference that any participant would agree to when probed on the matter. Despite the universal consent among employees in this state, the body of the corporation becomes enormous through the application of this division into daily operations. Arguably, the educational system preparing the future work force would have to be complicit in preparing such employees with such personalities for just this kind of "service".

Such "raw materials" can be inserted into the corporation using a kind of hierarchical platoon-based organizational bureaucracy. Some large corporations are constructed out of many such teams containing 6 or so people. Each of those teams of 6 reports to the same manager. Four of those managers might work together to report to a common manager between them, meaning that the 4 managers, 24 team members, and 1 second level manager form a group. There may be four such groups that report to a third level manager forming yet another grouping and then four such third level managers that report to yet another fourth level manager creating an even bigger institutional structure. We might call the first of these that contains the team a group, the second of them an organization, and the third of them a division. The work of the higher order managers in this hierarchy, the leadership team, is to bring together the disparate projects of the groups and organizations underneath them to present a large-scale set of offerings oriented toward executing on the business model encompassing all the work within the division. Only the fourth level manager, however, will have enough context to see the meaningful interconnections between the different interests promoted and pursed by the various employees underneath them. Only such a manager has enough perspective on what is going on and how it all fits together to provide offerings that further the corporation's interests while the people lower down in that hierarchy are less and less clear about what they are doing and how it all fits together.

A good corporation, one that is efficient and successful in carrying out its tasks, will not necessarily function exclusively through bottom up or top-down methods according to this structure. That means that the fourth level manager will not dictate the actions of everyone underneath them in detail, nor would we expect the "boots on the ground" at the bottom of the hierarchy to dictate action and policy to the people above them. Rather, what we expect is that everyone is expert in their little area, everyone can know the ins and outs of what is going on where they are located and where their focus is trained. The widget maker is an expert in making widgets and will send pertinent information on widget manufacture up the food chain to the managers above. And each will spin the information in terms of the larger context that comes into play with the project as the information moves up a level. Once the information about the problems and issues with the manufacture find their way up to the level four manager, the kinds of responses and solutions that might present themselves and the true nature of the problem itself will have gained more meaningfulness due to the varied perspectives that come into play as the information moves up and down and has been reshaped or translated while in flight.

Thus there is complexity at work in all decision-making in such a corporate setting, it is not as simple as one person calling all the shots, rather there are aspects and elements of every decision: deliberations are taking place not so much in the mind of one human being or another, but across the various roles and responsibilities at work in passing the information up and down while providing insight specific to the point of view that any contributor might have on the process. When interacting with others in the corporate setting, the key is to realize that you are working with someone who has some knowledge about some procedure or process that they are closer to than anyone else. They do not fully understand what that information means because they do not know what everyone else is doing and what sorts of environmental conditions might be going on in other parts of the organization impacting what can be done or seen in the information that they are working with. The questioning process that transfers information from one person to another is a cultivation process where a manager might be pulling relevant information out of the various contributors to construct a story in terms of the context that the managers are familiar with at their level. And they need to know how to question and cultivate information from others to form that picture properly so that they can then pass it along to someone else who will question them in much the same way so that still another fuller more complete picture of events can be formed. The same can happen as information works its way back down the organizational chain. The fourth level managers might have a solution for what needs to be done to address a growing problem discovered through the cultivation of information from the ground up. They can only present that solution, however, in the terms and semantics that are appropriate to the layer of the corporation that they operate on. As the directive works its way back down, each manager is responsible for converting the solution into terms appropriate to their context and their realm of familiar interests. Finally, the directive gets back to the individual at the bottom of the organizational chart, and that individual knows how to apply the solution to the raw material that is part and parcel of the work they do.

This is a non-life form, but it is an organism, nonetheless. And although the example stems from a large corporation and may be biased by my experiences in the technology sector where the corporations are efficient and highly optimized for maximum productivity, this form carries through to small corporations as well. Indeed, some corporations may be corporations in name only, meaning that they are closely held and are just meant to be liability havens for the ownership to act out their own personal interests and personalities, without the artifice of the corporate cognitive structure we have been describing. This phenomenon is easy to identify given the conceptual apparatus we are putting into place. Arguably, any intelligent human being would incorporate in a society where corporations are at such an advantage, and they would use that corporate status to protect themselves from socially coercive power and provide themselves with the augmentations of rights and subjectivity that come with it. Every corporation need not realize the "cognitive" structure that I have described. It is, however, a pattern made possible by the legal structuring of the artificial person and even a company made up of a dozen people will have sufficient division of labor to capitalize on this organizational optimization for structuring action to serve the corporate interests.

The employees of a corporation, big or small so long as it is well-run, cannot be completely reduced to cogs in a machine. They must be understood as intelligent contributors to a living process that goes beyond the perspective of any one contributor. The intelligence is required to allow the positions making decisions to be properly informed about what is going on in all the nooks and crannies of significance to the organization's mission. Dictatorial policy that may exist in some corporations is likely still getting information from all levels of operations, although it may be that the management chain is getting enough information from metrics attached to those levels of operation so that they no longer need any direct input from the individual contributors. Maybe they are watching the production line or measuring the performance of the employees in such a way that the measurements themselves are all that is required. Even in that case, there is some input that is coming from those levels and there is some feedback that is flowing back down the chain and to the bottom most layer. It may be that a machine expressing the rhythm or cadence of the work is the driving factor in the process, and the measurements work their way up the chain so that changes in the machine calibration can in turn work their way down again. The information in this case would be flowing up and down with the ultimate goal of local areas gathering the information and sending it upward to be processed before it comes back down in a new format that expresses whatever adjustments must be made to accommodate the analyzed data fed back to its point of origin in modified form.

This structural organization amounts to the basic processing and operational material of the corporation and the actions defining its personhood. The context is never sufficient for any piece to self-define the way it fits in properly with the whole, but the piece, the partial component, is not a dumb cog either. The telemetric understanding at work here is stark in nature, with the meaning of what is happening at any locale being obscure to the agents performing the work but becoming alternatively visible at other levels where the information synthesizes into a context that can feedback appropriate responses. The completeness of the picture is remote and the meaningfulness of what is taking place is happening in the office upstairs or down the hall. The organization is a collective structure no one piece of which has everything that is required to know the totality. We call these non-life forms cognizant because this organizational flow of information, this application of context and meaning in the office down the hall, is something removed from human cognition, but it is a process that approximates thought and computation nonetheless since human beings are some of the key ingredients and may have once been the model for such a leviathan. The corporation shares in an awareness or consciousness but bears no exhaustive self-reflexive position for it. Rather, there is no one single subject for that position, so it is an awareness or a consciousness that only approximates the human form. In transhumanist terms, the corporation is the AI and the employees are the nanobots.

We probably think of the managers of a corporation as the source of any real agency in it. There would be a tendency to dismiss any magical appropriation of attributes in the corporation for the supposedly simpler explanation that there are high level leadership personnel who are making the important decisions that drive policy and the orientation of the company in all the matters significant for realizing and benefiting its personhood. The information flow is critical and those people making those decisions are depending upon it, since their decisions are only as good as the information they have received and are acting upon. And there is more to it than that. As I suggested before, the leadership and managers themselves will be influenced by the culture of the other managers and leaders. The people at work in any corporate structure are guilty of all the petty human moods and traits that any people in any social setting are capable of. As a result, the communication that takes place across levels and among colleagues is not pure, it is not without power dynamics and coercive relationships that can impact what is said and how it is interpreted. There are people who know much more than they would ever be willing to say because they would fear punishment if they were to be truly honest about the details of their situation and the conditions they have seen first-hand. Furthermore, the information is often of competitive significance. If my colleague has some information and I have some additional information, it may be that the view each of us occupies conflict with each other. If my colleagues' viewpoint wins out and drives the project, my colleague will benefit from this more so than I will and that will be a disadvantage to me since it may cost me money or a promotion. The split between corporate interests and human individual interests becomes most clear in those cases where communication becomes coercive. The pressure is felt to get ahead, to move forward, and to advance up the organizational chart. To do that, one must compete with the others who are providing information up the same food chain and, most likely, one must appeal to the vanity of one's managers to ensure a competitive edge. This sort of thing happens in most any human social setting and so it happens within the corporation in so far as it –despite being a person– is also a social environment for persons.

Ironically, it is exactly this non-organic core of the corporation that most threatens the advancement of its interests and the promotion of its well-being. Because the people are socially oriented and because their personal interests find their way into the corporation's communication channels, the corporation itself is likely to suffer from it. Or at least this is a risk for it, that the human beings working to promote its interests will put their own interests ahead of it and cost it dearly at times. This structure is built into the corporation in so far as it is a person that relies on persons to do its bidding. Although the corporation is a person, it is formed out of persons as well, and *this* makes it very different from the human beings that work for it. It makes it very different from whatever machines work for it as well. In this regard, all the persons that contribute to the corporate collaborative environment may be ontologically different from the corporation. And this difference makes it possible for there to be a disconnect between the interests the person is fulfilling for the corporation and the interests that the person is fulfilling for themselves through the job that they do. Corporations depend on this distinction and they benefit greatly from it. First off, because it is the only way that they can exist at all. The dichotomy between the personal interests of the employee and the role-based corporate interests is what made it possible to form the corporate person at the outset. And then there is the matter of ongoing efficiency and optimization within the corporate structure. The corporation is made up of contributors and it may become necessary or optimal to remove some of those contributors at any given time. If the corporation does not need so many people with one set of skills, it can remove some of them. If it discovers changes in the market place, it can recalibrate its compensation structure. The corporation benefits greatly from the distance it maintains between its contributors and itself. And yet this same difference is a potential hazard to its optimal efficiency. The corporation would like to be made up completely of parts that contribute joyfully and intelligently to its aims without asserting their own interests. This could never be the case if the rudimentary resources that form the body of the corporation were human beings expressing human interests in the carrying out of their functions as private persons (aka people asserting themselves as the subjects of rights). To this end, corporations may deem it beneficial to align themselves with those social forces that coercively work to infringe upon the rights of persons that are most likely to wind up in their employ.

The corporate organization, cognizant and populated by human beings acting as persons fulfilling their role within that hierarchical order, runs aground on all the same patterns and possibilities as any society writ large. The state of nature/state of war reality that forms groups and factions, that produces identities and interest groups can exist within the corporation as a microcosm of the larger scale social macrocosm. We see repetition of the idealization and the disruption, the analysis and the bonds, all within the structure of the corporate operations and processes. The corporation as person is a super person, it is larger and more extensive than the natural persons that carry out tasks within it as its labor power; and, as a super person, it can be said to "have a culture" where that would amount to a set of common characteristics that tend to further condition the contributors. The way of life, the common way in which the employees will be encouraged to think, all of this makes up a set of identities that idealize agency within the confines of the corporate structure. And the personality of the super person is the embodiment of this such that its personality is formed in terms of these characterizations and qualities. As such, it is possible to speak not only of the personhood of the corporation, but we can speak meaningfully of its Germanic private personhood, its emergent way of being, its relationship to itself, and the characteristic organs and elements that emit that personality. Thus, these artificial persons can exercise control over how they are presented to the outside world and to the social world of other persons who do not work for it but are its competitors or customers, its suppliers or its government representatives. The corporate person is a private person in so far as it has a right to privacy among its other rights. And truly the right to privacy of a large corporation is a right to be envied. In so far as corporate personhood has been made the subject of a right to privacy, it has been given a great and powerful thing that is well protected by the laws of many lands.

A corporation that is engaged in data collection has the advantage of not having to justify gathering data on itself. The telemetric understanding as a way of gathering behavioral data for the sake of analysis that will lead into a feedback computation for greater automation and efficiency is not even experienced as an originating process distinct from the corporation itself. As we said before, this telemetric understanding for the corporation is the way in which it might be able to best get to know itself and best come to understand itself. It is its way of self-reflection and of being self-conscious. Telemetric understanding is put to excellent use by corporations to further their interests and promote their aims. What is more, these corporations are well guarded against surveillance by other corporations or by individuals who would use such means to find out trade secrets or proprietary information that is the sole possession of their targets. Corporations, it would seem, have a genuine and hard won right to privacy. They have information that they do not want to see leak out into the world and they are aggressive about protecting those rights. With all the data leaks plaguing corporate bodies, it is still rare to hear of a case where their bookkeeping data or their internal financials have been breached.

The legislative structure of privacy is almost completely oriented toward corporate personhood. Even the Germanic conception of a self-determined right to form one's own personality, to maintain control over the distribution and use of information about one's person, all of this is so much more powerful and so much more valuable when applied to the actions and boundaries of a corporate private person. Corporations routinely bind their contracts with non-disclosure agreements that place heavy fines and penalties on any who violate the terms. Furthermore, through their size, the bigger the corporation, the wealthier that it is, the greater its advantages in dealing with others. Individual human beings often engage in agreements with corporations as though they are one of many people making the same agreement. And that is because they are one of many people making the same agreement. When corporations enter into similar agreements for goods and services, they do not have such a disadvantage. Because they are unique, in a sense, the terms of the agreement must be unique to match with that singularity. The bigger the corporation and the deeper its pockets, the more leverage that it has in the agreements that it makes. And that advantage, expressed in its size and wealth, translates into greater respect from the other party. And by respect, what we mean is greater recognition for the uniqueness of the agreement and the uniqueness of the party making it. The corporation wants to be treated like a person, treated specially with added disclaimers and riders to acknowledge its ability to claim various damages or discounts if some or any of the terms of the agreement are violated. You will not see a large corporation sign a standard agreement on a par with an End User Licensing Agreement because their size and leverage mean that they do not have to sign such agreements. They can demand to be treated individually and thus use their size to their advantage in coming to an agreement catering to their interests in a highly customized fashion.

Continuing with the hypothesis developed throughout this book, this quality of agreements evades the telemetric understanding and thus establishes the corporation as a private person properly protected when establishing contracts. Such agreements will then better express corporate interests and enable them to better achieve their ends for the sake of those interests realized as profit, growth, or increased market penetration. As such, these cognizant non-life forms with their ontologically evident personhood and their legally supported and promoted private personhood, would seem to get the best of all possible worlds: employing the telemetric when it suits them, evading it when it does not. Because of that, I want to avoid judging harshly right from the outset and calling for some action to destroy corporate privilege or break down the advantages that they have. Instead, these conditions should be acknowledged as facts in the world within which we live, and then they should be studied in detail for a greater understanding of how that advantage works and what it has meant for the promotion and well-being of corporations, as well as what it has meant for the promotion and well-being of human persons. Corporations appear to be thriving in the world today and they would appear to have every advantage. What is to be learned from that? It could be that the community of persons that we find in a world made up of large corporations is exactly the ideal civic structure that the Enlightenment theorists were striving toward. Corporate persons may have achieved a

Utopian world here among us, in the midst of us, and through us; albeit behind our backs, or rather, telemetrically. We should at least consider the possibility that it might be better if all persons had it so good than that none ever should. The project of ending segregation between corporations as private persons and other human persons who do not enjoy the same protections, should proceed with a clear understanding of what those advantages and protections are and whether they are the true basis of private personhood rudimentary to all rights, regardless of whether the subject is corporate or human.

Artificial Intelligence and Neural Networks

Even Delphin Delmas did not argue that corporations are not people. In the case of Santa Clara County vs Southern Pacific Railroad, he restricted his argument to rejecting the railroad's claim to 14th amendment protection in their tax dispute with the county on the basis that they were not a natural person merely an artificial one. "Blackstone says, 'Persons are divided by the law into either natural persons or artificial. Natural persons are such as the God of Nature formed us; artificial are such as are created and devised by human laws for the purposes of society and government, which are called corporations or bodies politic" (Cf. *The Law Journal for the Year 1832-1849*, p. 21). Delmas went on to point out the absurdity of treating natural and artificial alike based on the consequences: permitting a will or a marriage among corporations or between a corporation and a person. The argument rested on the good common sense of the justices to see the absurdity in it. His outrage was doubly fueled by the fact that Southern Pacific Railroad was making use of the equal protection clause –introduced for the sake of protecting former slaves from discrimination– to justify their position that they should not have to pay a small tax on the fences that lined the railroad in one county in the state of California. The justices did not seem much moved by his argument. They found for the railroad, but not based on the claim to fourteenth amendment protection. The historical power given to the protection was based on a preamble made by Chief Justice Waite before rendering the Court's decision: "The court does not wish to hear argument on the question whether the provision in the Fourteenth Amendment to the Constitution, which forbids a state to deny to any person within its jurisdiction the equal protection of the laws, applies to these corporations. We are of the opinion that it does." (Cf. *Tax Cases Heard by the Supreme Court of the United States: October Term 1944 through October Term 1948*, p. 864). The court reporter wrote the following: "The defendant corporations are persons within the intent of the clause in Section 1 of the Fourteenth Amendment to the Constitution of the United States, which forbids a State to deny any person within its jurisdiction the equal protection of the laws" (Cited in Hartman, 2010 p. 23).

It came to pass without discussion or legislation, in lieu of considering the arguments, that natural persons and artificial persons were granted equal protection under the law. With the reconstruction of slavery, we saw the end of corporate servitude. Never more would corporations be exploited by and for the people, as now they were entitled to equal protection under the law. And I would gather that these protections would extend to any artificial person, any person constructed by social process or cultivation. The argument of this book might even suggest that the distinction between artificial and natural is merely convention since all persons are created out of the social soup that intertwines with the action responses of newly formed nodes for the sake of developing them into whatever their position in the world demands that they be. That is not a point that needs to be pushed too hard. The honorable justices of the 1885-86 Court spared the philosopher any effort necessary to untangle the knot separating the artificial from the natural person. Each in kind could be included within the society in all their glory just as they are. The causal chain which spawns the one and institutes the other is surely different, but in the eyes of the law it is a difference that makes no difference.

It is strange how inconsistent that same Court has been, however. For example, when instituting taxation against corporations and natural persons, there does not seem to have been any major outcry as to why the bodies of law that govern each are separate. What justified this segregation in tax code if all were persons under the law? Of course, tax codes are significantly older than the civil rights actions that tore down the Separate but Equal legislation of segregation among *human* people. Still, one would have expected the various landmark decisions to address the notion of taxation just as much as they addressed other laws that enforced the inequality of African-Americans within American civil society. When I, a humble natural person, employ a man to cut my grass and clear leaves from my yard, I cannot deduct this expense from my taxes because my home is for personal use and personal use is not tax deductible. What does 'personal use' mean for a corporation? It makes sense to say that personal use for a natural person amounts to usage that goes beyond any business-related interest that the person might have, but for the corporation there is no downtime, there is no separation of personal use and business use. They are business structures pure and simple and all their interests are tied to a business usage. That should mean that all such usage is dual purposed, both personal and business. And I thought the law is clear on this. At least as far as my home office is concerned, the dual purpose of that part of my house means that I cannot deduct expenses related to it from my income when computing my taxes. Corporations, even though all their expenses are dual purpose, both personal and business expenses, can deduct those expenses from their income computation and so bear no tax liability on that expenditure. How can we live with such injustice? History will no doubt judge us for this short-sighted discriminatory practice.

Because corporations are artificially produced, so the argument goes, they can be engaged in synthetic interest orientations which are deemed extra-personal, or so it would seem. That is, the consensus is that economic and business interests associated with corporate actions are not personal but constitute a new sphere of interest specific to the artificial creature that engages in such activities oriented toward achieving or satisfying those interests. 'Business' just is the name for this synthetic domain of interest. Clearly from a corporate person's perspective that domain of interest is intensely personal to it, it is the domain of its self-regulating determination of personality. When we provide equal protection for a corporation in terms of life, liberty, and the pursuit of happiness, we are specifically liberating it for the sake of pursuing its personal interests, defining itself and making a world for itself within which it can realize those rights and liberties. And the name of those purposes when perceived from the natural agent's perspective is 'business' purpose, or actions done for the sake of one's 'business'. For the corporation this is all personal business, these are all matters for it to take care of in defining itself as the corporate being that it is. All expenditures by corporations are thus personal expenditures. Paying employees, paying rent or maintenance on their properties, shipping, advertising, whatever they do to promote their way of life, their way of living, all personal activities that achieve personal interests. And therefore, if we were to use one single tax code and if that tax code were to be the code associated with natural people, then all these expenses would no longer be eligible for deduction from the gross income of the corporations. Alternatively, of course, it might make sense to place all persons under the corporate standpoint. As Marx was quick to point out, giving the employees a work life balance under conditions of capitalist accumulation is just a necessary cost of doing business. You cannot wear people out, they need to recharge their batteries, they need to replenish the population, and they need to reproduce their energy as a crucial portion of the means of production. And that could justify recognizing that all personal expenditures by natural persons are part and parcel of the business of living a natural human life. In either case, the point is that there should be one set of laws for all persons and that the interpretation of those laws should not be biased to serve the interests of one group over and against another. Separate but equal is a sham and Americans have already clearly expressed that through judicial precedent.

This consistently seems to be the case regarding corporations. Artificial and natural persons inhabit the same social space, live within the same social world, and come across each other in all sorts of social circumstances. The footing between them, however, is not equal. The sheer size and wealth of a corporation is a benefit because of the determination of principles that are permitted due to that size and wealth.

As we saw in the previous section, agreements between corporations can be unfairly tailored to their interests and needs because of this imbalance in power. And, to the extent that this imbalance is financial, it may be that any person that has a financial advantage can preserve their individuality over and against any other that they enter into agreements with. Anyone, corporate or artificial, who has the resources available to have an attorney on constant retainer will be able to negotiate better standing and more self-determination in their dealings. Among the ultra-rich, for example, one of the key discriminating boundaries is between being able to fly privately verses being forced to fly commercial. And at least part of the affront that is present in the latter but not in the former is the standard form of the agreement and service that the traveler signs up for. The private service is more heavily customized to the individual traveler, to their schedule and needs, whereas the commercial aviation system is on a schedule and runs on terms that are not negotiable by its customers. The experiences are thus widely different to the discriminating consumer and being forced to travel in the latter fashion can be a serious blow to their way of life.

This suggests that what gives the corporation at least some of its advantage is present in natural persons as well. What we learn is that it is not in virtue of the fact that corporations are artificial that they have this advantage, but that they have this advantage because they are rich. Wealth and the access to resources that most human beings do not have access to is not specifically a characteristic of an artificial person. What makes wealth so powerful in the advantages it provides is that it makes the natural person capable of being more like an artificial person. That is, the tailoring of custom agreements means that the personhood of the agent, of the actor, can be individually crafted for specific ends and interests. The negotiation team that represents the interests of the rich person is an added set of employees doing the business of that individual. The rich person has added advantage often because they have a staff that is working for them to achieve that advantage. And this achievement constitutes customization tailored to the circumstances, and benefiting from feedback, indicating the nature and quality of those circumstances when crafting the agency of the represented natural person, so that it best conforms to their expressed interests in developing their personality as they see fit. An artificial environment of interest realization can be created or simulated under these conditions. Thus, being artificial is an advantage because it provides the possibility of altering the real world for the sake of carrying out projects tailored to the individual's interests. And this happens on a much smaller scale to people of all walks of life now and again. People will be made to feel special in some random social encounter, as though the rules were being bent to their needs and adjusted to promote their interests. And in that moment, they are getting away with something, they are gaining some advantage. The nature of that advantage is that they are being treated somewhat artificially, where that indicates that they are no longer subsumed under the common categories and attributes that any prior analysis would have predicted for their treatment. Instead, they are getting a custom response to circumstances that were not adequately anticipated, they have become outliers. This artificial outcome comes across as luxury and privilege in the experience of its recipient.

When we resent the behavior of a corporation because it makes us feel subhuman, we are often reacting to the categorical application of general rules and principles that do not address the nature of our needs and interests in the exchange. Corporations, by and large, live by such practices and benefit greatly from treating their customers as generalized instances of a set of population characteristics, while at the same time being treated by their customers and suppliers as a unique and highly specialized structure. They are equated with their brand and their brand stands out as a unique construct with individualized meaning and context. The imbalance that often ensues between a highly individuated corporation engaged with a generic customer is the result of imbalance that always occurs when a private person comes up against a population at large. A private person demands individual recognition and respect whereas the population is a categorical generalization and a probability to behave in one way or another. As a result, the corporation can demand highly tailored action toward it while providing services that are generally and generically aimed (aka cheaper to provide).

In conversations with my boss during yearly performance reviews, I have pointed out this imbalance by suggesting that the corporation is receiving something infinitely more valuable from me than what it is giving me in return. That is, I am dedicating my life and energy, my passion and mental power to the corporation's projects and interests and all it is giving me in return is money. And, from its point of view, this money is a pittance. Hardly anything at all. And yet the imbalance is obvious because what I am giving in return is everything I have and am for some portion of each working day. And the working day includes the 'down time' required to shed this attitude and reacclimate to private life just in time for bed. This imbalance is on a par with that horrible imbalance we get when we consider that a corporation might kill human beings through its actions and, if they are penalized for it, need only write a small check as compensation to the victim's loved ones. The notion of incarceration and capital punishment is unknown to the artificial person not because it is inconceivable that such a principle could be translated into its ontology, but because the artificial quality gives it the advantage of bodylessness and thus makes it seem as if there would be something wrong with trying to perform the translation. The imbalance between artificiality as such and naturalness as such can be offset to the extent that natural persons become more and more artificial. Celebrity, for example, is an artificial construct that may augment a natural person and make them more like a corporate brand. The magic that this augmentation brings can have legal effects where it is no longer conceivable to treat them the same as other people would be treated in matters of punishment. How could you send such a famous person to jail? It seems like such an institutional violation of their enormously specific being. Individuality comes across as artificial in this sense because people, natural people, know that they are just like everyone else, that they are generic and part of an analytically determined demographic group that expresses itself through the same determinations and decisions. The celebrity, like the well-known corporation, has artificially broken free from this generic mold and converted themselves into a highly specific being. As a result, they could not be treated the same way and endure the same punishments in response to otherwise criminal activity.

It is specifically the artificial being of persons, or the extent to which a person can approximate such artificial stature, that gives corporations and others their distinct social advantage. The name we have been developing for this augmented reality is 'private personhood'. Here it begins to feel more like an injustice than the platform for liberation that we had hoped it would be. It is easy enough to set that aside by suggesting that the imbalance is the form of the injustice, not the thing itself. The nature of relationships and associations, the liberation from telemetry, that is at the foundation of corporate existence and makes it seem like an imbalanced circumstance because it is realized in a social environment where most persons do not exist in that way. We are mostly nameless and faceless to the corporations that we interact with, they do not know our names whereas we know theirs. The imbalance of power is that, on the one hand, you have this highly celebrated being known throughout the land and used by many for the services they provide, whereas, on the other hand, you have random joe or jane who is merely a customer settling up each month for their generic terms of service, the same terms of service that everyone else is getting. And this generic structure, this generalized involvement, becomes itself the focus while becoming the definition of fairness. The corporation or the highly specific person are bracketed out of the relationship, excused from the comparison, because they are obviously of a quite different kind and so the general population is deemed to be on an equal footing in their unequal relationship with highly specialized beings that come their way with goods and services for sale. Thinking that this status has been earned or that it is the logical upshot of true meritocracy indicates that the accumulation of power by those capable of expressing highly customized individuality is itself an acceptable and even desirable outcome of rights based civil organization.

If artificial persons have a uniqueness about them that is due in part to their artificial nature as persons, it might make sense to recognize that the cognition that goes along with such artifice is most reasonably called an artificial intelligence. It is trendy to think of artificial intelligence as belonging to a machine, but even our contemporary research in the area gives the lie to that. The intelligence associated with artificial cognition is often spread across many machines, living in a network between them, stringing the nodes together to derive notions and procedures and to invent processes for carrying out further tasks where those tasks may be spread across those multiple machines. Because such intelligence need not be localized in a singular robot that looks just like all other robots manufactured for the same purpose, but might exist in a unique singular structure that comes to know itself somehow and engage in projects in some fashion, we bestow this notion with a lot of imagination and extrapolation and think that artificial intelligence is somehow new or coming on the future's horizon. If artificial persons have been with us for centuries and if artificial persons are capable of rights and interests while carrying out plans and performing operations, then it would seem that artificial intelligence has also been with us for centuries and may amount to the manner in which the artificial person leverages all of its available resources for the sake of meeting its needs and furthering its interests.

If one were to go this route and recognize artificial intelligence in the context of artificial personhood, then a lot of the discussions that have been going on in recent decades might seem childish. We debate whether artificial intelligence will introduce new and different ethical problems that we will need to address one day before that fateful singularity comes to be. This is open to ridicule if the world where the questions are raised turns out to be a world already widely populated for hundreds of years with artificial intelligences running around unrestrained by moral principles. It is beyond obvious to point out that the corporation, as an artificially intelligent being, has no ethic other than the promotion of its own interests in their regard of natural persons. And in all due fairness to such beings, it may be the case that this is the best that anyone can do. That is, why blame corporations for not following the categorical imperative when it is at least plausible to deny its effective power over their form of cognition. The rational being to whom the categorical imperative is applicable is itself constituted by a specific form of cognition and no work has been done to establish that this form of cognition is common to the corporation's artificial personhood.

The corporation advances its interests which does involve concern for those with whom it deals. If the customers are upset about some business practice or other, or the quality of some product, the missing features in some service, then the corporation will respond and it will respond in such a way as to make sure the customers get what they want so that they remain customers and continue to purchase the goods and services the corporation provides. This real-world orientation is a legitimate ethical position. It would be wrong to suggest that corporations are without ethics. It may be, however, that their ethics differs from that of natural persons or persons with the cognitive structure of human beings. Not every natural person shares the same ethical insights or follows the same moral compass. And it may be that some artificial persons with their artificial intelligence are more hedonist than others, but that does not prove that all corporations are without ethics, only that some may cheat or that they do not include such ethics as part of their 'culture'. And even if the ethics that we find in corporations would seem to depart wildly from what we would hope to find in a natural person, that does not merit the accusation that they are without ethics at all. Any being with interests, any actor that performs targeted operations for the sake of ends in the world and which is governed by rules, may be characterized by ethical principles functioning as a kind of feedback loop. They make value judgements, they assert preferences for one course of action over and against another based on impact and circumstance, they recognize the existence of others within their world, and they evaluate how to treat those others based on the various projects and interests that they are currently trying to realize. These are all ethical concerns and indicative of an ethics at work. The ethics that a corporation or highly individuated natural person may use might turn out to be vastly different from the ethics we find in a generic or universalized person, but that is no evidence against the presence of ethics. Rather, it might merely suggest that various persons mean something very different when they use that term and that any human understanding of corporate ethics is always, therefore, partial.

Nietzsche might be a helpful guide in these matters. Imagine that the highly private person augmented and protected from socially coercive forces is a perpetrator of a master morality, whereas the generic others, those of us in the masses of the aggregate population, we are all governed by slave morality. This is a perfectly plausible narrative and it does not mean that the masters are without ethics or without morality, it just means that their morality is hostile to those who cannot live in a world with such standards: those who are not private persons and not capable of the actions reserved for such people. The lambs hate the eagles because they feed on them and the eagles love the lambs for the same reason. And in both cases, this functions as a set of values and as a set of moral principles that guide the actions of the persons involved.

I will have to defer any prolonged existential analytic of the corporate artificial person, but it is at least fair to say that artificial persons do not die or have a relationship to their death in the same way that natural persons do. If they are corporations, this might be because their life span can go on for much longer than is naturally feasible; and if they are natural persons who have attained individuality, then they will live on forever through their notoriety, they will take their seat by the gods and endure as immortals even if only in the minds of the public as opposed to some biological form. Make no mistake, however, they are working on that. To claim that such artificial persons are proponents of master morality should not be thought to indicate that they all share in the same form of ethical orientation. The point is that their artificiality enables them to produce their own orientation toward their interests, to perceive those interests as they will, and act to promote them on their own terms. They are not acting in response or reaction to something larger or more powerful that encroaches upon them. Rather they are acting from their own sense of overfullness, their own sense of self-determination, and the creative capacity to become what they truly are. And in some cases, this may be truly awful, vitriolic and hateful, manipulative and abusive: the purpose all the people in the world have in common is to serve at the beck and call of the master at work. It is likewise possible that they may evolve as a gentle giant filled with love and compassion for the needy masses who suffer under the weight of their collective lives, their lives dissolved in the collective sea of generic personhood, of a not non-private life as it has been aggregated and telemetrically understood.

The point is that either of these extreme approaches, or any of the variable spectrum that stretches between them, is presented through self-determination. The artificial person is able to make a meaningful decision on how to be and how to orient itself, whether it wants to act on behalf of the interests of others, become a protector of those interests (while serving its own of course) or whether it wants to realize the hedonistic life of the vampire who lives the more by sucking the blood of others. We do not need to theorize about the nature of ethical principles within the domain of artificial intelligence, we have living proof with centuries of history behind it. And what we have seen is that artificially intelligent beings will exhibit the ethics and moral behavior of a master, acting for the sake of promoting itself and what it interiorizes as its own best interests. Some masters may think the world of culture and civilization, the old Europe and all its sophistication, is the model for high living, and they may think that whatever can be done to promote such a world, to build it and develop the people to populate it, is for their own sake and benefits themselves because they get to live in a better world. Others might think the good life is what you find in the hotels on the sunset strip in Las Vegas and they do whatever they can to make the world that they are familiar with more and more like that.

Probably if you trace the histories, look deep into the conditions and narratives of the various artificial persons, you would see a yarn unfold that best describes how they came to have this master morality over whatever others might have been an option for them. You might find that the person who started the corporation, or some important CEO that they had at one point in their history, had a set of views or ideas that somehow found their way into its mission statement or guiding leadership principles. And in that case, we might see that artificial persons are susceptible to the influences that come from the natural world that contributed to their development over the years. This would be an important set of insights to cultivate. Not because we would learn how to change or impact the corporation's development, but because we would see that the artificial being comes to have its make-up in the same environment that we all have found ourselves in the middle of. It would suggest to us that the possibility of becoming an artificial person, of establishing oneself as an artificial person, is a process that comes to be naturally (sic). It is not necessarily the original state of any person but might be the result or outcome of a development process. This was Nietzsche's insight, you did not need to be born this way, you could come to be this way through activity, through guided action that aimed toward such an outcome. The slaves were capable of a slave revolt and could, through it, become masters.

Being artificial is not a radical departure from the natural state, as we have been describing it throughout this book. This is a focal theme of the essay: that there is something of the artificial that is always already at work in the so-called natural processes of human beings. Not only are our corporations the result of human action creating an artifice, but human beings themselves are the result of human action, socially organized and collaborative. We are products in the way that our corporations are products, even if we do have a different origin and even if the birthing process is wildly different for the two of us. As such, artificial and natural intelligence are not far apart from each other. What we see among the ones who are highly individuated and specific and what we see among the ones who are generic and just like all the rest, easily accommodated through the metrics and easily anticipated in need and desire through the analysis that we perform on them, is not so much a difference in artificial vs natural process, but a difference in the way in which the artifact is protected. How the creations are crafted, what resources are applied to the cultivation, and the social investments that are made in ensuring its quality, these are the differing factors. Some kids go off to public schools which are tailored to anybody and everybody. Some kids go to private schools which are exclusive, expensive, and offer education that is more closely tailored to the details and specific needs of the child or young adult. The methods of crafting persons may be oriented more toward creating specific and highly individualized persons or they may be more generally oriented toward creating one of many, toward a mass consumed service for providing all with a least common denominator approach oriented toward the most ordinary ways of thinking and responding to situations and circumstances.

The question that plagues us, that has probably plagued us for as long as there have been people, the question that we all know lies in there somewhere is twofold: 1) is not it better to be a master than a slave? And 2) Can there be masters if there are not any slaves? And the horrible truth of human being is that in some basic layer of our understanding we all grasp that both questions have obvious answers. The answers do not commit anyone to a plan of action, however. We can know the answers to these questions and nonetheless retreat from them, proclaiming life as unfair and humans as capable of rebelling against the inevitability of these depressing conditions. That response is all beside the point and we will not get anywhere if we do not start from a position where we recognize and see the brutal reality, so that we can understand how it divides us in various ways, splits us up, and puts us at odds with each other. It is better to be a master and masters must have slaves. Aristotle knew it. Kant knew it. Rawls knew it. We would not need any of the ethical theories presented by these philosophers if we did not all know it. I suspect that it is this rudimentary understanding of the basic conditions of human life that lies beneath the fear existing in the domain of technologically based artificial intelligence. People who realize this can only comprehend such a resource, such artificially intelligent beings, as either slaves or masters, and many do fear that the natural tendency of such creatures would be toward mastery. That puts human beings at risk (and corporations too) because it would make all existing persons into the slaves of the newly evolved artificial intelligences. Again, I am not nearly so worried about some future event that may or may not affect this state of affairs because this event that we are so worried about, that we devote such time and resources to asserting or denying (they will be our masters, no they will be our slaves, or, the incredibly absurd view developed with little historical awareness of human social organization, we will all be augmented masters and the technology will be our slave), that event has already happened. Corporations are those artificial intelligences and they have been with us for many centuries. Which are they? Are they the masters or are they the slaves? Have we all equally merged with them or are there only a lucky few that can claim that privilege?

The Rights and Responsibilities of People

Unlike the former slaves who were the prima facie beneficiaries of the Fourteenth Amendment, the corporations that gained protection from its concrete interpretations gained enormous value from the resulting equal protection. No doubt there were many who wished they could have had the same level of private personhood that the corporation achieved through whatever magical means were applied to the decision of 1886. When I go to work each morning, I leave whatever semblance of the legal right to privacy that I have inside my home. Or, at the very best, I leave it in my car in the parking lot at work. Although I am a little fuzzy about that since my employer does own the parking lot. The fact is that my employer can read all my emails, surveille my every act, demand drug tests, all for the sake of protecting their interests in so far as I am a labor resource for them to use in exchange for the salary they pay me. My employer, on the other hand, does not have to let anyone go through their books or gain access to their email servers without a court order. They do not need to let federal regulators come onto their premises without permission. Strangely, and in the name of protecting employee privacy, they do not have to share with me any aggregate data about my colleagues that might help me to evaluate the fairness of my own compensation. Truly, I am jealous of the extent to which their private personhood is secured from socially invasive actions and processes, even when they are carried out by the government in the name of the people. In their presence, I too feel like a fraction of a person. It is commonplace to rage against the injustice of this imbalance between the artificial and the natural person. What remains a curiosity, is whether this establishment of private personhood, this real-world example of a private person, yields such agents an ability to drive and control their fate relative to the telemetric understanding. Does this private personhood give them the potential to be ethical beings (in the natural sense) that can control their fate and act as they see fit independently of the social relations that might play with their actions as part of a larger scale set of operations beyond their capacity to change or control? Does private personhood yield the freedom of thought and autonomy of experience that we have been searching for all along? And if it does, who exactly has membership in its kingdom of ends or consideration in its computations of general welfare? Who are the citizens of the commonwealth under these conditions of telemetric understanding?

In the previous section, we seem to have come to the conclusion that it does have that capacity, that its size and power, its artificialness, gives it the means to demand special attention and treatment, and that this enables a master morality that allows it to "be as it will, live as it likes" without concern for potentially invasive social responses. Of course, that conclusion must be immediately tempered. We know that corporations cannot just do whatever they want, they are bound by some principles and constraints, the attitudes and impressions of their consumers or of the people at large can cost them dearly. One might argue that the technology companies have persistently had to navigate this love-hate relationship with the public. Over the last few decades, some companies that were once dominant in an area have fallen from that position due to their own arrogance and the public's reaction to it. The Microsoft Corporation, for example, once dominant in the Operating System market may have thought it inconceivable that there would come a day when their Windows application would not be the dominant product in that market. That arrogance drew the wrong kind of attention and the governments of the western world acted to remove what it referred to as monopolistic practices from the corporation's bag of tricks. This greatly slowed down the company in terms of both innovation and day to day operations. While it was bending under the publicly imposed burden of a consent decree that forced all kinds of compliance with documentation and public transparency for their protocols, other companies were able to race ahead in various areas and become larger and more influential worldwide. Corporations are not invulnerable, and their behavior can turn their customers against them. Their actions may not only bring about huge financial losses, they might cause a massive sea change within the company such that it is forced to reinvent itself to stay alive.

Larger corporations may be able to withstand the impact of such events and may have the resources to orchestrate the change, but corporations come in all shapes and sizes and the smaller ones may not be well disposed to weather such storms. They can be destroyed by a tragic event or a simple accident that so drastically impairs public opinion against them that they cannot recover quickly enough to continue making their payroll or keep their existing customers. The corporation is vulnerable to social forces and relationships, the smaller ones more so than the larger ones, and this vulnerability indicates that they must be social beings that live according to rules that make them acceptable citizens within those social relationships involving their customers, employees, and other companies. Asserting that a corporation must be socially conscientious, however, does not suggest that they have failed to properly situate themselves within their society as respected private persons with ample protection from socially coercive forces exercised around them by other corporations or by governments. The advantages of private personhood need not be established as a perfect cocoon of license yielding the ability to do whatever they want whenever they want. The point was to ensure that there is ample protection from the social forces so that the persons were capable of self-determination. We are asking whether each of them is protected as a person who can develop their attitudes and opinions, work out their conscience and act upon it regularly, express their creativity without threat of attack, develop their ideas in private without surveillance, and –ultimately– explore the universe and make up their minds (sic) on their own? Do they have the capacity to engage authentically with what is out there in the world and what it all means, without being bombarded by a set of interests moving them one way or another? Do they have protection from external powers that can make such bombardment appear as though it were simply an objective world being discovered as facts? All this, I take it, is the integrity of the private person and the question is whether the artificial persons in the world have enough protection to be so empowered even when their natural counterparts do not?

It is tempting to say that they are. And part of the reason for this is that they can escape their facticity or thrownness in ways that human beings are not able to do. Corporations, under the law, have the real and tangible rights that governments were attempting to afford natural persons, and this gives them an advantage. In addition, because they are not the kind of beings that natural persons are, the governments have been lax in assigning those corporations with many of the responsibilities and obligations that natural persons have in return for their rights. Every crime that a corporation commits, no matter how serious, is effectively treated as a misdemeanor since the courts have asserted that there is no such thing as incarceration in the case of a corporate person, because suddenly they need to be literalist about what incarceration means. The death penalty, common in many of the United States, is unheard of for corporations, since the courts are also literalists when it comes to what death might mean in such a case. The notion of personhood can be functionally understood when it comes to rights, but it must be literally understood in terms of a body in the world when it comes to the accountability and responsibility that are applied to the actions of the artificial person. The effect of this facet of legal application is that the corporation is always able to escape the implications of its past and has the potential to at least escape from the ill effects of past actions and recover from them. Human beings are not destroyed by misdemeanor offenses, they do not throw us "into the system" and force us to carry the burden of that for the rest of our lives. Misdemeanors do not need to be reported faithfully to anyone and everyone we seek to work with throughout the rest of our lives: felon or ex-con. A corporation is never weighed down by such a past: its participation in electoral politics is never withdrawn based on past crimes committed. They have a brand and that brand's history can be tarnished, but they can act against that effect, they can launch a campaign and try to change the mind of the public turned against them. The weight of such things is informal and loose, not legally applied, so it can be fought against and the corporation can hire smart people to figure out new and clever ways to reinvent itself, thus escaping the burden of its past.

The corporation's ability to develop and discover a fresh and innovative sense of the world and how to pursue and further its interests within that world is protected by law; and the restraint that governments exhibit in holding corporations accountable for their actions enables them to maintain a lightness of character that gives them the flexibility to maneuver easily through the obstacles and hardships that come to play throughout their lives as artificial persons. Having the ability to extract itself from its thrownness, to circumvent the burdens of whatever history it has had, at least formally, gives the corporate person an enormous edge and advantage. It must concern itself with its projected ends and it experiences limits of finitude, but the experience of its past and of its state of being in the world are such that its past does not weigh it down, and thus all sorts of future possibilities remain open to it without the encumbrance natural persons are so vulnerable to. Companies that have wiped out entire sea life populations with oil spills or shown complete negligence in how they prevent their cars from exploding and killing customers, are still able to get past these relatively minor inconveniences and endure through it to the next business cycle. This does not mean that the corporation does not occasionally get into trouble over its past decisions, but they will tend to be tied out to past decisions and actions performed by the corporation itself. A small corporation that takes on too much debt to try and develop into some exploratory area of concern might fail miserably and have to declare bankruptcy. The social circumstances in which they found themselves might have been to blame in some sense, but the corporation will have to face the fact that it made a bad decision and acted poorly in its efforts to understand those social forces in starting their new venture. In short, the consequences of bad action tend to be a direct result of bad decisions or lacking talent on the part of the corporation. The opportunity was there, but the artificial person was insufficiently capable of responding appropriately to the circumstances. The freedom of the private person is not the same thing as invulnerability, rather it is a freedom to fail on its own terms based on its own efforts, inadequate or exceptional as they may be. The fact that corporations can succeed or fail on their own merits is already an advantage over huge segments of the world's natural population.

We suggested earlier that the master status of the corporation and any other person with enough material resources to be treated with the special respect that private persons can command, may be dependent on the existence of those who must follow them or live as slaves to them. What I mean by this is that if the artificial person, the augmented person, can achieve the level of integrity that characterizes the truly private person, then it may be that this status is dependent on the fact that there are other persons who are not in this state. For the corporation to have achieved this position, there must be consumers and there must be employees. Both must be pliable to aggregate operations and targeting. A corporation that had to treat every one of its customers or every one of its employees as though they were private persons, so the conjecture goes, would be incapable of maintaining its solvency. The health and well-being of the corporation depends on the fact that it can have generically available employees and generically understood customers. Businesses that must treat their customers as though they were private persons are significantly more expensive to operate than businesses that can treat their customers according to generalized rules as generic instances of the demographically determined consumer. The service that must be provided to a singular person with highly specific interests and needs must be so closely tailored to those interests and needs that a very special attitude will be required by the corporation providing the service. Their customer base would not be very large because the market forces would end up pricing most people out of the services that are offered. And to some extent, this might be demanded by the customers themselves who may only be able to experience the highly customized nature of the service if they are able to see it as part of an exclusive offering. They may need to experience the service as highly customized and as such only meant for them. The notion of exclusivity would have to be baked into the specialized nature of the service.

Not only would the customers pose such a problem, the employees would too. The CEO of a corporation is arguably the only employee with enough status to be thought a private person. That may be an overly strict interpretation of private personhood. The presence of certain kinds of pressures to perform in the work place could potentially mean that even the CEO is not fully realizing their self as a private person. The CEO cannot explore and discover a world as they see fit, they are not free to self-determine their personality as the world within which they are living allows, rather they are highly constrained agents of the corporation who have likely reached that position because of their ability to adapt to its needs and goals. They have submitted their way of thinking to the corporation and even at the highest levels such employees may be so cognitively immersed in the power structure of the corporation that there is nothing personal or private left in them. It is entirely plausible that the individual contributor at the bottom of the corporate ladder is better off in this regard. It is easy enough to maintain one's status as a low-level cog and secretly think whatever you like about your job, your company, and your place within it. In that case, you must submit to the micro-management of those above you. The surveillance of your person might take place via corporate infrastructure and automated policies, and you are part of the feedback loop of corporate activity in the physical sense even if you can maintain your private thoughts for yourself. On the other hand, just as the CEO's mental state is what places them into the feedback loop of the corporation's automated predictive activities, the individual contributor could become just as governed by such concerns. Such a person needs that job, would be lost without it, would be impoverished, and their family would suffer greatly. Such is either really the case or at least perceived to be the case, and so could result in the individual contributor having to devote the better part of their mental energy to doing the best job possible. Employees who are so vulnerable to their employers may have to trick themselves into achieving an unrequited love of their jobs that is essential for properly applying themselves to the corporate mission while remaining –in some sense– essentially removed from it as natural persons living within the structure of an artificial person's life and serving its goals.

The flexibility and integrity of the corporation may therefore seem to rest squarely on the requirement that there exist employees who succumb to the artificial orientation of the corporation where they work, and that there are customers who can be aggregated into predictable behavior that can be targeted in various marketing campaigns or sales initiatives. The slave morality of the consumer with tastes that are sufficiently generic that the corporation can construct a regimen that produces what they will need now and in the future, is a necessary resource for the corporation as it asserts its private personhood over and against the social conditions that have brought it into existence. And what is more, the corporation, without any impetus for action of its own, depends on employees to initiate the embodied articulation of its interests and orientations. It must have these crucial ingredients in order to survive. They are its blood and its heartbeat. The master would not be a master without its slaves to do its bidding, to enable its mastery over the conditions of common concern to master and slave alike.

To the extent that corporations require this offset of power positions to do their bidding, to carry out their projects, and further their interests, they can be said to have no conscience. The most common state to discover in corporate-think is that this enforced master-slave relationship between employees, customers, and itself is providing a social service that is beneficial to humankind. Its conscience, therefore, is its assertion of itself as something of a savior in the social world, offering opportunity and livelihood where otherwise none would have existed. The cost of being an employee, even a well-compensated one, is dismissed. The cost of being a customer receiving a generic product usable by anyone with common interests or needs is also dismissed. The corporation sees the customers and employees as bodies that have no soul, no spirit, and no mind. In short, the corporation largely thinks of these stakeholders as pretty much like itself. It treats them as interest clusters with orientations that fit into such visions of a creature: it is fed, it is housed, it is clothed. What more could it possibly need? The notion that a being may wish to escape its past just as the corporation is able to do, that a being might wish for a genuine sense of privacy and security, just as the giant corporation that surveilles its users has attained, none of this surfaces to the corporation as a principle for action, or a daily concern that it must act upon for the sake of promoting the well-being of all in support of the commonwealth.

The truly large corporations, the largest among them, have worked hard to counter any possible emancipatory interest that such characters might have in trying to become private persons. Corporations have sought and received the legal status to participate in political campaigns to further their own agendas and work toward getting political representation that will steer the law of the land in their favor. There are no limits placed on their contributions even if natural persons have them. They have sought to reduce the capability of workers to engage in collective bargaining, they have worked to reduce the privacy rights of employees in the workplace, to reduce the privacy rights of customers as they wander through the digital marketplace, they have worked tirelessly to reduce the number of environmental regulations that are put in place to prevent them from polluting the world or irresponsibly consuming its resources, and they have spent a lot of money to ensure that their tax liability is as low as it can be. They have worked hard and tirelessly to try and build a social world that supports the integrity of their position as private persons in the world feeding off the coercive forces they enact upon employees and consumers that fuel their interests.

The schools that we have, the political position of the churches that we have, the media outlets that exist, the marketplaces that we have, all the major institutions of the social world, all those organizations and institutions that we said earlier had such influence in the creation of persons, are working with funding and influence from corporate sponsors who seek return on that investment through conditioning from those institutions that will further the interests of the corporations themselves. We do not learn random things during our young lives, we are not learning to fish and hunt and farm to stay alive, we are not learning the tribal rituals of our people. We are learning what is the best thing to buy for this or that need, we are learning to be comparison shoppers, we are learning to have preferences that can be measured and used to better market to us in better support of our needs in predictable ways. And furthermore, we are being chased into STEM-based education not because this is where the secrets of the universe lie, but because this is where the corporations need us to be. We are corralled into such vocations because they are where the innovation is required for producing a better battery with more life per charge enabling people to use their smart phones for longer periods of time without needing to find access to a charging station. The private personhood of the corporations has been achieved and it has been achieved at the expense of the entire civilization of natural persons.

I have read an argument that suggested that the cephalopod is the closest that human beings have ever come to meeting with an alien race here on earth. Just as I previously sniped at the artificial intelligence community for failing to see that corporations have long been an artificial intelligence here at work in the human world, I also think that the cephalopod's alien nature pales in comparison to that of the corporation. To the extent that historical ethicists have always bound the action orientation of ethics to the rational structure of the agents engaged in such action, the alien nature of corporations suggests that they are bound by a different set of rules and driven by an alternative set of directives. The critique of corporate reason entails an independent critique of corporate practical reason and the principle of corporate utility entails an alternative conception of generalized corporate well-being. To the extent that ethical relations are bounded by social relations evolving throughout the passage of history, alien beings will have been participating in a distinct set of such relations and would have developed a very different set of practical responses and normative orientations. Living among us, they have had alternative lives grounded in a radically distinct form of experience. The telemetric understanding connotes a distinct phenomenology and a distinct commonwealth at work conditioning the private persons fostered within it and by its constitutive reason: emission, analysis, and feedback.

In the early 20th century driven by the innovation of Henry Ford, human beings may have been just as much at the mercy of corporations as they are in the early 21st century, but there seems to have been a difference. Ford's revolutionary notion that his employees might also be his customers at least gave human beings a chance. When considered as an amalgamation of customer and employee it was possible that the corporation might treat such stakeholders as private persons and grant them the respect that they deserved. Only a brief 100 years later there is a growing sense that this is gone and that corporations no longer see it this way. There is a clear indication in the corporate motivational literature that we are at the dawn of a new era in corporate profits and productivity. We are about to be freed from the burden of human error in the production process. Machines and algorithms are about to spawn a 10x (or more) event that will make it possible to produce and innovate like never before. All the imperfections of a human being operating a fork lift, or driving a car, or writing a piece of computer software, all these imperfections are on the verge of being removed from the process, are on the verge of extinction. The corporation teeters at the precipice of a singular event, a moment in time in which it can sustain and promote itself, possibly at all levels of administration and activity, without the assistance of human beings. This may not be fantasy, it looms as a real possibility on the horizon. The talk of it, the fantasizing about it, is in any case very real and shows the true colors of the state of mind behind it. The extreme advantage of the corporation as private person is insufficient, its private personhood must be enhanced even further, and that means purging it of any possible impurity. The impurities are the natural persons living in the works.

One standard response to this observation is that, if it is true, it is short-sighted of corporations to move in this direction because they are killing the goose that lays the golden eggs. The employees do not have any money, no real money anyway, as the discussions about income inequality and wealth inequality are pointing out. The real wealth lies in the hands of very few people. As long as a corporation continues to feed its productive interests, it can continue to grow. There is a universe in which the expanse of machines is done solely for the sake of the consumer capabilities of the owners of vast wealth, where the mere consumer who is dependent on a wage-based income is no longer of any use to anyone at all because they are a drag on the demands and interests of such private personhood. What with all their gawking and their chattering away about the comings and goings of the rich and famous. Eventually, the rich and famous, functioning artificially, may come to realize they only need the gawking and they will liberate us from our servitude as both consumers and employees. To continue to see the economic dynamics in the way Henry Ford saw them, where the citizen customer employee was the king or queen of the corporate agenda, is to live in an outdated world that does not comprehend the way in which private personhood has evolved in the last 100 years. If there were no people behind the stock digits that rise and fall, if the "value" of a corporation had no human beings participating in the equation, the Dow Jones industrial would still be capable of rising and falling. The value of corporations is not bound to human agency, it is bound to the markets and the markets require less and less human involvement in any direct fashion. An algorithm can buy and sell, and another algorithm can buy and sell from it. Together they can form markets where prices go up and down without any human involvement making that happen. Machines can slap communication devices onto other machines to cause a demand for communication devices. There will be markets for steel and silicon, markets of all different sorts, and very few of those markets will require that a human being participate for the producer of that part to be successful or to increase their productivity.

Insofar as natural persons are just part of a larger order generic structure, part of the telemetric understanding, the artificial persons can manipulate that larger scale understanding for their purposes. It is a mistake to think this is something "we" are going to have to worry about in the future. The indicators are that this has been going on for centuries and that we are right in the middle of it. The telemetric understanding would appear to be at the beck and call of the private persons that inhabit our world. It would appear to be climbing up its growth ladder through the process of using up human beings to achieve higher order ends that make it more and more powerful as it secures the integrity of those private persons, the ones that really matter, through the instrumentalized agency of the persons who function as generic consumers and components of the corporate machines. I have suggested that this is something of an alien among us. That is because the private person does not appear to be on the same planet as the natural person who cannot achieve a perspective that grasps the telemetric understanding at work in the lives and social relations of these artificial and corporate persons. It is tempting to suggest that corporations are not like people and do not have experiences in the way that people have them. This is to fail to recognize that aliens will always appear radically different. This is what makes them alien. The corporate experience, at least with the "smartest" corporations, the ones that can achieve awareness of the structural dynamics that surround them, is of a world belonging to the telemetric understanding. The biggest players, the giant 5 that are worth 4 trillion dollars between them, are the true leaders of this world, and their experience is an experience entirely alien to any human being, beyond our comprehension in its details and nuances. This understanding does not have the best interests of the employees / customers in mind. This being will emit whatever it takes to make the natural persons compliant in their roles furthering the goals and interests of the artificial persons. The telemetric understanding is the closest we can come to articulating that alien experience. Any project that attempts analysis of this phenomenon is vulnerable to Kojeve's criticism of Spinoza: how could you see this; how could a single natural person gain the insight needed to tell the tale? Not so much a cracked pot as a broken clock.

The condition of natural persons relative to artificial persons reminds me of the story of the cows being monitored for their stress levels as they are led to slaughter. The insight was not that the cows are stressed and maybe we should stop leading them to slaughter. Rather, the genius of Temple Grandin was that she saw how to lead them to slaughter without unnecessary stress. It does not matter that her insights were driven by compassion, whereas her employer was interested in avoiding the ill effects of stress on the meat. Leaving aside whether this reveals that there is greater autism in her employer's concern for the meat than in her concern for the cows, we can observe a deeper cruelty at work in this brilliant solution tailored to a narrowly defined problem space. It looks like a lot of problems characterized out of customer "obsession" when working within a corporate environment. No matter how you slice it, in the end, the cows are still being slaughtered and these innovations have not been for their sake. They were right to be stressed, it was an entirely correct and reasonable response to what was happening to them. Fuck them if the meat tastes weird: low all you want guys, stampede if you must, do whatever it takes to stay out of the slaughterhouse.

Chapter IV: The Residual

The residual of telemetry is privacy as expectation. The residual of phenomenology is the telemetric understanding that undermines any notion of privacy as self-determination within a socially coercive world order. What is the residual of commonwealth if the only private persons that remain are corporations fit for ethical action only in so far as they have achieved a state of mastery over and against an employee and customer population absorbed in the objectives and operations of the telemetric understanding? Privacy, in the residual sense that follows, suggests the power of the person to freely associate protected from coercive and invasive forces that shape and cultivate the actor in an opaque manner. Such protections are the source of this power and provide support for enabling the true individuation of the agent. It is the commonwealth that must maintain such protection as its residual. A social order structured according to the rights of corporate persons, for example, will provide such corporations these protections for the sake of their invasion of the personhood of those upon which they depend for their livelihood. Infantilization amounts to the appropriation of the product's associations for the sake of the producer. Not found in nature, such privacy must be fabricated.

Every day I get up and go to work for a large corporation, every day I have submitted my energy and cunning, my intellect and passion, to efforts that have nothing to do with me. To excel, I convince myself otherwise, I make its goals my goals and its ways my ways. I make its projects my own and put myself into the outcomes as though my life depended on it. And for this submission I receive a salary, generous by most standards. What do I do with it? I consume, I buy this or that necessary item to keep the household going or I pay for a brief entertainment to help me unwind and recover my spirits so that I am rested and refreshed when I return to work the next morning. What in my life is not employee or customer? When I spend time with loved ones, we consume, we go here or there to engage in some fun activity that helps us collectively unwind for whatever trials await us on the following day. We develop routines that orient ourselves toward the pattern of week days as scheduled pieces of the work week. The very notions of work week and weekend have been completely colonized alongside the notions of consumption and employment. Even in those moments when I convince myself to extract my frame of reference from this rationalized world and direct it simply and wholly to the people that are close to me, I sense that even this respite has meaning only in so far as it references the primary dynamic driven by these cycles of employment and consumption.

This is a yarn that I have been spinning and it has been a reasonably long one, demanding a lot of patience from you. And this would appear to be its final objective, its upshot and goal. Where is it? Where is what remains? To me, the story suggests that human beings have committed a great crime. Not that each of us has been a perpetrator, but that we have been collaborators in the sense I described long ago. This is the world that we have made, that we are making, that we have agreed to, that we have taken up residence in, and which we are either comfortably or uncomfortably spending our lives for the short while that we have one. Look around, it has all come to this. I do not understand the explanations trying to penetrate this bewilderment, they make no sense to me, strike me as originating in the same omnipresent and coercive understanding that I have been describing. Some would have it, for example, that there is a reasonable apologetics for the corporations and the role we have allowed them to take. Replacing our gods and our kings and queens with corporate power and its seemingly omnipotent private personhood royally situated in the commonwealth, dominating the marketplace of ideas as well as the marketplace of goods and services, all of it justified somehow. To the extent that the American Revolution was a war against the East India Company and the influence it had come to have over the lives of the colonists, and to the extent that other revolutions like it have been aimed at wresting power from some overlord or dominant being that ruled according to their whim, it would seem that the spirit of a truly human being is long gone, co-opted by the advance of corporate interest and influence over all the conditions of life, the societies within which we organize ourselves, and where we live our lives. In search of an emancipatory interest, in search of a sense of self and privacy that might provide a happy ending, I am struck senseless by telemetric forces slamming against the naïve and idealistic nature of my project. In the face of interest structures that were above and beyond anything that I could perceive from whatever perspective I have achieved here, I sense that the apologists do not mean what they say and may not be aware of what truly drives them. There is a lurking beast and we are prey to it.

What book is going to say anything at all that is helpful to members of a civilization who live under the guise of corporate power expressing telemetric understanding? There have been great books in the history of human beings, there have been great messages as well, and they have proliferated and taken root among different peoples over different times throughout history. What chance could there be for such an impact now? The global telecommunications that belong to the machine of telemetric understanding is a constant messaging, a constant amplifier of messages, of requests and responses, emitted from everywhere and bouncing around here and there, catching on for a moment in pockets around the globe only to disappear to all memory moments later with some new idea or thought or brain fart that carries the attention off into another realm with another purpose. And who does this serve? Who benefits from this absence of depth, this impossibility of impact? The deeper social relations that have already been established can only be reinforced by such a dynamic, they are never challenged by anything that originates within them. Even revolutionary politics is just a momentary blip on a feed that is constantly moving past any such messages and onto other messages, presenting everything side by side and continuously replaced by yet more. Even this message mourning my own meaninglessness and conflating meaningfulness with telemetrically focused attention, trite and often heard in variable form, is just like that. I told you I was going to issue a warning and then immediately succumb to it. What could possibly remain in such a case? What is left when words, messages, and learnings themselves can no longer have any kind of impact on a person who has succumbed to the total all-consuming power of their role as employee, if they are lucky, customer, when they are lucky, and center of attention, when they are loved by the gods. And it is the hope for such luck that still rules over us more so than any experiences might. We fight to be so lucky. If there is anything like a radical point of view, it is aimed at making sure that all people have an equal opportunity of being employed, of being served as customers of the global artificial understanding, and of having a few moments to shine at some point in their lives. These last moments, if they occur, are technically anti-experience in that "shining" suggests others see the light while the person at the center experiences the sense of being an object of attention. Each of the three moments in the lucky human's life under the telemetric are themselves telemetric: the employee's experience is in the employer, the consumer's in the seller, and the star's in the audience.

If we are lucky, there will be a job for us. If we are lucky, we will enter the market place to acquire everything we need, and maybe even a few things that we want. If we are lucky, everyone will look to see who we are. This is truly a slave morality where we are transfixed in listening to the arguments of the politicians who would tell us that we deserve our crumbs as they fall from the table of their donors, the people who matter, the artificial persons with their staff of assistants, accountants, lawyers, and the customized lives that such employees are able to construct and maintain for them. They are the courtiers of corporate power and they have enough cushion to weather whatever changes are in store as the process of production morphs into whatever forms the new world order of advancing capitalism has in store for us. And these folks love it when we fight among each other, when we point at other employees, generic customers, and whatever celebrity is the flavor of the day, and say they are the problem, they are the source of my condition and all that seems to be missing from my life. When we do that, we sacrifice our personhood and obtain an aggregate self that is incapable of digging deeper into the nature of the telemetric logic at work everywhere, moving us this way and that with the sea of others likewise isolated by their abstraction into the statistical sum total of opinions. We remain rooted in our cages and unable to take on the ones who have, in plain sight, locked us in them. Identity politics and the fear of it, globalization and the fear of it, all these ideological positions in so far as they are socially driven fail to point in the right direction if they do not understand that whatever their point of view, it is realized as the perspective of a slave whose personally identifiable information has been masked to better aggregate their personhood for proper use. Such conditions deprive them of any awareness of their chains because these conditions are those very chains fastened to their bodies, keeping them in place, and forcing them to focus on the entertainment that has been mass-produced with exactly their demographic in mind. Thus, they are lost in images flashed before their eyes as though they were shadows on the cave wall so far from the origins of truth and the light of the sun that no one with any sense could see them.

And even this cynical and resigned position, completely ineffectual, has long been caught up and captured within the telemetric's characteristic dynamic. You know already exactly how to categorize me, you know what analytical group of persons I fit into, and you have an array of canned responses that you are ready and able to make against these observations and tales. Where did you get them from? Where did you get these notions, how did you learn to so quickly and efficiently categorize every idea and notion that you hear so that it immediately fits into and reinforces the view of the world that you already have? Do you think that is the natural condition of human beings, acting and living as if every idea or point of view that they could come across is already part of a grand scheme that is clearly mapped out and understood? Mapped out by whom? Understood by whom? The indication of ready-made response and rapid interpretation of any point of view that you hear in the sea of messages that are out there bombarding you, should inform you that something is going on, something that you are not locked into: something alien to your human consciousness is taking place. If all the new and dazzling ideas that you perceive are products made by corporations, then this should be an indicator of a persistent state of affairs that is alien to you. This alone should clue us in to the fact there is something going on that we are not fully privileged to, that there is a logic at work which may be only partially visible or partially meaningful to the human beings living and breathing within that logic.

I have given the name of 'telemetric understanding' to this thing that shoots down every option I can think of as soon as I think of it, that appropriates every notion into an order that has already categorized it and made it part and parcel of a well-established set of daily behaviors and predictable responses. I can immediately feel that name slipping into the soup of taxonomies that are intrinsic to that understanding. I do not get the sense that this name will communicate anything new to anyone, rather it seems familiar as soon as anyone hears it, as though it were a notion that someone else had already said. Despite the fact that there are hardly any telemeters who were formerly educated as phenomenologists and despite the fact that this notion –as I have developed it– could only be the confluence of those two disciplines and the experience that stems from those two areas of expertise, it still feels like something we must have seen before, something we must have already figured out how to fit into the table of categories and theories that human beings have available for use in their application store of knowledge. With the panopticon effect, not only is everyone being watched, but it seems as if everything has already been seen, that everything is already known and understood. Nothing can get through to the foundations of these conditioning orientations and the fiber of our involvements in so far as they are guided by those conditions. All the lights are on and everything that can be seen is being seen and has been seen already. This condition itself hides something.

I can imagine that there are apologists for some piece of the system that matters to them who will argue about this. They will give examples of discoveries and inventions that have come along, they may even point to some radical new ideas in some best seller which seem to have changed the way everyone sees something or other. Or maybe they will point to a political movement, a green new deal, that they optimistically hope will sweep across the land. When the hummingbird hovers around the honeysuckle, something new is happening, even if it repeats a pattern that it has executed every day this summer, this day it is happening again. I just wrote that down, just presented that image in a story that I am telling. I could take a photograph and post it on some social media feed where my followers will see it. Some may like it because of the detail my photograph captured. Does it make sense how these forms of representation are taking place within a conditioning process? Does the conditioning process appear as one of regularization, normalization, or rationalization? Does it appear as some such procedural smoothing out of experience into something that fits in with everything else, that is just like what has already happened? It is not that the world is not an extraordinary place with adventures and extraordinary emissions happening all the time. This idea is not new, so of course, I am aware of it. This is immediately interpretative, it immediately succumbs to this global conditioning effect. A radical book on income inequality breaks onto the scene with an immense amount of data to show a trend. The association of that trend to historical movements and conditions would appear to be clear, everyone reads it. Something new has truly happened. The excitement is in the air, so many people are energized by it. The spinners spin, the apologists apologize and soon this new thing is nothing at all, is absorbed into the flow of patterns, the set of categories that are already understood, have already been explored, and now the next new thing must come along. It is not that there are not any new things, rather that the meaning of newness is constantly recaptured by the logic of the systematic understanding that is at work in the world and this logic fits it into the slot where it belongs, shows that there was already much that has happened which anticipated or expected such a thing, and so we can all just move along. Nothing to worry about here, no adjustments to be made. Everything is as it should be. The interpretive mechanisms that are at work have run afoul, that is the point. And that is the telemetric understanding at work in every aspect of human experience. And you already know this. And it has already been seen, boring, what else do you have for me today?

And the matter itself (I almost wrote 'problem' but even that is overly eager and ambitious), is already one we have taken note of, the matter has been registered and everyone knows it as a set of conditions that gives the human lifeworld, in its broadest sense, a *meaning*. For no good reason, I assert there must be something amid this light, amid everything that is so clearly seen that there must be something that remains beyond it, something that cannot be taken hold of, that is not in view, that does not appear as such. And this is something that I sense for no good reason, I call it the remainder or the residual, following the standard laid down by some of those phenomenologists that I have mentioned. Even this notion has already been hashed and rehashed, thought of and categorized, placed in the order of things and comprehended in impact and relationship to everything else that there is. No, that cannot be. And this is more rage than reason. Think of Kant in the opening sentence of the first Critique: "Human reason has this peculiar fate that in one species of its knowledge it is burdened by questions which, as prescribed by the very nature of reason itself, it is not able to ignore, but which, as transcending all its powers, it is also not able to answer" (Kant, 1965 p. 7). But what if that were not a principle of pure reason or knowledge but a principle of production under certain material conditions? And so as not to bias my intent, what if it were a principle of action under certain physical conditions? What if that something or other that remains is a product of human being as we are, in so far as we are, what we are? An existential category? And that very same principle (if you will) could be the origin of the telemetric understanding itself. Make no mistake about it, we have been referring to the corporations that seem to realize this telemetric understanding, that seem to "experience" the terms of its order and dynamic, they are the socially protected artificial persons that have been constructed out of the social conditions and relations that likewise have made each of us. They are our brothers and sisters coming from the same parents. We have built this understanding that conceals something from us, we have been selected as the best suited configuration of living matter for building it, and we are the sorts of beings that are best suited for coercion into acting on such principles and under such conditions. The bees must first build the hive so that they can realize its operations as their daily lives.

My point, probably a familiar one, is not that since human beings made it, we can unmake it. Because I do not think human beings made it in the commonly understood sense that this suggestion might have. We are siblings with rudimentary elements in common, the origins of each of us, persons one and all, are the same mold, the same source; and this gives us at least the chance of having an insight that might reveal what is hidden inside the creations that we have "made". This undermines the point immediately upon making it. Even if there is something hidden and at work in the telemetric understanding, the principle of reason converted into a principle of experience or action is such that the revelation itself, the shedding of light, the act of reason, are all acts that hide in the moment that they reveal themselves. This Heideggerian trope may not get us very far, but if you prefer you can think back to Descartes' pattern in the *Meditations* when he suggests that the thinking thing is presupposed in the act of doubt. The act or the principle actualized is itself a process unfolding and as such hides something of itself from the unfolding act itself. There is a process at work that is intrinsic to the result, implicit to the outcome, that is part and parcel of the activity performed. And that something or other is not revealed, it is not an outcome. Transcendental deduction, as Kant showed, can get at it as a condition of possibility. That cannot be it. For something to be hidden in the process and procedures of experience, it must be truly hidden to view. It must be impossible to see, impossible to make apparent, and impossible to take hold of. This is a paradox, a limitation in the mechanism. Reason itself, or whatever name we hope to give to this set of insights that realize, that comes to know, that becomes aware of whatever *is* in the way that it *is*, is victim to a paradox, to an impossible state of affairs that makes it such that there is always something missing from its point of view in the expression of it, in the experience of points of view per se. Reasonable beings cannot escape this partiality.

What is that? What is it then that remains in human cognition or consciousness? What is it that establishes the partiality of the partial? This kind of questioning may not get us there, as it attempts to invoke exactly that phenomenon that is intended for the sake of clarifying that which cannot be clarified because it is residual to clarification. Such questioning then embodies the paradox that it questions as it questions. The relationship to the telemetric understanding is an opportunity hiding in this regard. There is something hidden to us, we cannot ascertain the experiences of such alien artificial intelligences and the way in which they apprehend and understand the world. There is something beyond view here, something that cannot be viewed at all. Whatever that is, it is the hidden thing of the understanding that makes it be whatever it is, that gives it the flavor and form of its character. And yet I have claimed to see some evidence to this effect in the very notion that nothing is missing from view to such an understanding. At least in so far as such an understanding is occupied by human beings acting out as consumers, employees and the celebrated, locked into such a lifeworld and experiencing things through their place and position within the telemetric understanding. Because it leads you to believe that you have seen it already, heard it already, and know how to classify it, there must be something about it left out of view. In the very fact that we think things are so clear, there is a hidden something or other at work. And if that is the case for the telemetric understanding, then it must also be the case for any human understanding that any one of us embodies in our lives and experiences in this world.

Throughout this work, I have made many different points about many different things, and –despite the fact that many of the pathways and roads that were taken came upon failure as they were spun, analyzed, and understood– I do not wish to let go of anything here in the residual moment of the commonwealth. I want to keep everything close by and, as a result, I do not wish for any of this hiddenness mumbo jumbo to appear as an appeal to a deity or something divine. Have not I had done with that? Has not that been swept away clearly and distinctly along the way? This is not something exotic and strange. Rather, it is the sense that human beings are finite creatures, not just in the sense that we are finite in time and will die, but finite in space since we have position and location. This finitude is a limiting factor in anything and everything human. It characterizes everything that we are and everything that we do. It is what makes it possible for us to have involvements and be conditioned, to enter relationships with some things and not others, to have a point of view, to see things at all. Everything about us that characterizes our experience of the world is due to these boundaries that are conditioned into us as personalities and which we seek to protect from onslaught as we age, become more accustomed to ourselves, and who we are. This finitude is extraordinary to that effect, but it is a serious problem as well. It is why there is always something missing, always something that cannot be seen. We are limited in our point of view and so cannot see everything all at once or even in endless sequence. The telemetric understanding is alien to us because it does not have this characteristic, it is not bound by such conditions, it can experience many things at the same time in many locations and places around the globe. Its finitude exceeds our own even if it is not infinite. It can have arguments with itself in a way that no individual human ever could, even if we cast those arguments as inner conflict and wrote a great epic novel to that effect. There is a space that opens between these patterns of understanding and experience; and that space would seem to be the scope of operations and the breadth of insight that can be cultivated by such artificial persons.

The closest we come to breaking out of this finitude is our perceptions of self in the aggregate, and the closest the telemetric comes to breaking out of its form of positional being is when it articulates itself as a private person. Human beings are locked in bodies and breach the boundary through aggregated statistics, whereas artificial beings are locked out of bodies and breach the boundary by articulating a private personhood that gives it the semblance of having a body, of being an incorporated individual. There is a residual at work in each of these types of persons and their common ancestry is what makes it possible to expose this fact of hiddenness. Exposing the fact of hiddenness is not the same thing as exposing the material of that hiddenness. That, I assert, is something we must live without, something we will never transcend no matter how much it pains us or aggravates us to be incapable of doing so. The vicious mastery of the corporation may very well have its roots in this ability to go beyond its limited form of being, just as the slavery of the natural person may have its roots in this yearning to grab hold of some universal point of view, some viewpoint from the perspective of eternity. The human being wants to see the world telemetrically and the artificial person wants to see the world phenomenologically. And the telemetry engines, all the enormous data pipelines, are there to open two-way traffic enabling the one to see through the other's eyes. Subject yearns for substance and substance yearns for subject.

The temptation in trying to articulate this maneuver is to attempt a confessional. The human being may be interrogated to discover its deepest urges and fantasies, to learn where it wishes to go and what sort of a life it imagines for itself. We look within and discover there are two powerful drives: to understand what others are like in the aggregate so that we understand how we fit in, and to transform ourselves into something completely custom and unusual, someone who must be treated in a highly specialized fashion, a celebrity. The former is the desire to realize the position of servitude in its most complete form and the latter is the drive toward mastery. Despite being equally illusory, the two push and pull throughout any individual human life. Social media services play on this push and pull and offer opportunities to feel and express both urges, both drives. All offered up in a pretend setting that makes the experience safe and controlled. Until something carries it away and we are caught up in too much individuality or too much universality, either of which can be destructive or produce a pathology. This is a confessional form because it is a pyscho-analyzing of the experience, it is a psychologism inhabiting the phenomenology. Structures no longer formal become material, become a play of interrelations that need to be worked through, that need to be endlessly turned over and over. The telemetric understanding can insert itself into such psychologizing of experience and take over its interpretation, provide a ready-made and easy to consume order to apply to this material. And no doubt there is some medication that can be prescribed to reduce the symptoms of whatever name this disease may be given.

It would make sense to posit a similar dynamic at work in the corporate body, the order of the telemetric understanding as socially gigantic beings, would be expected to exhibit a similar pathology, a set of symptoms where the extreme possibilities of its orientation become tangible. The corporation has human envy and it has an extreme form of its private personhood, it may be seen to swing between these options as a kind of de-customizing of itself through attempts to fit in, to be a good citizen, and one of the upstanding members of the community that is here to help. This amounts to an egalitarian urge where the master thinks it is in it with us and needs to help the good along by contributing to community programs or having meaningful advertising campaigns during events of extreme human need and crisis. The corporate person is not just playing at a role here (although it is playing at a role here), it is expressing an urge, an urge to fit in, to be part of the social setting in just that way that any person would situate itself relative to the society at large. And then it has the moments of extreme narcissism where it asserts its mastery in bald and obvious fashion. It expresses an interest in tax reform regardless of the larger social impacts, it acts out in hostility to government or other types of organized efforts to tether it to the nation or the city or the community. It asserts its desire to be rogue and wild and threatens to pick up and move to another place or punish the offending organization if it does not get what it wants and what it knows deep down that it deserves as the only real and valid creator of value in the world. For the corporation, this too functions as a form of the confessional as it is an admission of its deepest urges under the pressure of some intensified observation whether that comes when turning toward it during crisis or during some legislative session where it focuses its actions among the people.

I do not think the confessional moments are essential components of either of these dynamics that we are positing in the residual of the commonwealth, and I do not intend to try and build some emancipatory interest off them, some strategy for how to purge the human psyche of such things or to constrain or restrain the corporation. Again, as I have said, any such platform would be immediately recuperated as something within the telemetric system, something coming from some categorized point of view that reflects a categorially apprehended agenda. The confessional may tell us more about what each of the persons are, it might give greater insight into the nature of private personhood and its opposite, the aggregate personhood of the servant, but it does not offer up anything residual to that system itself, it does not escape its logic or offer anything emancipatory for the flesh and blood humans that find themselves baked into the order of things.

Beyond the confessional, it may have been (or so I thought at times) that some form of ethics might emerge as the upshot of the dynamic at work in the commonwealth: an ethics that combats this urge to be seen from far away realized as confession. Thus far, however, the approach I have taken only establishes that ethics is not possible among natural persons in their current state, that human beings without private personhood are not capable of properly ethical action because they do not yet satisfy the preconditions of being able to act otherwise. The conditioning at the hands of the telemetric understanding prevents any ethical uprising, only the artificial intelligences of the world would seem to have sufficient standing to develop an ethical position that could move the world one way or the other in any such time of crisis. And I sense that such is exactly what is taking place in the world as I write this. The concerns of the masters are not the well-being of the slaves, of the customers and employees, not even the well-being of their own fans and supporters matters much to these beings. Rather they push for their own protection from social circumstances, they seek to instill a set of directives and rules that will maintain their mastery and safeguard themselves against whatever crises and conditions the future has to offer. They are afraid of losing their position and they have an interest in growing their status. The only ethics they understand is the exertion of effort to acquire more for themselves, the empirically verified ethical principle of the artificial private persons in the world today is: get more, be more, make more, all for the sake of protecting ourselves from slavery.

And truly it is not like there is any major innovation going on in the field of ethics. The major positions have been established, most humans have a sense of right and wrong, obligation or duty and so on. They may not be exactly utilitarian or exactly deontological, but they have some sense of these notions. They probably have some sense of the variations on those themes and even more so than any philosopher has ever been able to describe. What good has this done? Ethics and morality are no mystery to human beings and yet here we are, neck deep in the telemetric and subject to its mastery. And human beings were instrumental in making this world. Human agency succumbed to social pressures and social conditions and understood the origins of corporate power so much so that here we are right in the middle of it. Why would anyone think that ethics is the key here, if only we had the right set of principles, the right set of ideals, or if only we acted on them? This is not new, ethics did not just suddenly come into the world as an offering that can cure us of whatever ails us. And yet there are still political movements that justify themselves based on ethical principles, thinking that somehow people just did not know they needed to express certain values in their behavior toward others or in the institutions that they build or the policies that those institutions enact. If there is anything that even vaguely looks like an ethics today and that might be residual to the commonwealth, residual to the socially scoped telemetric understanding that drives the civilization in all its norms and principles, then that would come from an effort to make ethics possible. And reason has not cut it. Reason has been the theoretical basis of ethics for millennia. And here we are. Empiricists of the world, so scientifically minded, consider it falsified. Try something else. What is the foundation of human ethical capability? What makes it possible for human beings to make the world a place that is good to live in, to be in? Even if we could think it, this essay has tried to make the case that we have no way to make it happen. And therefore, the priority in ethics would have to be the real and material conditions for the possibility of human ethical action at all.

It may be suggested that such a material set of conditions is the aim of activism and organization. That is, to properly change the material conditions of this world that prohibit human beings from living as private persons among each other, that prevents the existence of privacy among individual natural persons to scale, there would need to be a movement. There were many times in human history where that would be spot on true, but I worry that it may not be the case here and now. Despite the hope in the name of a "green new deal", activist movements are largely operating in a classical universe in which the agents are all human beings struggling for relations and conditions, involvements and orientations that benefit them alone and in groups, we have not yet figured out how to be activists in a world with super beings, with omnipotent private persons that can spin and twist messages with such agility and ease that they make it cool to have a pussy hat or be part of an occupy movement. People wear out, they must go to work, they need to buy food. The corporate spinsters and cultivators are working all day every day to achieve their ends, they are the best and the brightest. Activist, they are smarter than you and they are wiser than you and they have more resources than you do. That is because human beings are willing to do their bidding and have an interest in doing so. The artificial persons among us have figured out the best way to recruit and the best way to cultivate the best talent and best minds that the civilization has to offer. And all the activists have is good intentions and a concern for all. As noble and beautiful a story as this might be, the attack on the East India Company as a mythos that energizes us and inspires us to rally around the causes that need attention, it is nonetheless vulnerable to recuperation and translation, to being swept away in the sea of messages that relay through the telemetric scroll of ideas. The impression was that the Arab spring took place on social media, at least the organizing of it did, but they learned the lesson from that one. The conditions for the effective establishment of ethical capabilities will not happen on Twitter.

One direction that it might occur to some activists to go is to organize through the apparatus that oppresses the natural persons of our civilizations, the corporation. Government may be too outdated to offer a real opposition structure to corporate interests, it is largely in their pocket at this point and we should not expect much in the way of massive reform coming from that direction. There are certainly many who think that the social democratic wave could sweep through the world and save the day, but democratic socialism has been around for a long time and the new hope is mostly just an American thing. The government structure in East and South Asia shows no vulnerability to this approach and nothing in the current movements across the globe suggests that there is any such impact on the horizon. Instead of thinking in terms of government change, someone might propose that the residual for the commonwealth lies in its incorporation. We the people hereby issue the following common stock to each and every one of us: a single share is granted to each natural person in so far as they are a natural person. This share is allocated at birth and bound to the body of the person, it becomes the socially awarded token of that person's privacy, so to speak. The token is returned to the kitty at the natural person's death. Immediately, such an incorporation would become the largest and most valuable in the world, it would dwarf all the other existing ones and so it would be able to immediately demand highly customized treatment from the other artificial private persons that live alongside it. It would be lovely to fantasize about all the interesting power that this would apply to the commonwealth of people so incorporated, but the fact would remain that this residual effect would ultimately amount to a surrender. It would establish private personhood only at the cost of losing it for every single one of the token holders. The stockholder is not a private person per se, rather just an invested party, an interested party who gains some benefit from that relationship. Surely, it would make us something other than employees and customers with the urge to attain celebrity, it would expand human horizons to an extent and it might even improve the lives of many by asserting this leverage over some of the coercive forces that occupy the social order. It would do so at the cost of just what I posed at the outset as a critical end: the integrity of the natural person as a meaningful finite boundary condition, worthy of self-determination and development to its fullest potential.

By its very nature, the residual effect is elusive. Each attempt we have made previously was ultimately recuperated by the dynamic of the structure that it implied. And commonwealth would seem to be all the worse because it does not yet imply some higher order or greater dynamic that can be discovered. If we cannot build this together for the sake of ourselves, then where is there to turn, what else is there to rely on? Under the conditions established in the order of the commonwealth, the assertion that "Only a God Can Save Us" is tantamount to saying that nothing can. Should I leave it at that? What remains to be said? Bear with me as I step into and step through the architectonic movement itself.

Metric Emissions (Regional Differences)

Here is a metric emission: I deferred writing this book for longer than I needed because I thought the crucial ingredient, voice, was missing. I thought there was an artifice that had to accompany the tale and give it its power and depth. The deferral ended the moment I decided to be simple and clear in a way that social philosophy has not always been. I found Durkheim to be most inspiring because I saw a simplicity and straightforwardness in his work that set it apart from the philosophers who were performing related investigations. The germ that started the formal planning of the project was the sense that on the one hand privacy is the most important thing that there is to a human being and on the other hand that there is no such thing as privacy, it is a fiction. Simplicity of voice is a form of intimacy that verges on being a violation of my privacy. Here is what I assert, not adorned in any way to impress or complicate matters, not hiding behind any terminology that might make the point of view seem artificially more intelligent or informed than it really is. What could be more private than one's wisdom, or whatever it is that passes for wisdom in these matters of experience that have been the mainstay or substance of a life lived? Whoever said it and however it goes, I remember it as "Remain silent and be thought a fool, speak and remove all doubt." This simplistic notion was further fed by Heidegger's image that only someone with something to say can remain silent. It is the pattern of these sayings that was so compelling, the sense in which they suggested that something held close, held dear, was existentially bound to the being who held on to it. There is nothing universal here, it is a metric emission. And what that amounts to is something that comes from me but is not just about me. It is categorized and meaningful, familiar in some way to those who can read it. And to cloak such an emission in technique or decorate it with exorbitant style is a form of fancy dress and pretention: it amounts to an urge to speak without being understood for the sake of holding on to a sense of intelligence without being discovered as the fool that one is. And how do you know whether you have anything to say if you remain silent? And how would you know if you were saying anything at all if you cloaked it in style that worked harder and harder to hide in the open whatever came out so that it could be both seen and not seen simultaneously?

There is no accusation in this. If you think I am accusing Heidegger of something, then I will have been misunderstood. I found it generally clear how to go about understanding Heidegger during the extensive period I spent studying his work. You read the prerequisites so that you have a strong background in the problem space that he is working in, you do your best to grapple with the whole of his work in its original language to try and meet it on its own terms, and then the work yields. Heidegger and Habermas, Husserl and Hegel are doing the best they can with what they have to say, whatever they are trying to say. I am not accusing them of being overly obscure, but I am suggesting that it is possible to work in the domain of metric emissions in such a way as to try and hide behind various artifices to trick or fool others into believing you have something of greater substance to say than you do have. And I decided that I did not want to do this, even at the cost of making it painfully clear that all this education and all this time spent studying the work of others had, in the end, amounted to very little. And that is an extremely intimate act, it is a decision to emit messages in such a way as to let others know what is going on in here, to remove all doubt about the foolishness that lies behind the silent façade.

There are people who only sing in the shower and those same people might not do it if they had the sense that they were not alone there. When do you have the sense of being alone? And how do you respond to it? What are the truly private metric emissions of your life? I have tried to establish that you are not truly alone in those moments and I meant it. The social conditioning, the crowds and generations, the entirety of the species has produced you so that you could get into that shower by yourself with the door closed. Still, holding onto that and feeling the impact of its meaning, you feel that you are alone in that moment. And are you lonely there? How do you occupy yourself? You may have removed all distractions and that may be why you sing, you have nothing else to do? In those moments when it is only you, how much of that is patently artificial? Sometimes I would sort through a work problem in the shower, another way of chasing the solitude away. I'm not alone, I'm here with the code before my eyes, working through the problem that the team needs me to solve so that my component can perform its function integrated with the rest of the product. Or maybe there is a family matter that needs working through, or some friend who I have been concerned about. Even that song I sing, I did not write and arrange it. When am I alone if not in those moments standing by myself under the falling water and away from anyone else's view? People say that it is when they cannot connect to anything, when they cannot feel the impact or conditioning of any other person on themselves, that is when they are truly alone. To that extent, we realize that "feeling alone" is a pathology and that it is based on a lie. Probably even in that moment there are connections and involvements at work in the way we are thinking about ourselves and our relationships to others, but they are not the connections and involvements that we want, that we fantasize about having, that we are thinking of as the ones we need to have or that we should have. You can only be alone if there is a disconnect between the conditioning at work with the emissions taking place: there must be a disconnect between you and your environment. And phrased this way, it is not a good thing, not a pleasant thing, it is a problem and there are entire industries that aim at making the feeling more manageable or making it disappear entirely. Contemporary wisdom says it is reducible to an issue with one's brain chemistry. And that makes sense since the social reality is that you are never alone, the language you use embodies you with a history and a set of social connections that you can never turn away from, so that only when the neurons are misfiring and leading you to draw conclusions that are self-contradictory is it possible to feel alone. The feeling is a lie, you are not and cannot be isolated. That voice in your head is the voice of everyone else, all of history, and all the generations that have ever been. That is the secret of human consciousness.

This is likely why we have adapted so easily to the panopticon effect in our society. In just a few years' time, we allowed telemetry to take over the entire surface of the planet and there seems to be a growing realization that we have done so, but the reaction is remarkably staid. We do not seem to be panicking or angry about the fact that there are eyes everywhere, cameras, listening devices, trackers measuring activity and inactivity. I am surprised at how easily human beings have accepted this new reality into our lives, and I suspect it is because at some level we were already aware of the constant inner workings of such pressures and powers. We knew that we were filled with our history and our society, that we were immersed in a domain of conditioning that threatened our natural personhood with the constant penumbra of the artificial, of the crafted, and of the socially manufactured lives that come so easily to us. Not that the lives are easy or that the adaptation is easy. I do not know if we really are political animals, but we surely have become social animals in the deepest possible sense that this might have. And our social nature makes it hard to be natural persons, hard to find the natural personhood in ourselves without realizing it as a pathology.

Considering the matters at work in this book or any matter at all that relates to some vague sense of a love of wisdom, I realize of course that I am merely aping the work of others, mushing together the many books that I have read for the sake of presenting the hodge-podge that is the outcome of what I have been able to make of all this. This is how any project works when it is not data driven. Peppering it with specific experiences, trying to integrate it with stories others have told me, it is not an act of solitude. Or at least I can intellectually recognize that it cannot be since it involves so many connections to others even if not in a rationalized form. All that comes together here and now in its articulation and in that moment of metric emission is personal to me. Partially so because of all the errors I have made, all the misunderstandings that inhabit my interpretation of events and how those events fit together. These errors, these misfires and pathology relative to the actual things that conditioned me, or that should have conditioned me if I could remember them better or if I had paid more attention to them when I was working through those events or experiencing them, these are my solitude, this is my personal point of relay, the node that I am. The smooth pathways of the network are traceable but not worthy of attention, unless there is a misfire. This is standard telemetry practice. Sure, we like to measure the positive quality of the service, its availability, but the measurement is largely for the sake of the errors, the subtraction from 100% reliability. The errors and mistakes are what is significant to the measurements. I cannot see why I would be any different, how I could possibly have escaped this dynamic that I inhabit through no choice or act of my own. Appropriating something as one's own is not the same thing as being the origin of it. The very notion of appropriation suggests that something that was not mine has come to be mine through an act of taking ownership over it. Any such appropriation amounts to a kind of error, a mistake based on a set of faulty positions that do not adequately reflect the pathways executing throughout the network within which the appropriate is merely a single node. I turn that around and suggest that not only is appropriation an error, but error itself is an appropriation. When you make a mistake, if you are adhering to the highest ideals of your civilization, you "own" it. Or own up to it. This need not be a distinct response, there is a sense in which the ownership lies within the mistake itself. In the very commission of the error, you circumvent the conditioning, you ignore the training, you fail to realize the pattern that should be applied, and you take ownership over the process through some act that no one else could have possibly assigned to you, but which you take for yourself.

Surely there must be common ways in which people make mistakes, common ways in which these errors reflect a failure to live up to the norm or match with the principle. Of course, and that is why I call them metric emissions. This is a domain before ethics, before the establishment of good and evil or good and bad, this is the domain of human beings acting out in the context of their social conditioning. The "mistake" is already a categorized notion, the history of the people already contains the wisdom that such and such errors are typically made under such and such conditions so that even when we appropriate, we appropriate like numerous others who were just like us under the same conditions. There are many who have aspired to wisdom and managed only trivial musings, there is nothing world historical in that, it is not an extraordinary act that makes the agent committing it into some strange and unaccountable beast. Rather, it is a binding function, it connects the errant one to the rest, it assigns a set of dimensions to their action, and yields the possibility of measurement. It is not that the act is one's own so much as the acting is owning, a process of appropriation more so than something appropriate that has been done. It is decidedly not appropriate, it is in error, a mis queue, but as action orientation it is a procedure as opposed to a quality. In the metric emission there is an act of emitting, of doing, the act itself. And that is the appropriating action. And in some sense, there is no pristine instance of any such acting operation. The human being is an error machine, a manufacturer of mistakes. Nobody is perfect, they say. And what they mean by that is that no act is perfect. Of course, if you narrow your scope sufficiently enough there may be moments when a highly isolated sequence of operations performed by a person or groups of persons seem to perfectly reflect the principle of operation that they were aiming at and which may resemble perfection. And these moments are truly dazzling, but they probably do not bear up to intense scrutiny either. There may have been ever so minor mistakes embedded in the operation and what makes the overall act extraordinary may be that the actors were able to accommodate the tiny misfires for the sake of the act itself. In an athletic performance, the gymnast tumbles to perfection. The gymnast might be aware of ever so slight miscalculations that are at work in performing the tumble, but is trained enough to be able to respond, to adjust in midair to give the overall appearance of perfection even while making a series of micro errors that only they can perceive.

The intent here is not to build an ethics or value theory out of the error condition. Errors can be gruesome and have horrifying and catastrophic effects. No one wants to make a mistake, but the point is that the act of making an error is the most personal experience a human being is capable of. When we make a mistake in our metric emissions, we extend the boundary of the categorization, we mold the dimensions to ourselves, and we step outside the boundary of the appropriate response in an act that appropriates a response. It is a huge leap of logic to make the comparison between the act as it should be done and the act as it has been done. Where does that right come from? That assertion of privilege that says this is how it should be? What are we referring to when we say this? Is it something as simple as "that is not what I meant to do" or do we mean something more like "that is not the way people expect that sort of thing to be done" or better still "that is not the way this conditioned act has been historically conditioned for execution"? And are these different formulations merely different ways of saying the same thing? Somewhere there is some standard, some set of metrics, against which this sort of thing is measured and the act that has been performed, by me or another, has been performed in such a way as to fail to meet the bar, fail to exceed the minimum measured requirements of that action. The prototype for the action has an acceptable allowance for errors up to a point, exceeding that threshold puts the act outside the realm of the normal and turns it into an error. Error itself would seem to be a telemetric notion, but only once the comparisons can be made. And it is extremely difficult to imagine a form of constructing error that did not already instantiate the notion of the standard against which the act has been measured and has fallen short.

It is precisely this pre-normative state of the error that I am trying to point to, trying to evoke in the very moment that I am making that same error, making a series of huge mistakes, each sentence itself a further token of the type. The act appropriates the dimensions of its performance and does so in such a way as to fall short of them. In a child, for example, away from the watchful eyes of their parents, such tumbling is not yet a mistake, not yet interpreted as such. And I want to both call and not call that a mistake, an appropriation or act that emits measurable data, but does not yet attach it to any precise validation process or reflective recuperation. The act is done from a conditioned sense, but still not yet fully comprehended to make the necessary comparisons that would reveal it as an error, or as laden with error, as mistaken in its effort to properly realize the form of the act, the tumble as it was meant to be. And this is something of an animal performance, or there is something not quite human in this form of the action. It is valueless and potentially horrifying, possibly the raw and brutish operations of a sociopath. It exists in a domain before there is measurement, before there is ethical judgement and determination. It is the origin of human being as the raw material of the conditioning acts of civilization and, as such, is uncontrollable and dangerous. If anyone other than a child was capable of such experience, that person would be a threat to all the rest of us. That is the pre-mistake form of the action, the action not yet understood as either done correctly or not. I am trying to suggest that the error, the mistake, the appropriating act that owns up to a sub-standard operation, is the closest a civilized person comes to this raw and childlike state.

If possible, it would be insightful to try to time slice a human being's activity so that the action is carried out with focus on the conditions required by the action. The agent is responding to conditions, trying to say what needs to be said, for example, in getting a point across. In that moment when the act of saying is being performed but before the person can reflect on the import or significance of what has been said, that is the moment just before the mistake has been "caught". That is the closest we come to this animal being that beats in each of us. The catching always pushes us past that state, puts us back into the world where we live with others, and lets us know that a mistake or error has been made. Some people may not catch these errors. Most people do not catch all their mistakes all the time. Others might point them out in some cases, in other cases no one may ever point them out at all and the person might live their entire life not recalling the event or considering it as a mistake. Presumably, this would hold for every human being. I am sure that there are mistakes that I have made and that there are mistakes that I made that I do not remember. Furthermore, there are mistakes that I have made that I never considered to be mistakes. If there were all knowing telemetry on every event of my life and if that telemetry could be collected in a data store somewhere, it is entirely possible that a machine learning algorithm that would be trained on that data would, for all intents and purposes, be me. This is probably a joke. Such telemetry with its metrics and dimensions would indicate the mistakes in all their glory, whether there had been anyone there to point them out or not. The telemetry would be there, would have been there. The telemetry would be able to function as conscience and as constant observer. Better than any husband or wife there ever was at witnessing the smallest details, the tiniest of errors, in all that we have said and done throughout our lives. Such a perfectly trained mechanism would capture my essence in a way that not even the narrator seems to have been able to do. The mistakes that I have made that I did not recognize as such, that never registered as mistakes, would have been captured there in that store and so even if I could not bear witness to those events, those mistakes, then at least the telemetry could. If god is dead, at least we have telemetry.

This is an extraordinary paradox and the cleverness of it is really the only thing that has led me to break my silence on the matter and proceed with all the mistakes and errors that I have made here in this project. All the words may be nonsense, but this idea seems righteous. We are looking for the most quintessentially private being of ourselves, the own-most purified and singular personhood of all our conditioning, we are looking for the freedom to be or not to be moral, to be monstrous again, the rudiments of our ethical being, in the mass surveillance of the telemetric system. We believe that privacy will be established in this exhaustive collection process and that aggregating out any data that is personally identifiable will protect our privacy. And by 'we', I am not trying to say that there are not resisters. I am sometimes one of them. At work, I go on routine rants about the nature of the telemetry systems we are building and the damage that they are doing. By 'we' I am referring to the social conditions within which we live, the factual world that displays what we value, and what we strive for, not what we are claiming to value and claiming to strive for. This new god that we have posited, even in the most Germanic regimentation with regulations aiming at self-determination (especially there), the drive is toward a complete digitization of the person and this 'voice of the friend' that can be established in the giant data stores will be the truest expression that could ever be constructed of our private personhood.

If that algorithm could be trained and if there were one for each human being, each of us would be fascinated by it, drawn to it, wonder if it is the truly real self that we knew all along that we were trying to protect and defend from those social conditions that were exercised to construct that algorithm in the first place. And that is the paradox of human consciousness, that it is prone to define itself through an immanent and ongoing observation process that presents its inner workings in complete clarity as a process to be completely comprehended and understood. And in that complete and total understanding lies the truth of the person in its most essential and private way. What I am deep down is the complete chain of events that not even I can exhaustively grasp and actively take hold of in comprehension and awareness. If only every single emission could be captured and saved off somewhere, that would be the real me, the private me, the most personal representation of who and what I am in all its detail and throughout all its developments and changes. Heidegger's "Gestell", the way of Being that is expressed in the epoch of technology, is realized in human beings as a yearning to find the truest sense of oneself in the dominant ontological paradigm. The drive to uncover the private person, to isolate the metric emissions in exhaustive form, is itself an act of the telemetric understanding emitted in measured events. For why else would we emphasize the revelatory need in the capturing of these idiosyncrasies? Why not just hold that the mistakes made, known or unknown, recognized by others and communicated to you or discovered on your own, are with you in so far as you are as you are all your life? How you think and feel and how you react to things, what makes you laugh and cry and so on: all this is just the embodiment of your errors, your attempts to appropriate and your repeated failures to do so properly. Why do the "mistakes" need to come to light, need to be comprehended as the revelation of truly private personhood? What pushes us in this direction?

The problem of privacy may very well be the matter of determining to what extent errors need to be comprehended. I do not mean to romanticize the error in the way that some philosophers have done. I get that an error could be a catastrophe for an entire community and the errant individual hated as a result. When is it reasonable to permit the mistakes of a person to remain in the dark, if such is the case, and when is it unreasonable? The question of ethics and social theories of value seems to reside in this general vicinity. Up to what point does the society have an interest in letting errors remain beneath any social level of comprehension and when does the society have an interest in bringing those errors out into the open for all to see? And this again is one of the great pulls of the telemetric system. Would not it be great if all the emissions could be captured by an impersonal collector, a gathering of events that judges and determines the good and evil, what needs to come out into the open and what can remain hidden? But the knowledge of good and evil is biblical in origin, why turn over the right to make ethical evaluations to an impersonal system? Whatever possessed human beings to turn that task over to a deity, for that matter? Might it not be the very essence of our liberty as human beings to take ownership over this process ourselves? The slave morality begins by turning over the right to judge to some other origin, to a deity or impersonal telemetric data collection mechanism. And this right to judge amounts to turning over the question of life or death that is our privacy to something that, by its very nature, must violate it to establish it. That is the paradox that we are up against.

These are my mistakes, these are the errors that I am making, and I do not need anyone to point them out to me. I know deep down that a book will not make anything happen, that a single person cannot just mutter something or other and make it so. Proscriptions are ridiculous at best. I do not think it is possible to instigate a social change all at once. Anyone who makes such utterances can only be a part of a larger set of ideals and operations that take place over years and decades or generations. Changes to the social conditioning of human beings are slow and they are cumulative. My mistake then is not an isolated one but may contribute to a larger history of this error, a sequence of instances of a mistake in these emissions become critical in the conditioning of responses to events. And the fact is that there is nothing that any single person can do to circumvent this. You never get to tell the whole story. When you spin a yarn, you are just playing a very small part in it. If that part remains small because no one else makes the same mistake, or exacerbates your mistake by adding to it, then it will be wiped clean by human history and no one will remember your mistake or your contribution to it. If the mistake is massive and lots of people do contribute to it, reinforce it, then at the very least the mistake itself will have been remembered and will persist even if no one remembers that you added some tiny piece to it. Metric emissions, but without the exhaustive comprehension, stay with us, are incorporated into us, and that process can be isolated in your body as private to you, or it could be widespread among many. And even in the latter case, when many share in the contribution, there is still a sense in which those errors that you contributed remain your mistakes and are retained privately.

Analysis (Federal Needs)

If it has not become clear yet, let me say it again: ethics originate in social conditioning and ethical arguments are one set of social conditions "arguing" against another set. And by 'ethics' I am not referring to the categorical imperative or the principle of utility, but to the social order that either leaves acts in the dark or exposes them to the light. The paradox at work in this definition forces acceptance of an omniscient principle to bring it all together: some events must be exposed, and some do not need to be. The social order has an interest in exposing and pursuing some actions and has no such interest in other cases. The paradox of omniscience here is that the determination and the analysis of the act in one fashion or another already requires that it be exposed to determine whether it ought to be. The fiction of the moral agent is meant to resolve this paradox. The moral agent, guided by a universal and rational principle, can act on behalf of the civilization as its agent, arbitrating the revelation or concealment of the "mistake" according to the embodied principle. If the agent is conditioned through and through, then the rational foundation of this arbitration is lost, and the probability of trustworthy agents will be increased or decreased by the presence of various social conditions in the formation of each person. This is unacceptable, for it means that the more the social conditions become perceptible to the agents acting on behalf of the society, the less effective is the ethical nature of those conditions. The people figure out how to game the system, to lie, to hide, and to take exception to the rule. These violations become widespread and embedded in the fabric of some forms of advanced civilization. The persons can no longer be trusted to police themselves, their conscience has been weakened, and so must come mass surveillance via the panopticon and all the behavioral nuances that it introduced gradually throughout the years.

If it is the case that whatever this rudimentary ethics coming from social conditioning was, it had no resemblance to what we get from the works of Aristotle, Kant, and the Utilitarians, then it is still an open question whether we find in those works a sensible set of notions to apply to something with the socially conditioned nature of human beings. Are we the sorts of creatures that can follow rational principles based on autonomous decision making? The first in all the history of the earth and in all the history of the known universe to be capable of doing so? And with seemingly very little empirical evidence behind it, it has none the less become a popular tale to tell. I do not mean evidence of doing good things that promote the general welfare or make the doer more excellent or comply with the conditions of the universal law, that could happen regularly and still be just a coincidence. Evidence in favor of the moral grounding of these actions would come from data demonstrating the presence of an autonomous will of some kind, a deliberative capacity that is free from the natural causal chains. I'm not trying to venture into the realm of classical questions of determinism, rather I'm merely pointing out that there is a heavy orientation toward scientific explanation with a basis in theory, and I do not see where there is yet anything like an authoritative position concerning the autonomous capabilities of human beings relative to deliberation on a course of action. What is the science of that? If it is Psychology, does not that self-contradict the very notion under consideration? That is, if truthful explanations come from scientific approaches to analysis and discovery, and such approaches are based on the discovery of principles that enable predictions to be verified or falsified in such a way as to lend credence to the hypotheses used to formulate the predictions, can there be a science of autonomous action and deliberation? Or, is not psychology more likely to fall into the area of statistical aggregates and probabilities based on past observations of the outcome of "deliberations" or "autonomous action" where those are merely empty place holders for the impetus to act? The science itself does not get at the inner workings of the process, it gathers and collects data, looks for patterns in it, and produces guidelines that describe the likely events. It is all correlation and no causation.

A scientific investigation cannot reveal the righteous and necessary foundation of ethical action. All we can get from science in these matters are patterns and aggregate practices, we will only ever get factual statements: this is what happens when this happens before it if the following set of minimal conditions are in place. Is it the position of the great philosophers that Ethics itself is such a science? And if so, is it a science that acts on the preconditions of ethics? Kant did not think so, he emphatically thought otherwise, he thought he needed to get his epistemology and his metaphysics straight before he could ever venture into the ethical domain. And it is arguable that even the great Kant never quite succeeded in this project. Rather, he left the preconditions as something of a toss-up, maybe we can be ethical, maybe we cannot, it is not altogether certain, but we surely must act as though we can. He does not mean "act" as though we can because that would beg the question, what he must mean is that we must pretend as though we can, we must affect a state of mind that assumes we can make ethical decisions and base our actions on autonomously determined intentions and action orientations. Arguably, even in that case he has begged the question, since the notion that we must do so fails to account for the possibility that this necessity has been given to us, has been conditioned into human agents through the social world from which we each emerge throughout the course of our lives with a growing sense of self and a growing sense of the errors and mistakes that we have made and which thus form us into the persons we have become.

Conscience is the urge to expose one's mistakes in those cases where the social interest is that the mistakes be exposed. What if we do not know that we have made a mistake? What if a person's conscience is incorrectly conditioned and so they are unable to perceive their mistake as a mistake or, even if they do, they do not think they are compelled to testify against themselves by exposing it? The approach that seems to have been around for many thousands of years was to work backwards on these matters. Something horrible would happen, or just something bad. Someone would suffer an injury and the injury itself would become known. The provocateur of the injury would leave a trace, witnesses, evidence to the fact that such and such an act had been performed by such and such a person and the resulting injury incurred. The society determines that it has an interest in protecting against such injuries and so allocates resources to following the evidence trail and exposing the crime and the criminal along with it. Some civilizations deemed it perfectly acceptable to protect the individual person against being coerced or forced to testify against themselves in such cases, they saw that this would be a huge conflict of interest and besides, in most cases, it would be possible to accumulate the necessary testimony without relying on the accused person providing any of the relevant information. This approach seemed to work for a very long time: procuring hidden details for the social concern, exposing what needed to be exposed when it needed to be exposed even if that exposure violated someone's privacy.

The first indication that this approach was insufficient was when people started to wonder how to prevent crimes before they happen. And by 'prevention' I do not mean something like condition the educational environment within the civilization such that the society procures an aversion to such criminal acts. That sort of cultivation of prevention had been going on alongside the much older strategies in the form of norms and taboos, the threat of punishment or even death, excommunication or banishment. This was a different form of prevention. A prevention that likely stems from our orientation toward scientifically predictive methodologies for determining patterns of human action and the likely directions that it might take. If you look at the metrics describing the past crimes of any population, you can learn all sorts of things about it, you can discover the attributes and characteristics that tend to accompany various criminal behaviors. It becomes possible to build a profile of a likely criminal under a given set of circumstances. With other methods that track the circumstances in the environment and the details of those people who will not commit the crime but who might be its witness or its victim, a scientific investigation can determine likely events that will lead up to a crime before it takes place. You do not need to be a mystical precog in a Philip K. Dick novel to see the crime that is about to happen, you just need access to the data and a good algorithm for sorting through it to determine the patterns and probabilities of people doing specific types of things and the circumstances under which they will do them. If those circumstances can be determined, then no magic or mysticism is required, precognition of criminal behavior is just empirical science driven by the accumulated data.

Tools for gathering data are only part of the story, the experienced detective has the data and the tools up to a point, they are just too limited in their scope to be able to focus their attention on the society at large. They need the criminal act to trigger their attention. What if attention were always possible in all places and on all things, the all-seeing eye could attend to all possible conditions, gather all pertinent data, and effectively render precognition of those crimes? It may be that the events of September 11, 2001 were exactly the world event that was needed to trigger this shift. Or it may be that the tools had already been developed and were already in use, but this was the event that made it acceptable to go public with the procedures, or at least made the risk of being exposed using such procedures a reasonable one for some of the interested parties. A terrifying event of epic proportion is enough to convince many people that whatever it takes to bring about such precog awareness of crimes would be well worth whatever sacrifice it requires. And following upon that, it is just a coincidence that the civilizations impacted by such events started a massive technology race that involved more and more surveillance capabilities, social networks, tracking devices installed on nearly every human being (aka smart phones), and a general sense of the necessity of such mechanisms for living a modern life properly protected from the catastrophic harms that threaten us. And by coincidence, I mean part of the Zeitgeist, or maybe they had just become baked into the set of social conditions that everyone was thinking about, everyone was focused upon and attending to? It was only natural that there would be so many innovations and advancements in that same general vicinity, many mistakes made by many different people all culminating in a multi-year project to institute a set of changes in the direction of a worldwide telemetric understanding. No conspiracy theories needed, it is enough to say that we all lived amid the same world when certain events struck and so we were all singing the same song in the shower afterward and the result was the seemingly calibrated effort to institute the most massive network of surveillance mechanisms that the world has ever encountered. Although there are big players in this mass surveillance, there are multiple of them, and we need not think of their results as intentionally collaborative when understanding that they all emerge from a common social fabric made up of a set of interests being served at the expense of others. How else could one account for the enthusiasm with which these offerings have been accepted by all and sundry? Would you have it that we are all co-conspirators or merely that we are all complicit in these conditioning events?

Oddly, this precog behavior rests on a dependency on data science that binds together even very diverse interest groups. It is not uncommon to find people who doubt the authority of science on various issues like climate change based on fossil fuel use, for example, but who also think that using deep analytics to determine patterns of behavior leading up to terrorist attacks is a necessity for which any allocated common treasure would be well worth it. And this is so even when the reliability of the data in the former case is significantly stronger than the reliability of the data in the latter. It is important to note from such circumstances that the core of the argument is not whether data analysis is a valid form of knowledge acquisition. There is no culture war on the value of science when it comes to formulating possible causal chains and their relationship to desired or undesired outcomes. Rather, the issue is a simple battle of interests. Where the conclusions do not fit with the desired outcomes they are discarded, and where they do, they are pursued. Simple as that. Ockham's razor. This explains why you cannot use analytics to justify the hypocrisy of your opponents in such arguments. Either they already have an interest in your conclusion or they do not, the data will not convince them one way or the other. Or rather, any such analytics will immediately meet with opposition that points in another direction and offers an alternative story. The recuperative capacity of the telemetric understanding prevents discussion. And, so the theory goes, always has.

What is new, what did not exist before as far as such detailed analytics is concerned, is the megaphone that makes the results of the analyses available for browsing on the worldwide network of computers dedicated to the rendering of data and the analytics that can be run over the top of it. This megaphone makes the commonality of attributes visible unlike ever before. When we invade someone's privacy for the sake of the society at large (and in the surveillance state that means all the time), the output of that exposure is louder than ever before. The characteristics of people, their perversions and quirks, are everywhere open to view. People have learned that they are not freaks for having the orientations and involvements that they have. They need not live quietly and invisibly without showcasing whatever characteristics provide them their personalities, rather they can boldly bind together with the rest of the segment of the population that is just like them, and they can recognize themselves as part of an interest group. Mass surveillance has made analytics ubiquitous to the extent that "mistakes" are not even perceived as mistakes any longer, they are life style choices and preferences. This is perceived in a liberating experience as something that may have once been thought to make one a monster or a freak. Now, there are freaks everywhere and we have many tools at our disposal for finding each other. Surrendering privacy had a normalizing effect. And so long as no one is directly harmed, then the people who had been living under the shadow of 'guilt' need no longer harbor such feelings. Seeing this as emancipatory, however, may fail to account for what made the characteristics 'freakish' to begin with. That is, something that amounts to an accidental change in social conditioning from one season to the next can go another way just as easily. Backlash can occur, reordering, new social interest groups emerging in opposition to one set of interests for the sake of another. The circumstances where the analysis led to liberation can easily snap back into analysis that leads to persecution and assault. Now that all the 'freaks' are out in the open, it is easier to round us up.

The analytics is critical to the ethical agenda that we are exploring. The way the mistakes of individual human beings are exposed will show up in the analytics. The metric emissions of those mistakes are flush with attributes and characteristics. As such, they can be aggregated in the analytics that pulls together all the metric emissions of all the contributing agents. This is how the ethical is put into practice: not made possible, in the sense that we previously discussed, but conditioned into human action. There is nothing transcendental about it. Analytics provides a digested way of conceptualizing the things that make up the world, attributes them with universal qualities, and so binds them together into interest sets. The interest is crucial for determining the exposure or non-exposure that constitutes the boundary condition of social involvement. Whether to expose it in the aggregate or in the detail will be the only necessary determination. The feature of "drilling into the data" has critical social value. When the aggregate indicates a pathology, then we must drill into the aggregate to examine the details behind it and the contributors who make up that detail. These are the deviants, the criminals, the ones that need to be pursued and incarcerated. The ethical determination of good or evil is derived from the analysis and so constitutes something that is itself decidedly unethical in that sense prevalent among traditional 'lovers of wisdom'. Rather, ethics becomes the judgement of others, pure and simple, it becomes the cudgel that can be used to pound an action or characteristic out of existence when it threatens the desired order. It becomes measurement at a distance. This practice is not attached to an autonomous deliberation of any kind and it is not bound to a decision-making capacity. It seems that whatever mistakes Kant may have made, he was clear that the ethical was not some set of judgements made about actions after the fact by observers. He was in search of the origins of the deliberative act. His solution to that riddle may have been flawed or missing in various of its features, but he at least knew that the question of acting ethically was a personal matter, a set of personal considerations that governed the origin of an action, and not merely a matter of applying the principle to the action after the fact as a form of criminal forensics.

This point explains why I tend to ally the sense of personal dignity and autonomous deliberation –or whatever it is that we hope to call the origin of the ethical– with the notion of privacy, the expectation of privacy, and private personhood –or whatever thought heuristic we use to get at this seemingly possible impossibility. That is, it is in the moment of determination, during the event of appropriation, when things happen or continue to happen, that the import of the ethical must be at work if it is to exist at all. If I cannot interrupt myself mid act with a set of concerns that bring pause and a change of course based on my concern for what the world I am making here and now is like, and how it shall emerge from me and for me, then I am not capable of acting ethically. If I am not able to draw from the power of deliberation and orient my actions as I see fit and as it is appropriate for those actions to be, then I am not a free being. Somehow, I personally have come to be conditioned to think that this ethical impetus is both necessary and impossible. Necessary, because without it I am unable to realize myself as the sort of being that the history of human beings suggest I should become, and impossible, because I am a product of my social conditioning and there is nothing magical about me that would suggest that I am capable of such a feat of power and determination. The more analytics that there are at work in sorting out the good and the bad, the categories under which we fall, and how we align with each other based on those properties, the less privacy we have (whatever that might mean: telemetrically or phenomenologically). And that suggests that analytics are diametrically opposed to this potential to be ethical, to be your own best guide at preserving the boundary between what should and should not be exposed, what mistakes and errors can remain hidden and which ones must be revealed. Under this approach, the burden of proof remains on the detective and the detective must begin with a body, something that originates the investigation or launches the concern over whether there is probable cause to violate someone's person for the sake of pursuing a criminal case.

Under these conditions, the precog culture will be thwarted and the predictive power will be surrendered for the sake of the personal privacy of the human beings that populate the social domain. The form of analysis that the detective traditionally employed is not based on large-scale surveillance, but on specific reconstructions. Big criminal acts do not come from nowhere, do not emerge fully formed from soupy connections in the social world. Rather small crimes are committed to prepare for larger crimes and there are cases where traditional detective work has been used to uncover a larger scale criminal ring headed toward some diabolical act based on the evidence trail left by those smaller crimes. Respecting detective work of this kind, encouraging, and recognizing the analytic behavior that is carried out by such experienced professionals, was once valued appropriately and not merely as a pale substitution for mass surveillance. People use analytic capabilities in their lives all the time, these capabilities can be cultivated, and the skills they involve can be mentored and improved with collaboration and experience. Rather than spending enormous amounts of social treasure to build the machines of mass surveillance, civilizations may choose to invest in the analytical capabilities of the people that inhabit those societies and develop their abilities to think critically and deeply in areas that may protect the society from harm.

Of course, I cannot help myself when I speak of these things as though they are available as a conscious choice that people or societies can make. I realize how ridiculous that sounds, but I have not made the choice to observe it either. Nor is it a choice to point out that it is possible to pay more attention to the causes of the developing circumstances which give tacit permission for those enormous crimes that we now wish to prevent. What brought these conditions about, how did the world become such a place? We can use the analysis of historical events and past governmental or military action to come to some degree of understanding on why the crimes that are being planned seem desirable to the perpetrators. In traditional crime-solving activity this is known as the motive. To understand the future actions of the person, our analysis may focus purely on the aggregate characteristics and correlations, but if we try to understand the people and how they came to a point in their lives such that their actions seemed like the appropriate set of errors and mistakes to commit, then a different kind of preventative measure may be formulated. It may be that our precog orientation is too late in the process. If we had properly understood the motives involved in the forming of the criminal intent, then maybe we could have changed larger scale actions and conditions that continued to promote those criminal responses. Maybe the criminal is responding to some very specific set of events and circumstances, and maybe the society needs to evaluate whether those circumstances are crucial to it even at the cost of the crimes that they inspire. Unless of course it is the case that such crimes promote a set of interests occupied by people who have become collaborators after the fact by capitalizing on the promotion of those interests.

Just as metric emissions can have a more human meaning, can be situated in a specifically human context, so too can the analytics. It is shocking how much effort is placed in the analytical infrastructure that aims at data analysis and correlation when so little emphasis is placed there as a valid form of human cultivation and education. Surely, if it is worthwhile for the entire civilization to depend in every aspect of its existence on the analytic capabilities of machines, then it should be safe to say that teaching human beings to be analytical, or at least to have the capacity to do it better than they do already when the circumstances require it, would be of some interest to the society. Unless, that is, there is some reason why it would be contrary to the society's interests to be populated by human beings with strongly cultivated analytical skills. And drawing on all the work we have done throughout this yarn, we should draw conclusions about interests based on what we see in the world rather than basing them on anything that interested parties say. Whoever's interests are running the world's civilizations at this time, they do not appear to place emphasis on the capacity of human beings to think and act analytically and critically based on discovered circumstances and involvements. The phenomenological training of the citizen is left in ruins and the only thing that matters in our real-world emphases is the marketability of whatever skills a person can acquire through their education. This marketability is rarely ever conceptualized in terms of the micro skills that go into that employment marketplace. That is, we emphasize engineers or mathematicians and scientists, rather than emphasizing the cultivation of cognitive capabilities that make one a good engineer or a good scientist. We pay a lot of lip service to analytics, but we do not have a very deep performative association with those skills and the extent to which they should be embodied in living people. And most notably, the interested parties would seem to be clearly aware that the likelihood that a human being might repurpose such skills to areas where they do not belong is strong, thus guiding their orientation against promoting such basic capabilities. It may not be possible for analytically minded agents to initiate revolutionary acts that upset the balance of power among interested parties in the social domain's status quo, but history has clearly shown that such agents are significantly more expensive and difficult to manage.

Just as metric emission ends up being a moment in the telemetric logic that need not be surrendered to the telemetric understanding, so too analysis. The basic components of the telemetric understanding, the systematic maneuvers or patterns that guide its operations, may themselves be residual to the commonwealth that articulates that telemetric understanding. In following through on our insights into metric emissions and analysis, we have gained the sense that it might be the case that what remains in the telemetric understanding is the telemetric logic appropriated by human beings for use in working to catch a partial glimpse of, or insight into, telemetric operations.

Feedback (The Constitution)

The residual climax to the story structured by the logic of telemetry is feedback. The aim that inspires any such understanding, even that of the artificial persons best suited to experience the world that telemetric understanding exposes to corporations and other highly customized private persons, is the feedback of the results of analysis into the action world of the socially conditioned persons. The critical insight here, if there is one, would be that feedback orienting personal involvement is marked as the love of wisdom expressed ecstatically in etiquette because it is the practical navigation of experience that draws together all metric emissions as the mistakes that they are and spins them together with the years and years of pattern discernment that has been collected for the sake of comprehensive engagements exerting themselves upon each socially conditioned member through that very experience. To the extent that the analytical is a phenomenological engagement with the domain of others and the environment that we inhabit together, it is the truly scientific philosophy achieved as love of knowledge. Husserl's dream is realized, but not in the way that he hoped. Philosophy as rigorous science is not to be realized as phenomenology. Rather the phenomenological, the human-centered engagement with the conditions of possibility, with the world that is there to be understood, and that conditions the human being in the course of all action orientation, that is the application of scientific, and even analytical, rigor to experience rather than from it. Such a telemetric maneuver is residual to phenomenology and is not done for its own sake, it is not the be all and end all of a rigorous science. We do not engage in such analytical pursuits in the domain of our errant ways for the sake of gaining greater knowledge about the objects and subjects that populate that world, rather the work is done for the sake of continuing the action in a feedback loop that makes further mistakes. And such ecstatic commissions break down into the emission of more metrics with the impact and conditions of the analysis in view to better adapt and better comprehend the conditions that present themselves in experience. This translates into better anticipation and expectation of what is coming for the sake of producing a better situation for oneself among those others with whom one is collaborating to collectively establish that world.

Since the analytical phase is often oversimplified as 'philosophy' (where everyone with a PhD or equivalent is a philosopher) and spun into a telemetric endeavor, we can distinguish the feedback moment from the analytical moment by referring to the former as the 'love of wisdom' embodying ecstatic forms of know-how over and against telemetric forms of knowledge bent on coercive acts. This repeats the notion from earlier on helping to recall that this is an unrequited love of wisdom as it is, marked in a further pursuit of that same bound and conditioned set of emissions and characterized as errancy or the personal ex-propriation of social conditioning as personality. And such personality has been established already as transcending the bounds of the natural person through the various social augmentations and etiquettes that are part and parcel of the analysis to be done in the process of feeding these patterns back into aims and an understanding which can only be realized as an unrequited love of wisdom.

This is pure fantasy, of course, because spinning such a tale about the love of wisdom as the "automation" of the telemetric logic that unfolds as informed error would easily be recuperated by dismal mechanisms of the telemetric understanding that work to transform such a painting into something familiar, something well understood, and something that has already been categorized in accordance with a set of principles providing a standard taxonomy of human possibilities and situating any potential approach that might be discovered or formally presented there. We would be cheated out of our ecstasy. And the fantastic nature of the offering is due to the apparent proximity of automation to autonomy. This proximity is very confusing. When we suggest the automation of the process, we are suggesting that it be made into a self-governing process, a process driven by its own rule, but this driving force is externally defined. The automaton is the opposite of a free being, it is the setting in motion of a chain of events that cannot be interrupted because once set in motion it is under its own command. It did not launch itself into this motion, rather it was launched. Autonomy as such is supposed to suggest the capability of self-rule that is commonly associated with freedom and the emancipatory interest of the rational being. Such autonomy, however, is an internal conflict where the one connotation suggests that the other, machinic connotation, is substandard; whereas this machinic connotation suggests that the liberal connotation is impossible. We can complete the fantasy by suggesting that under the conditions of telemetric understanding the human being functions as an automaton, whereas in the residual, the remainder of that understanding, the human being may function autonomously. And the latter meaning must have already been infected by the former suggestiveness. That is, the fantasy of autonomy is made fantastic by its roots in automation, in an automatic procedure that is subsumed under the higher order systemic logic of an understanding that is responsible for its context. The human understanding, if there is to be such a thing, must come to be amid this originally automatic state, cast within the boundaries of its finitude, unable to completely escape it, and thus unable to realize itself as the free and autonomous being that it has always hoped to be.

The love of wisdom, as such, is always something of a pathetic state, an orientation that is carried out in the context of impossible odds, statistically speaking. The wise person understands that there is always too much behind them preventing them from attaining any complete grasp of whatever it is that is in front of them. They merely love wisdom, they do not necessarily have it but remain forever in its pursuit as something deferred. Suppose each step forward reveals a greater understanding, but that understanding includes recognition that the wise individual is further away from what they were pursuing than they realized. The lovers of wisdom are more like Tantalus than they would have liked to be, would have preferred to be, with their beloved always slightly out of reach. At a distance, but only just barely. An unrequited love is a love from afar. Because the beloved is not turning toward the lover, the relationship never achieves the desired intimacy and remains a telemetric maneuver. The residual needs to be understood as a residual to the telemetric. That means that it is always in a relationship with that logic but not quite reducible to it, rather it is what remains and is left over from the systematic determinations with which this logic operates.

It is possible to spin this love of wisdom into something of a stalker dynamic. Wisdom does not want the lover's attention or does not know about it at all, but the lover is oblivious to this and pursues it nonetheless. Such a lover acts more like a hunter or ranger. And to this extent, it may be critically challenged whether such a dynamic truly expresses love. And no doubt, the actions of any lover of wisdom may, on occasion, degrade into such perverse expressions. There are ancient names for the use and abuse of the love of wisdom and there has been a constant concern by lovers of wisdom that they are falling into the trap of sophistry. Instead, we may be lucky, there may be a bright spot in all of this, although it is difficult to set it out in a guarantee. The end of this pursuit, wisdom itself, may be more aware of the attentions being lavished upon it and may have some 'feeling' in response, although that feeling does not itself match with the intensity of the lover's feelings. This may be the vision of parental love that some philosophers have made use of. The love of the lovers of wisdom may be a parent's love for their child. The child is grateful for the attention, needs the attention, but often turns away from it. The purpose of parental love, in the philosophical use to which it has been put, is to shape an independent being who will be able to set off on its own at some point. The parent wants to create an autonomous person that will no longer need them in the way that it once did. And this asymmetry is a direct result of the form of parental love. To the contrary, the child –if they are equally devoted to the parent– would be incapable of ever properly receiving this intended benefit. Luckily, or so the story goes, nature has not crafted us that way and the child loves the parent differently than the parent loves the child. And this difference amounts to being able to turn away from the lover and make use of that love to launch themselves into the world on their own for their own aims and ends.

And we might call this love unrequited only in a loose sense to suggest that the return of the love is not in the same form as its investment. Even in cultures where the return is devoted geriatric care, the imbalance remains, since what was in one direction the fostering of independence has been returned in an acknowledgement of growing dependence. This kind of a return is a deferred return and not part of the same venture made when expressing love while raising the child. In this immediate expression of love acted out in actual parenting, the parent does not get back exactly what they give and so there is a kind of pain in parenting that is not there when the conditions on either side of the relation are symmetrical. To that end, it is the best of all possible examples of unrequited love. It is not that there is nothing that comes back, it is rather that the return is something very different from what has been ventured and that gives the appearance of an imbalanced dynamic. What is returned, however, is the pride of an artist in the outcome of their creation. The parent, in so far as they sought to produce an autonomous being, one able to stand up against the social conditions and navigate its own way (which is really the best that can be hoped for in matters of autonomy), enjoys the ever-changing surprises that come with that creation. Such is the metaphor.

To place that vision into the realm of wisdom, it would have to be meaningful to characterize wisdom as one's child. In so far as you are experiencing the world, the experience you have accumulated, in this picture, must be formed into a thing, or rather a creation, some conditioned output that you have been endeavoring to produce while analyzing all your mistakes and errors over the years. And that is not a common approach in matters of wisdom and might be characterized as something of a horribly twisted version of it, a fetishized view that alienates it from the lover of wisdom in their act of producing it. And yes, that is the connotation I am aiming for as something like that must be at work in the process if there is this distance in the creation that puts the wisdom pursued at a distance from the pursuit. The wisdom that human beings have to "achieve" in the course of their lives, under these conditions, is not something that they are, it is not an existential state that they come to occupy, rather it would be a pursuit that they become better and better at performing as it continues and their tactics are refined through feedback. And yes, it is perfectly reasonable to think of this as nuanced in some difficult to manage space between the stalker and the parent that we have set up as the key character development in such a lover of wisdom.

This strange casting of the love of wisdom is meant to explain how human beings can be endlessly engaged in such a pursuit and yet may seem to make nothing of it. For some, the pursuit of wisdom would seem to have had no fruitful effect on themselves: they are tiresome and boorish. The possibility of this lies in the action orientation toward wisdom that the pursuer takes upon themselves. Such a boor thinks it is the trinkets they have produced and accumulated over the years that marks their stories with riveting content such that any who hear them will be enthralled and will likely identify with the narrator. Such a boor is contrasted with the lover of wisdom who knows that it is more of a skill than a body of knowledge and so every tale that they tell, every yarn that they spin, will evoke the way this skill has been won and which now can be advanced in the processing of the story. The listener understands that they are in the presence of a wise person because they see that the story is not about what it seems to be about. It is not a story that is limited to the characters and events that comprise its content. Rather, the listener learns that this tale tells them more about *how* to be wise, extends their own pursuit of wisdom, and their own efforts to love wisdom and develop its skills. They embody this as they feed what they have learned back into further attempts to err and appropriate their personality in the events and operations that characterize their lives among others, involved with them, and –importantly– mutually engaged in this common pursuit. Aristotle may have called this friendship.

And this indicates a crucial piece. Because the love of wisdom can only be characterized in the context of the residual that remains in the dynamic of the commonwealth, the love of wisdom is usually conceptualized as a cognitive capability that works as a faculty in the subject of the individual human being. Rather, such a being is always an agent for the commonwealth in which that agent lives, an extension of its hive mind. The subject is not autonomous in the classical sense. Rather subjection is the outcome of conditioning and social involvement so much so that the love of wisdom can only be realized in the course of it and as a moment of it. Only as conditioned beings living among each other and collaborating on this world that we are making day in and day out is it possible to love wisdom. Only under the conditions of life as a social being can this skill be developed, and this know-how realized. It is only in so far as we are in the world together that we can feedback metric errancy appropriately analyzed in the actions that we take to further articulate additional error conditions that have been conditioned by the collaborative efforts of others who are likewise engaged. It is not so much that there is this way of behaving that is common to us all because each of us is individually engaging in that behavior and so there is analogous operations at work in each node of the system. Rather, the collaborative commonality of the love of wisdom is because it is the operation of a networking principle and it is common to us all because it is the logic of the conditioning process relaying across that network. It is embedded into human beings in the process of becoming autonomous, of being parented. And as such, this functions as an imprint of the type upon the instance. This means that there is something of the generic at work in the procedures, but invisible as such because it marks what it is like to be the instance of a type: "the class instance," says the lens grinder.

Consider a moment when a group of human beings are together in a room collaborating for the sake of some end. The end is of a type that we have described before as being in some sense external to the human beings in the room, but crucial to the enterprise for which they are working. They are trying to realize some corporate aim. And they are struggling. They work together to come up with some solution to the problem and they all know it when they hear it. This happens sometimes when the conditions are just right. There are no disputes or arguments in the end (although there may have been plenty of them along the way), when they finally come to a suitable approach. And it may be the case that any observer to the meeting would also agree that they have managed to work together to come to some perfect conclusion to whatever issue they were working on given its specific parameters. Concluding the matter in this way does not make the result any more personal than it was before they started. These are still, in the end, human beings who are acting on behalf of the corporation in so working together to solve this problem. They each might experience something deeply personal and fulfilling in that moment. The residual dynamic, the operation that they have carried out for the sake of the corporation, will have left a mark. The participants will have sensed that they have somehow grown wiser in this outcome. They believe that their pursuit of wisdom may have been furthered along by this successful event of collaboration. The residual is this love of wisdom there in that telemetric moment where the participants are brought close to themselves when they most righteously express their distance from themselves in realizing the corporate objective.

And this story in the corporate context need not be corporate. It is a common story in many different types of group dynamics. The co-op board may come to such a conclusion, the activists planning outreach might come to such a conclusion, and the baseball team trying to win the game may come to such a conclusion. There is nothing specifically corporate about this pattern, but it is important to highlight that the pattern is possible within the corporate domain and that even there it is the case that the outcome of the operation still contributes to the love of wisdom born by each of the participants. I suppose this could be interpreted as a kind of apologetics for spending one's life as a corporate stooge and surrendering any claim to autonomy in the process. Quite to the contrary, it may suggest that wherever human beings are, they are capable of the residual and of appropriating something personal in experience even if that is immediately recuperated as wizened job experience by the circumstances in which it has been elicited. Even under the most dehumanizing of conditions, there is such a possibility. This is indeed the lure of such megala-domains of human action, especially when there is a hook that binds one's interests to that megala-domain. And, to be sure to maintain proper perspective, let's be quick to point out that working for a corporation is not the most dehumanizing of conditions. A peasant's tragic life may only be exacerbated if they were to achieve complete clarity in apprehending their condition within the global machinery that requires peasants so that others might thrive.

Appropriating something personal in such telemetric contexts is not predisposed by some value judgement. The pursuit of wisdom, as I have run up against it throughout this essay, does not appear to be a skill that is itself the embodiment of goodness, of the ethically bound agent who is now acting for the sake of the good. Both Dalai Lama and dictator may love wisdom. The love of wisdom is potentially something quite dark. There can be wise butchers and thieves, a pirate can –through their wisdom– become one of the most powerful people in the world and use that wisdom to feed their coffers with the stolen goods taken from their victims. There is no guarantee in this love of wisdom, it does not orient the human being in a way such that they will be determined to do the right thing for the right reason or serve the common good and promote the ends that best realize the principle of utility. The lovers of wisdom are the embodiment of the best of human being to the extent that, for them, ethical action is possible, and we are entitled to claim nothing more than that. In biting the apple, any such human being becomes capable of right and wrong, good and evil, good and bad, but the love of wisdom is not itself wisdom, not itself the differentiator and the judge. Pursuing wisdom may never result in the personal accumulation of wisdom, rather it may accumulate in the circumstances providing the context for one's actions. In the telemetric understanding at work in such action, even the lover of wisdom may be a tool with a partial role in the instrumental dynamic. The residual of such exploitation, fortunately or unfortunately, is more of the same: continuing to love wisdom as an agent that can be proclaimed guilty as a collaborator or guilty as a revolutionary.

Just as Kierkegaard's commentary on Abraham suggested that he never got any further than faith, we might say along with him, with the same emphasis, that even if human beings were to never get any further than the love of wisdom, that itself would be extraordinary. Being properly oriented may just be enough even if the automaton never gets anywhere. Emergence in autonomy is not a final state, it is an entrance into a kinetic state that amounts to little more than a condition for the possibility of something or other, the potentiality to be something or other. The monstrous possibility of the individual. This marks a departure in what might be codified as the logic of analysis and distinguishes the feedback moment from any such knowing-that which characterizes the second phase of the residual maneuver. Knowledge as traditionally contemplated, by some at least, has the characteristic of seeking to grasp its object and contain it somehow. The absolute grasp of certainty may have been relaxed for the sake of degrees of probability, but the paradigm shift is not so very radical after all. Ultimately, there is still an idealization of the grasping itself, a notion of exhaustive visibility that guides the analysis and that provides the bar for probability. It may be that 1 is never the correct probability of a future event, but 1 still provides the boundary condition for the scale that is used for that projection of probability. In the love of wisdom, the effectiveness of such a spectrum is set aside. It is not so much that knowledge of that sort does not matter, it is that it is recognized to be an ingredient, a necessary condition for proceeding but not sufficient. And the necessary further step in the autonomy of action that haunts any human being like a poltergeist, playful and taunting, is to engage forcefully in the love of wisdom, which is always a state on the verge of disappointment peppered by the affective hope of better things that will propel the lover forward, or sideways, or backward, or whichever way wisdom lures it during the pursuit as though it were a toddler being chased by its parent.

If such a feedback pattern is the third moment in the residual to commonwealth, could not it be said that there is a residual in the residual? Does not this architectonic, with its overly clever mechanics suggest an ongoing processional that is a residual of the residual which can only be interpreted in the process of a further recuperation: another phase, another movement in the architectonic? Might this further recuperation resemble the spirit that follows upon reason and which is followed by a religion culminating in absolute knowledge which, truth be told, can only be the end in so far as it is the beginning of another rotation, of a re-entrant attempt to recuperate everything that has happened thus far in the light of such an ending?

The residual of commonwealth cannot be advanced positively since there is no determinate moment to resolve it. This may be the problem with Absolute Knowledge in the Hegelian architectonic. There is no supporting it within the so-called systematic design because its function as End makes it impossible to articulate within the system itself, thus forcing another cyclical movement, a re-entrant initiation back into the process unfolding. Or so goes the conjecture. Each of the residuals (Telemetry and Phenomenology) were only fully realizable in the following part, telemetry's in phenomenology and phenomenology's in commonwealth. That makes for an awkward requirement in this final chapter. It occurs to me that the only plausible residual (the only statement of what's left) that could somehow be evoked –without systemic and telemetric recuperation destroying it as residual– is the architectonic itself.

I do not know if the architectonic has been successful, but my intent was to repeat the pattern of the thinking thing, the act of thinking that asserts itself in the moment of doubting itself. Or to provide a repetition of the pattern of the transcendental deduction. Or of hermeneutical circling/understanding. Repetition as architectonic reinvents 'privacy' in the form of a lover of wisdom: open to association and constantly vigilant against moments of recuperation and partiality in the feedback of analysis gathered from metric emissions. And I do not know if the architectonic accomplishes such a goal in anything like a plausible or well-established theoretical sense. No matter, I assign that impulse and pattern to the telemetric and do not care much about succeeding in such fashion since it would be partial and immediately recuperated. What ultimately –if anything– may come across is that we do not yet know if we can become lovers of wisdom and thus ethical beings. We have not yet done the work to establish that and so are not yet entitled to such claims, to such "a way out" –whatever that might mean.

In the end, I see the desirability of attempting that very maneuver, I see the necessity of it, but not its possibility. It is in keeping with the residual of the commonwealth to engage in a collaborative wide-scale endeavor to prove that I am in error. Continue.

Epitaph:
Here lies TPC: recasting Pascal's wager, lover of wisdom.

Appendix

There is a loud and international mantra going on in the world today in which many different people from different walks of life are calling for greater protection of their privacy. They are concerned about corporations that are harvesting their personal information and using it to profit, they are concerned that governments are collecting their personal information and using it to secure party agendas, and they are concerned about black marketeers who are stealing their most confidential information for the sake of criminal activity and identity theft. If we were to judge from these concerns, we would conclude that 'privacy' amounts to restricted access to personal information that has been published by parties named in the information. The tension in this concern is very real. People are publishing more information about themselves than ever before. And this rampant publication has made the people doing the publishing vulnerable to attacks, subject to analysis, and better known to external organizations whose interests are served by getting to know people better. We run the risk of thinking something absurd is happening in the world around us, in thinking that people who are putting their lives on open display are fiercely concerned that they no longer have any privacy.

Simplistically speaking, I might revise my original contradiction and say that we must beware the urge to be seen from far away because it might cost us our privacy. It would seem that –en masse, at least– we want to have our cake and eat it too. We want to be seen from far away and we want to maintain our privacy. What could that possibly mean? By hypothesis, I suggested that this is not merely a matter of some people wanting one thing while others want something else. The claim is not meant to be an indictment in the aggregate: that there is a higher order phenomenon taking place where "civilization" as such has spawned drives that want various things that are contrary to each other. Instead, the proposal is that there is something at work such that there are specific people, individuals, who both want to be seen from far away and want to preserve their privacy. The two aims appear to be collocated in the same behaviors. Consider, for example, a person who repeatedly publishes information on a social media platform proclaiming how great their life has turned out. They love it when their friends bubble with adulation, but they withdraw in disgust if any of their nameless followers call them out as phonies or attempt to creep too close with lurid direct messages. This person both publishes content in the hopes that their audience will attend to the message and then becomes outraged when that attention manifests itself outside their pre-conceived notions. To which one might ask, why not just stop publishing at all? What drives you to have distant relations with friends at the cost of being an object of interactive voyeurism by acquaintances?

And such a person has a simple answer to this. They do not want to be left completely alone and they do not want to live in isolation. Privacy is not the same thing as isolation, it is a relationship that they wish to cultivate with some set of persons of their choosing. Their friends and the friends of friends, making up a network of followers on the platform, are trying to build some form of community. And they want to proceed with this project without concerns that opportunistic secondary usage (another name for the interactive voyeur?) might be made of the materials they present to community members. They want as many followers as they can get, they want to associate freely with their peers without anyone inappropriately interjecting themselves into the communication channel, and they want to be arbiters of what counts as propriety in the responses they receive. They may not have a clear vision of the distinctions they are making here, but they want to build communities and have a say in how the scope of those communities is defined. As counterintuitive as it might seem, such community building may be coherently advocated as a privacy right. I will not be trying to build a case for or against this, I'm only claiming that the way in which the issues and concerns are framed matters in that some ways of doing so will give the impression that private relationships are at stake, whereas other ways of framing it will suggest that private property and control over it is the central concern.

The distinction is important because it is what allows us to make sense of the perfectly reasonable co-existence of the two acts: 1) people are putting more content about themselves into the hands of other people than ever before. 2) people want greater protection for their privacy. If putting more content about themselves into the hands of other people is understood as an act of publication, then we might think that they want to maintain copyrights on that data and establish a fair use policy for it. This is a way of thinking about the publication of personal information as a property matter and the demand for privacy as a demand to maintain control over the use of it. If, however, we understand putting all that content into the hands of other people as behavior that makes use of a networking platform to cultivate communities of connected persons, then the demand for privacy is a demand to avoid surveillance while making those connections because of the impact it will have on those connections. This is a way of thinking about the dissemination of personal information as a matter of intentional association and the demand for privacy as a demand to avoid being subject to unreasonable searches of one's person while exercising one's rights to freely associate. It may remain an open question at this point whether search and seizure are distinct categories of rights or whether both are intricately intertwined. Is constant surveillance merely a search of your person or does it also take something from you? Is a search always for the sake of a seizure?

Thinking in terms of publication and maintaining control over the published materials, however, is idealistic and far too narrow in scope for the subject area invoked by my warning. Sometimes the attention trained upon you is surreptitious or clandestine, sometimes it is aimed at capturing your attention without you realizing it, and sometimes it aims at drawing your attention away from something that should be your focus. The warning may be appended with a further warning to be careful what you wish for since you may get it in ways you did not anticipate. Let us consider a variety of different scenarios and then some points of alignment. The point here is to vaguely gesture in the direction of a coherence view of privacy. This initial gesture will lay out some of the terrain that we will need to navigate as we explore the contradictory urge and the dynamics within which it is exploited by various interested parties:

1. Every time you go to your favorite restaurant, you order Pinot Noir. The sommelier remembers you and starts offering you special vintages whenever you come in.

Note that in this case someone is observing your behavior and remembering your past selections to guide future interactions with you. They are not taking anything that belongs to you, you are not giving them anything to store for you until you need it again. Rather, you have merely visited them at their establishment and made selections from their menu. We might think that such a sommelier, in this case, is doing their job thoroughly. Many people might have increased loyalty to such a restaurant and be more likely to return given the excellent service. This sense may continue to hold even if, at some point, the sommelier was to offer you a Malbec that they think you might like. And suppose you do like it? You might find yourself migrating ever so gradually from your devotion to Pinot Noir to becoming someone willing to alternate between Malbec and Pinot Noir depending on the circumstances. In this perfectly innocent act of excellent job performance, the sommelier made use of your personal preferences to impact those preferences with alternatives that you initially resisted or showed no interest in trying but later came to enjoy.

2. You keep a diary filled with all kinds of personal reflections. Passing thoughts. Half-baked ideas. You die, and your father comes into

possession of this diary. He publishes it and it becomes one of the best-selling books of all time.

This suggests the historical case of Anne Frank who left her diary behind in the secret annex in Amsterdam when the police raided it and took all the inhabitants and some of their caretakers into custody. Miep Gies, who had not been arrested, gathered up Anne's diary and then held onto it throughout the remainder of the war. When she first saw Anne's father after the war, she did not give him the diary, because at that point she did not know whether Anne was still alive. Once Miep received further information confirming Anne's death, she sent the diary to Otto Frank. We know that he then published it and it has become a widely read testimonial to the events that took place during the Nazi occupation of the Netherlands. What gave Otto Frank the right to read his daughter's diary? The actors involved appear to have treated the situation as a simple case of property rights. The diary belonged to Anne. Miep intends to hold onto it until Anne can come to claim it. Anne has died and, presumably, did not leave a will so Miep gives the diary to Otto as the rightful heir of all Anne's worldly possessions. Otto reads the diary and decides it is of historical significance and then publishes it. On the surface this does not appear relevant to Anne's privacy or the violation of it. Maybe dead people do not have any privacy rights? Some reductionists might say that this sort of reasoning can be broadly used in privacy related scenarios and all such cases can be analyzed to show that privacy can always be reduced to some other right or set of rights.

3. You are a child and your parents teach you all sorts of things by example and by direct instruction. You learn how to count money, that nickels are worth less than dimes even though they are bigger. You learn to go to the store on the corner by yourself and pick up milk for your mom.

Each child born does not start over from scratch. Thankfully. What we have learned over the years, what each of our civilizations uses to assist in the struggle for survival, is passed on through parental guidance and instruction. Here we focus on parents, but there are many facets of a civilization that are aimed at introducing children to the accomplishments, wisdom, and errors that the group or species at large has achieved. As such, guardians of children make all sorts of decisions about how to develop and influence the behavior of those children. The child's temperament is quickly absorbed in the personas of the adults. Class centric behaviors may be passed along, whether the child comes to love learning or money, honor or fame, may develop out of the example and guidance set down by that child's parents. The parents, by the sheer force of their constant proximity, do not leave the child free to develop and decide for themselves what kind of a person to be, what sorts of things to like and what to dislike. The parents are influencing their child's behavior right from the very beginning of that child's life. And –at least until the child gets a little older– no one will complain that their privacy is being violated while being raised. There is no question that this scenario is all about influencing behavior and not about property. It is no longer commonly held that children are the property of their parents and yet the parental right is widely held to include (and may require) the modification of the child's behavior for the sake of giving it a proper upbringing so that they can go on to become upstanding members of the community and admirable representatives of the civilization.

4. You make a telephone call from a telephone booth and there is a listening device that picks up everything you say. You were placing illegal bets and federal law enforcement uses this recording to prosecute you for the crime.

In Katz vs. The United States, the Court held that the plaintiff had an expectation of privacy when using the phone booth. The Court also held that this expectation was reasonable, meaning that the society at large recognizes the expectation as a legitimate expectation given the circumstances. It therefore found that the use of the listening device constituted a warrantless search of Katz and that the evidence gathered by the search was inadmissible in the case against him. As part of the decision, the Court spelled out that privacy was not about things and places, but about the subjective expectations of the actor. When we act, our action is decorated with a variety of different expectations. One such expectation is whether we are under scrutiny or being surveilled. The zone of privacy that is guaranteed persons is therefore guaranteed via their expectations and the social recognition that those expectations are reasonable. Conversations taking place in an open field may not be reasonably thought private, but a conversation in a closed telephone booth can and should be.

5. You are sitting in your house and someone video records you through the window.

Of course, this peeping tom is violating your property rights. They are trespassing, but it is not just about trespassing. Following on the scenario in Katz, it has been determined that you have a reasonable expectation of privacy within the confines of your own home, bringing to bear the common conflation of property with privacy. One of the privileges of ownership is that there is a socially recognized expectation of privacy that goes with it. Because something is yours, you may reasonably conclude that the law will protect your ownership and your expectation of privacy while enjoying your property. The suggestion made in calling it private property is that we are intentionally binding together these closely-knit notions.

6. You shop at the local store using a 'club' card which gives you discounts on groceries in exchange for providing registration information that includes your name, date of birth, address, and phone number. You often receive mailers from the grocery store with coupons for items you have bought in the past and for competitor items to those goods.

This would appear to be a transaction or exchange where carrying the card and using it at the grocery store provides certain advantages: being able to get discounts on some items, being able to get bulk discounts on others like buying 2 for the price of 1, while the memory of your purchases and the ability to associate those purchases with your address gives the grocer the ability to send relevant coupons in direct mailers. You give them information about your spending and purchasing habits and they give you cheaper access to the products you are buying. Many people are happy to make this trade for the advantages it provides them and even in the simplest case where the grocer only uses the data to make stocking and purchasing more reliable and predictable, they gain a clear benefit from the exchange. They may also gain greater customer loyalty because once you have the ‘club’ card you will be more likely to use it. The detail that has been snuck in, of course, is that this exchange also gives the grocer the capability of impacting your purchases by sending you coupons that are not exactly aligned to purchases you have made in the past but might offer you discounts for competitor items. There is an opportunity here, as in the first case with the sommelier, to slightly change your purchasing decisions and shopping behavior. This capability can become a revenue stream for the grocer if they allow providers of the competitor goods to access the data or share in determining the details of the targeted actions.

7. An agency you never heard of keeps track of every 'credit' transaction you ever initiated. They fail to secure your transaction history and all data associated with you and that history is leaked to criminal hackers.

Whenever a party extends terms of credit to a buyer, they both check the past credit history of the buyer and make updates to that history based on the transactions committed. The creditor checks to see if the buyer’s credit is reliable based on past transactions and if so, extends further credit, then the creditor relays ongoing status of that relationship back to the credit bureau. Anyone who has bought anything on credit establishes a credit history. The data is descriptive of past behavior and is enough to uniquely identify the buyer via attributes such as social security number and historical trajectory data like addresses, phone numbers, energy service accounts, car loans, mortgage or rental agreements, etc. Access to a person's credit history provides detailed data about the life history of that person. In the United States, nearly all people are enrolled in such credit services indirectly through the purchasing and credit decisions they have made throughout their adult lives. Since utility usage is considered a credit-based transaction, nearly all people are included in the category of debtors. They did not directly enter into agreements with the credit bureaus, rather they entered into an agreement with a creditor who included a waiver or rider allowing this information to be shared and the credit check to be performed so that access to this information and permission to contribute to it was a condition for receiving credit (viz. a condition for becoming a customer of the creditor since nearly all purchases nowadays tend to include some form of deferred payment). In actual events, one of the 3 major American credit bureaus, Equifax, was hacked and the data for 100s of millions of Americans was stolen, exposing the credit history and all the identity pointers contained in it to criminal elements. The data that has been breached is not the personal property of the individuals described in each history, they are not the authors of that data, but it is a transactional history of those people's lives. Equifax was not fined for their carelessness in letting this happen, they were given a new set of risk mitigation steps that they must follow in their future business practices.

8. You store documents and photos in an online repository. The owners of the repository routinely scan your content –in compliance with federal law– to ensure that none of it contains child pornography.

On the one hand, documents and photos may be and remain the property of the person collecting them in online storage. I take this picture, it is my property and then I store that property in some location that is the property of another person, the storage provider organization or enterprise. Terms of service agreed to with the storage provider may grant an easement allowing them to scan those documents and photos. The storage provider is only interested in complying with the statutes in the countries where it is operating and, let us say, one of those countries demands that the storage provider use a search algorithm to scan the content for child pornography. Should the scan trigger a positive identification, the storage provider may be required to notify law enforcement and give them access to the user data given all the proper warrants and procedures. The photos and documents provide a transactional history of the person's actions and behavior and are not merely the property of that person but represent a record of that person's life such that the scan –if performed by law enforcement directly– would have been in violation of the user's rights. The routine scanning of the photos and documents is a warrantless search of the user's possessions, but since it is not carried out by law enforcement directly and was included in the terms of service established at the outset, it is legal to use the evidence obtained by that scan in any subsequent criminal proceedings.

9. You are arrested after being stopped for a traffic violation. As part of the arrest, the police search your person and, finding your cell phone, perform a search of the information accessible on it.

In Riley vs. California, the Court found that a cell phone, although near at hand during the arrest, could not be treated simply as a piece of property functioning as a container of data. It ruled that established precedents regarding arrest procedures did not apply and the officers were not authorized to perform warrantless searches covered by their sworn duty to secure the criminal and prevent the destruction of evidence. Rather, the phone was recognized by the Court as a window into the transactional history of the owner extending well beyond the parameters of any arrest. The ruling was that law enforcement could not conduct a warrantless search of the cell phone because of its broad range of content and the expectation the owner had that this transactional history is not subject to arbitrary or warrantless surveillance.

10. A law enforcement agency gains access without a warrant to data produced by your phone which transmits your whereabouts to cell towers as you drive around the city. They use this data to construct a narrative describing criminal activity and you are charged with a crime.

In Carpenter vs. The United States, the Court recognized that the data allowed law enforcement to construct a transactional history of the actions and behavior of the driver. This, likewise, constituted an unreasonable surveillance upon a domain in which the person had a socially recognized expectation of privacy. Because the emission of data was constant and ongoing as part of the internal workings of a locator application and not the result of a deliberate submission by the user, the ubiquity of the transmissions placed them beneath the user's awareness. The expectation of privacy cannot be reduced by machine-based operations and so warrants are required to gain access to the data from the cell phone provider. Passive telemetry, because it is ever-present and non-intentional, was ruled extraneous when deciding whether a person had entered into an arrangement that knowingly surrendered their expectation of privacy. The data stream was deemed to exist below the level of awareness required to establish an active surrender of rights even if it was included in the terms of service with the cell phone or application provider.

Notice that in several of the core scenarios the reasoning for why some specific behavior constituted a violation of a person's privacy was that the act of surveillance was being performed by a law enforcement organization without a warrant. The findings in those cases were not that there was an inalienable right to avoid scrutiny, but that a warrant must be secured to do so. The officers do not have the authority to surveille actions in cases where the agent has a reasonable expectation of privacy. In such cases, the officers, should they wish to surveille the agent, need to bring their probable cause argument to a judge and obtain a warrant for conducting that surveillance. They can still gather the information if they have enough cause suggesting substantiated suspicion of criminal activity, but they cannot do it without following due process. Due process trumps any expectation a person might have, no expectation is recognized as reasonable when due process has been followed. Furthermore, the findings of the Court only relate to the relationship between law enforcement and the individual agent. In cases where the person has entered into an agreement with another person, the notions of surveillance and criminal activity are no longer relevant. In those cases, what becomes crucial is the presence and content of a service agreement between the two parties. Prima facie, Equifax may not face criminal prosecution for leaking the data of 100s of millions of credit-seeking Americans, they are only open to civil suits since they have not violated anyone's rights by failing to properly secure their data. In the case of the storage provider who routinely scans digital content for child pornography, the direct relationship is covered in the terms of service, although the impetus and result of the surveillance is notification sent to law enforcement including enough evidence for establishing probable cause to secure the necessary warrant.

Because violations of privacy in the United States are often framed as acts of surveillance, the constitutional foundation for privacy has been found to rest in the Fourth Amendment protection from unreasonable search and seizure. And because search and seizure are characterized as a component of law enforcement's use of due process in enforcing the rule of law, persons in the United States have been left to fend for themselves when it comes to safeguarding their privacy within civil society. In the so-called "private" sector, each person is free to engage in whatever terms of service they wish to secure the goods and services that they need or want. Expansion of the notion of person to include associations of persons and other intelligent artificial non-life forms (like corporations and NGOs), extends the rights of privacy to those organizations in such a way as to produce a domain of secrecy that enables ruthless behavior away from the eyes of government surveillance and regulation. Since the privacy domain is constructed in the protection of the person from government surveillance, relationships within the privacy domain gain an effective protection from capricious government action. This leaves the privacy/secrecy domain to the private negotiations of persons entering into agreements and fulfilling their terms. Let the buyer beware. Civil cases may ensue if one of the parties believe themselves injured by the way the other party fulfilled or failed to fulfill their end of the bargain. In such cases, the relative strength of the different persons involved will become part and parcel of the resolution of the dispute. The 'little guy' will have less resources at their disposal when taking on the big guys who have wronged them. When the damage is done in aggregate to large segments of the population, there may be government-imposed restrictions on how the suit can be brought. The domain of privacy runs the risk of becoming a bellum omnium contra omnes with the courts standing idly by, or actively assisting some segment of the actors, as if to proclaim, "enjoy your privacy domain for that is where true liberty lies: the freedom to be eaten, to be exploited, to be victimized by fine print fraud and opportunism, and –if you are one of the chosen– to excel and acquire the good life."

This cynical comprehension of human life in the United States only covers the scenarios described in a few of the items above. The others are very different in kind. I would wager that many who read this list might think these do not belong there since they are obviously not anything like the others. They are ordinary examples of everyday human practice and encounters that do not say anything interesting about our rights and responsibilities in legal matters of privacy. To be a good sommelier is not just to serve your customers properly but suggests going above and beyond the call of duty and remembering their tastes and preferences to better serve them. The 'club' card example might work the same way, but there is a larger scale business involved in providing the service. Maybe such scale makes the scenario look more suspicious than the case of a single employee at a single establishment performing similar actions. Why consider them to be so different? In both cases, someone remembers the customer's preference and –for the sake of argument– does not share that memory with anyone else. The provider then uses that memory to give better service the next time the customer comes into the restaurant. Many people would not object to either of these scenarios, they might even view them as services worth paying for. After all, not every restaurant is so elegant that it has its own sommelier, the people who go there are at least in part paying extra to be able to get service like this. And maybe people appreciate the special treatment they get when their local grocery store makes the effort to keep them aware of relevant savings and promotions for items they purchased in the past.

As to raising children, who could possibly be so polemical to think that child-rearing is an invasion of privacy? Parental love is the most expensive bond a person can have, it costs the parent so much to devote themselves to the well-being of their children. Parents care for children and express this love by instructing and correcting their behavior for their own sake, to promote their well-being as they grow into personhood. This might explain why people do not generally think it unreasonable that an online service surveilles all content for child pornography. Because parenthood is such a common experience, we are quick to generalize its qualities into an aggregate attention. We have an interest in taking care of children, of raising them properly, and we do this for the sake of the civilization and for the sake of the children themselves. All well and good, but none of this reasoning means that when we do raise children in these ways we are not infringing upon their privacy. We put baby monitors on them, we put parental controls on their internet activity to restrict what they can do online, we insist on meeting all their friends, we keep tabs on where they are at all times when they leave the house, and even when we put away their clothes, we might be performing a spot check of their bedroom to make sure there are no red flags that their lives are in disorder. All for the protection of the child. We are looking out for them and they need this care. Because they are children, we do not think it reasonable for them to have an expectation of privacy in such circumstances. The rights of passage of all children include the necessary fact that they will be surveilled by adults. Provided the surveillance is done by the right adults. One reason they are being watched so closely is to ensure that they are only being watched by the right adults. At school, at home, in public. People are watching them, keeping an eye on them and they will come to learn and even depend on this. Children grow up to be adults and the whole point of raising them in certain ways and teaching them certain lessons is so that they will carry those lessons into their adulthood. If they are accustomed to being surveilled as children, why would not they remain so as adults? And if their expectation of privacy is diminished by the way that they are raised, then their tangible actual privacy is diminished. To the extent that privacy just is the expectation of privacy, wherever that expectation is reduced, privacy itself is reduced.

Before we get to a proper introduction ("Have We Met?"), I would like to provide something of a thought experiment to describe why I think the question concerning privacy matters ("A Father Reads His Daughter's Diary"), and then make a few remarks on method ("What is a Lens Grinder?").

A Father Reads His Daughter's Diary

Otto Frank not only read his daughter's diary without her permission, he published it. Not surreptitiously, but publicly with the intent that everyone on earth should see it, read it, know where it came from, and exactly who wrote it. I do not believe that anyone challenged his right to do so, at least not publicly. The diary may have been treated as a standard piece of property. Otto Frank was the girl's father, she had died, the diary came into his possession and –as her guardian– he was entitled to dispose of that property however he wished. He did not need a release from Anne, did not need her permission, did not need anything at all. He could just publish it as he saw fit. It may not ever have been Anne's property to begin with. Not really, not legally. She was a child, her parents fully controlled her life, what she learned, what she studied, where she lived, who she interacted with, and –presumably– what she possessed. The diary never belonged to her and the act of writing it was only permitted because her parents permitted it. Obviously then the decision to publish the diary is completely up to Otto Frank. Upon learning of Anne's death, Miep Gies returned the diary to Otto. This was an error. She could have returned it to him immediately, as soon as she met with him after the war. She did not need to wait until she learned that Anne was dead, the diary belonged to Otto Frank and should have been returned to him as soon as she saw him. No doubt Otto was a liberal man and may not have been the kind of authoritarian father who demands his due in all things as a matter of principle. He likely accepted that the diary was Anne's diary, that it belonged to her. While she was alive, he never would have read it without her permission. He loved his daughter and respected her privacy. That aside, he published it. He read it and he published it.

The justification for this is widely known and discussed. Oh, but of course he published it. And Anne would have wanted it that way. She spoke of it to him, she wrote of it in the diary itself. She got all excited when the Dutch official on the radio announced that they were looking for writing that reflected the Dutch experience of the war, she set to work in getting things ready, they said. Obviously, it was Anne's intent to publish the diary. Well, at least she meant to publish something. Some revised version of it, maybe it was the Secret Annex story that she meant to publish, something that had been scrubbed and prepared for publication, for other eyes, for strangers who maybe would not understand her raw emotions and reactions to her living conditions and situation. Aside from the fact that a good portion of this justification could only be derived by someone if they were to read the diary (a catch 22, you have to read the diary to confirm that she wrote a couple of entries indicating that she intended to publish it), it is hard to imagine that Anne would have wanted all the material in the diary to be released. She writes some harsh stuff about her Mom and Dad and includes some very private descriptions of her first menstruation, kiss, and other such things. It is not at all clear that her fantasies about the diary becoming a public artifact would have included all the details that she had thought to put in there when writing it. And the fact is that she never did get a chance to scrub the diary and prepare it for publication herself. Any conclusions that are drawn are merely guesswork and follow upon a reading of her diary.

I am not asserting that it was morally wrong of Otto Frank to read and then publish his daughter's diary. It may have been. If it were, there would still be the question of whether it was wrong for me to read it. I did read it. As an adolescent and then again as an adult. The first time through I could claim innocence. No one ever pointed out to me that this might be a problem. I was assigned the book in school, my teacher told me to read it, and my parents approved of my reading it. Everyone around me, all these people that I trusted, were unquestioningly telling me that this was what I should do. I did it. It made an impression, many impressions. It was a great learning experience. Yet I never once recall raising the concern to myself or hearing someone else raise the concern that it was something I should not have done out of respect for this young girl's privacy. And then there is the matter of the second reading. For this, I have no excuse. Not only was I an adult when I read it, but I was explicitly aware of the moral dilemma itself. I knew that I was reading a diary and I knew that it was commonly held that one should not read another person's diary without their permission. I read the diary anyway. To make matters worse, I had previously concluded that I would not read Franz Kafka's diary. I owned a copy of it, having bought it out of curiosity. Shortly after bringing it home, I realized it was a diary and all this entailed. I researched it and discovered that Kafka had written to the executor of his literary estate and told him that everything that he had written that had not been published by him should be burned. My interpretation of Kafka's literary will was that I –were I to remain true to Kafka's wishes– should not only refrain from reading the diary, but I should burn the copy I had purchased.

With all this background, I nonetheless read Anne Frank's diary. And I cannot ask her forgiveness, so the burden of this moral dilemma must remain with me. And now I am running the risk of deepening the violation of her privacy by subjecting it to analysis, by examining it under a microscope, by using it as an interesting anecdote to increase the marketability of my book. There are so many great passages and entries in the diary. Many of them would provide rich content for my project. She talks about things that are very much relevant to the ideas I am trying to explore here, so it would further my aims to quote her, to use her words, to make my points. These passages are from her diary and everyone can read them on their own, everyone can find a copy and pore over it in whatever way they choose: on a bus, in the bathroom, on the beach. The secret is out and certainly nothing I say here will make the diary more public than it already is, but still I hesitate. Because not only is there the question of reading it, and publishing it, and publicizing it, there is also the matter of profiting from it. In complete confidence, I might divulge that Anne does say that she would let someone read the diary if they were a true friend. This is crucial. She would let this happen. She would sit with this true friend and show them the bits she wanted them to read. Or maybe something else, but when she said it, I do not think she meant that her diary was simply a message in a bottle and anyone who found it could read it so long as they were a true friend to her. She is thinking in very intimate terms, she is thinking of someone who is with her, who *she* can appraise, someone who *she* thinks is her true friend, rather than merely someone who has come to think of themselves as a true friend to her. The decision would be hers, the act of permission would be at her initiative. She would let a true friend read her diary.

I do not think I will find justification for careless and opportunistic actions in her words, in the diary itself. This is an intimate work, she is in the middle of it. She is on every page in every way. Her whole being is present in each of the entries. Anne, dear Anne, is so alive and so involved in her life in everything she relays in the pages of that diary. This is her book, this is *her* self, this is everything that she thinks sees and understands as she goes through it, as she thinks it, sees it, and understands it. If I stand far away from her, if I produce a critical analysis that captures her meanings, abstracts them into universal principles, and applies those principles to case studies on the notion of privacy, what kind of a friend will I have been to her? When Anne says she could only let a true friend read the diary, she is –maybe– thinking of some shared event that is just as intimate as the writing of the diary. Maybe she would let you stand over her shoulder and read it like that. Read it with her while she writes or while she narrates the passages that she is sharing with you. No, she does not mean she would let you read it by taking it off into some corner somewhere to make of it what you will, she means that she would let you read it with her. In no way shape or form is Anne Frank offering herself up to telemetrics: to measurement or analysis carried out from a remote location. Anne does not want to be viewed and studied from far away, she wants to be up close and in the middle of her life. Being a true friend means being there with her.

Let me be clear: Otto Frank violated his daughter's privacy by publishing that diary. He did it because he loved her. And he did it because he wanted all the world to share in her loss. People are not likely to blame him for his actions, I certainly do not. People who read the diary are violating Anne's privacy by reading it. Each of the diary's readers needs to be aware of this and needs to come to terms with it. It is unlikely that there would be much public outrage or condemnation at this violation. If you have read it, you have violated a little girl's privacy. Admit it and come to terms with that fact. No one is going to fine us for this, no one is going to forbid it. There are still people who encourage this moral violation. And that is fine so long as it comes with this warning: beware, you are reading someone's diary, you are violating her privacy, be very careful with how you tread here. I already violate the spirit of this notion when I put it that way. I am not really talking to you or any of the people who will read her diary. Truly, I may be talking to myself when I say this. I need to tread carefully here. The last thing I would want to do is exploit her story for my own benefit, to stand far away from her and render some distant abstract learnings that promote my own theory on the foundation of some overly critical interpretation of her diary, its contents, or its context.

The Franks were forced into private, into secrecy. They hid themselves away instead of responding to the summons to appear in public. If they had heeded the summons, they would have been sent off to a concentration camp in mid-1942. Instead, they went into hiding and remained there almost two and a half years. Most of Anne's diary was written while she was in hiding. Her diary is a history of the time spent in hiding, a log of all that transpired in that little annex, what they ate, how they bathed, how they passed the time, the mishaps that took place, the help they received. Arguably Anne was confident of the fact that her writings were private. She writes as though she had every expectation that no one else would read this work. And that is what makes it so wonderful to read. That is why diaries in general are taboo for the reader. They were written with every safeguard of privacy and with every assurance that no one else would violate the sanctuary of their pages. To the extent that a diary reads as forced or with an awareness of public consumption, it becomes less interesting to the reader. The desirability of reading a diary is that you are getting access to the unedited and uncensored thoughts and feelings of the diarist. The pull is the very taboo, the allure is that the act is forbidden.

And that helps us to separate some of the critical facets here. The diary is not merely an artifact that can be discussed and treated as property, as a container of thoughts that can be consumed by whoever has them in their possession. Rather, the draw of the diary is that its emissions were meant to be private, it has been crafted in private, and it is something other than a simple thing available for public consumption. The interest in reading the diary is that its content is raw emission not reducible to an artifact or commodity, but something privately produced, something not produced for consumption and which was never meant to see the light of day, to be in the hands of many, and possessed by all and sundry. The diary is published and so accessible, but it was published without the author's permission, even against her wishes, and that is where the extreme desire to read it comes from. Somehow, when we read this we are getting underneath the object, the thing, we are getting directly to the exposed subject expressing itself as itself in the pages of the book. Objects are largely available to remote examination, to remote inspection, they are the subjects of all telemetry. As powerful as this vision of things can be, human beings find it lacking as well. We find that there is something missing from such a point of view. There is some great lesson to be learned in how this human being lives her life, how she navigates its stormy waters, and how she muses on the daily events not from far off and at a distance, but from up close. The diary both presents the contents of a life in a way that is accessible to measurement from far away and as a presentation of the viewpoint held up close.

Still, knowing this does not safeguard the observer from the moral violence that must be kept in view as one sifts through the lessons to be learned in this girl's diary. The Nazi occupation of The Netherlands was only possible because the Germans were able to manage distances, communicate across the European continent, telegraph and telephone their policies and directives from position to position. The growth and power of methods for dominating public communication not only made the occupation possible but made it more and more difficult to maintain a private existence away from the occupiers. Rationing coupons and identification cards, address registration and public rolls providing an account of the population and where they lived and what attributes they had: Jewish or not, etc. All these techniques and mechanisms of communication and cataloging were essential to the occupation of Dutch life by an alien power. Yet somehow Anne Frank managed to live in secrecy for 2 and a half years before she –and her family– were discovered and sent off to concentration camps. She died in those camps most likely from Typhus. This is a disease commonly associated with over-crowding combined with filth: too many people in an unhygienic situation and living together too closely. It is hard not to think that she died a very "public" death. She died from being too close to too many people. Disease can easily spread from person to person when there are so many packed together so closely. This is a terrible irony.

The point of view that I am taking from my reading of this private diary is the point of view that is most detrimental to the essay that follows, and I assume this point of view knowing full well that it exposes me to moral condemnation since it is extracted from the reading of a diary that no one ever gave me permission to read. Remote measurement, gauging and evaluating things from far away, is at odds with the human tendency to be in the middle of things, to be involved with things, to be tied up in passionate biases and concern for the events and objects that happen or pass through one's life. And what is more, human beings tend to vacillate between taking up a position that is intimate and attached and one that is distant and removed. We take a telemetric view of ourselves, we look at our friends from far away, we look at "people" as though they were far away and then we turn right around and interact with groups of people who are as close and as dear to us as any ever could be: our friends, our family, even the person you unthinkingly give up your seat to on the bus. We constantly vacillate between a position that puts us right up close and a position that puts us far away from the smells of the world within which we live. We are constantly moving back and forth between an intimate phenomenology of experience and a telemetric one. And although I have cultivated this narrative of vacillation from a reading of Anne Frank's diary, I have only been able to do so because that reading itself is undergoing a constant vacillation. One moment I am close and thinking of the world as she saw it, the next minute I stand further back and think about her from a distance. I easily maneuver back and forth between her position and the distant image of it that I get from my living room so far away from her in time, space, and meaning.

Fair enough, but why is it wrong to violate someone's privacy? So far, we have said a lot about the variable nature of privacy as bound to expectation and amounting to freedom from surveillance and seizure, either of property or of one's person. The Franks were driven into secrecy because there was a political force in power that wanted to seize their persons, to take them and place them in a concentration camp and, ultimately, a death camp. Why is it wrong to kill people, why is it wrong to round up groups of people deemed enemies of the state, and why is it wrong to violate the privacy of a person by reading their diary, or analyze any information that they have emitted without the expectation that this information is being surveilled by other people? The fact that I have not yet made any kind of utilitarian excuse on behalf of Otto Frank suggests that these questions have been strung together as though they gain their legitimacy from a common deontological source. I am not willing to excuse Otto's violation of Anne's privacy because the publication of the diary served some higher end. He violated her person, and this violation of personhood can be viewed on a spectrum that includes brutal incarceration of people who have not committed crimes as well as the brutal murder of incarcerated people whose only crime was that they personify some attribute that makes them an enemy of the state.

This is a morality of distances and proximities, of the telemetric and the intimately phenomenological, bound to private personhood emerging from a set of social conditions as opposed to a pure practical reason. This book could have been called "Ethics" in the same sense it had in an earlier book with that same title. The task at hand will be to try to show that both moral violation and moral permission are pursuant to the right use of remote measurement and phenomenology which therefore function as ethical preconditions. People are murdered, people are incarcerated, and their privacy is violated. We live in a world where this happens, where governments, groups and individuals are making life and death decisions that impact the fates of their fellows. Our focus is going to be on privacy because the way in which the telemetric and the phenomenological interplay in the realm of our privacy deliberations is what enables and excuses our decisions in other matters concerning life and death. There is a spectrum of violence that our ethics are meant to protect us from and the basis of that protection lies in the boundaries established between person and world. Shatter the glass at your own peril.

What is a Lens Grinder?

Put simply, lens grinder was the profession of one Bento Spinoza, 17th century Philosopher of Sephardic Jewish descent. Initially, Spinoza made his living as a grinder of lenses and did his work in philosophy on the side as an amateur. This notion of practicing a trade while working as a philosopher stimulates the imagination. How did he fit it all in? Where did he find the time? Why would one continue to carry on the rigorous life of a philosopher while unable to practice one of the professions more commonly associated with that calling? He was not a professor of philosophy, nor did he undertake tutelage. There is some indication that he thought there was something inherently wrong with taking money to think, and so one of the primary drivers that led him to lens grinding was that he needed to pay the bills and it would have been wrong to do so by exploiting his study of philosophy. It is wrong to take money to think. That is how the story goes at least.

If this is the case and if he were pure in his practice of the separation, then there is nothing in common between a lens grinder and a philosopher. "The writing of the Ethics, and thus the contingency (or necessity) of the circumstance that Spinozism as a doctrine made its appearance in history by way of a lens grinder, leave no trace in any of the master's books." (Althusser, 1997 p. 102) If one were to try to probe into Spinoza's philosophy, or into his method, there would be no trace of any contact between the two. Lens grinding and philosophizing, so this line of thinking goes, have nothing in common and there is no point of connection or influence between them. Lens grinder or shepherd, it would not have mattered, neither would have had any impact on Spinoza's way of thinking or violated the principle that it is wrong to take money to think. There is no philosophical thought in lens grinding, no philosophy is being done there, and so one can live as a lens grinder, philosophize on the side and be confident that the principle is being upheld.

As someone who spends the better part of his week working as a software engineer and who spends his spare time practicing philosophy, I find it hard to imagine that Spinoza would have been able to keep these two parts of his life separate. Rather I imagine that the one would have crept into the other, that the concerns of the one field would have continually triggered ideas and insights in the other. And this not only in the direction of his trade infringing upon his philosophy, but I imagine his philosophy would have found its way into his trade as well. I do not have evidence here, I do not have a clear story to tell about how the geometric method, or the pantheistic equation of God and Nature, would have found their way into lens grinding. There is no detailed data or telemetry driving my suspicions of how those notions could have impacted someone's work as they polished lenses. People, when considered up close, typically have not been very good at separating and isolating the parts of themselves that they express at different times. People love to try, there is plenty of discussion to that effect, but there is so much discussion on the matter because it is so difficult to pull it off. If you are fishing in the morning, hunting in the afternoon, and doing critique after dinner, I should expect to see your fishing habits have an influence on your hunting, and I would expect your critique to find its way in there as well. In software engineering, problems come up all the time that force one to carefully evaluate conditions and implement solutions to problems without introducing disastrous side effects or regressions on other functionality. Philosophy comes in handy at these times. The root of many software problems is that the model for an action is incorrect and a consideration of that model in terms of what it presupposes about the world can greatly impact how one redesigns the flow of instructions to achieve the desired mechanical effect. In a data pipeline, for example, a query is distinct from a platform specific command and a command is distinct from a connection to the store against which it will be executed. A set of problems can occur if these ontological distinctions are not modeled into software that constructs and executes queries. To whatever extent lens grinding and software engineering may be similar, I can only imagine that similar such overflow must have been a part of Spinoza's professional life.

Lenses, whatever else they are, make things that are far away seem closer (making things that are small seem bigger is just a variation on this as is any magnification). The lens grinder is regularly and repeatedly interested in giving the telemetric a new spin. The very notion of remote measurement is filled with contradiction. If it is far away, if the thing that you are working on is indeed remote, then what meaning can the act of measuring it have at all? Unless, that is, the measurement itself brings the thing closer. Spinoza's philosophy starts from a great distance away from the things that make up our lives. Substance and the modes of it, these are abstractions that do not concretely appear as such before us. *Ethics* progresses in such a way as to bring the thinker closer and closer to the phenomenological upshot of these abstractions. Substance is no longer this purely abstract notion, it becomes affective throughout the course of *Ethics*, and its affectivity lies under all human emotion and feeling in the end. Human cognition provides the middle way and becomes the movement and interaction of elements within this larger context. *Ethics* does not remain at the level of abstract sets of insights into the underlying being of all that exists, it attempts to reel that into our daily lives, into things that are up close and familiar. And maybe this is the lens grinding at work in Spinoza's thinking? This is Spinoza realizing that looking at the world through a lens that brings the stars down to earth, or the microbes up out of the microsphere and into human awareness, is what the philosopher is meant to do as well. Philosophy takes on the role of that which brings the far away close, appropriates it so that it is affective in the human being's life.

Or it might be objected that the movement is the other way around? Does his philosophy take what is up close and maneuver into position to see it sub specie aeternitatis? The upshot of the analysis is that I, here and now, am substance, am the modes of it. Nothing could be more intimate to my being than the substance and modality that I am in each breath that I take, thought that I have, or action that I perform. The philosopher demonstrates this intimate status in seemingly detached theoretical representations. He puts what is close-up at a distance so that it can be seen, so that we can get some perspective on it. Either way, the facility to make the switch, to focus the lens, must be operative in the process. Does the heavenly body come to earth when we look at it through a telescope or does the eye venture out into the heavens? Do the microbes get extracted from their position when viewed under a microscope or does the miniaturized observer venture into the micro-world? The lens and its effect on the subject relative to its object draws the one down and pulls the other up, or vice versa. The lens grinder, as near as I can tell, is everywhere in the work, there are all kinds of lenses everywhere working in all directions making *Ethics* into an optics of epic proportion.

It is tempting to compare this dichotomy of near and far with theory and practice. It may be my point that the lens grinder is trying to work out something that is both practical and theoretical and that his philosophy reflects that fact. The distinction between these viewpoints, although helpful at times, is far too sharp for the distinction that I am making. Rather, the distinction is more spectral where even at the opposite ends the other side is still at work. There is always an eye looking through the telescope to see what is far away. The measuring maintains an intimate experience and experience is always self-transcendent. When human beings are least themselves, looking out into the cosmos, they will still find themselves out there. And when they are not looking at all, when they are so deeply engaged in their actions, they are –in a sense– quite distant from themselves. Whichever way we turn, there we are whether we see it or not. Failing to see it, no doubt, because the focus is on what we are seeing and not the fact that we are seeing it.

It might be even more tempting to make a comparison here to the ready to hand and the present at hand. A thing is present at hand in so far as we hold it at arm's length and consider it as an object, whereas it is ready to hand when we take it up and use it as an extension of ourselves in carrying out some plan or action. Or maybe we would compare them to lifeworld and system, suggesting that things nearby are never systematic or rationalized and that things far away always are. These notions will be helpful when working through the matters that lie ahead. At this point, however, there is a huge distance between the two sets and mentioning them in the same breath is already to anticipate what is to come. Both sets involve ontological commitments that inhabit action orientation, but they are distinguished through notions of scope and scale of operations. Lifeworld and system introduce large scale notions of action summarizing involvements in generalized form, whereas the ready to hand and present at hand are small scale and positioned in contexts relative to those larger scale involvements. A micro analysis of action exhibiting an expectation may inhabit various positional orientations that are present at hand or ready to hand. The eyeglasses, shoes, and keyboard may be the latter whereas the works of Spinoza may be the former. As I grow tired, the status may reverse. These are micro declarations because they do not stretch out to contain the entire surface area of my endeavor as I carry out an action, they are momentary. These ontological commitments are variable as I act, regardless of the actions. Some aspects will withdraw from clear vision into the nexus of relations making up my activity embodied in muscle memory, whereas other aspects will be contemplated as my focus searches for opportunities to direct the next step.

These micro inhabitants are very much different in kind from what goes on when I move between lifeworld and system. These larger scale categories are modes of rationalization, degrees to which the span of all momentary involvements have been committed to social norms and explicit values. A string of maneuvers orchestrated in ready to hand orientation may realize a rationalized or systematic structure of action that has developed through collaboration over long periods of historical time. And then, to the extent that I operate in a realm free of large-scale rule directed structures, I may be operating free of system in a set of relations that are not yet clearly articulated within the world at large: I may be absorbed within a lifeworld that is free of or limited in categorization demonstrating the long-standing analyses carving out the terrain. Any aspect of thinking or acting may be systematic and rationalized in this sense conforming to standards and established procedural rules constituting a domain. Likewise, it may be possible for thinking and acting to execute free of such standards and beyond any such domain. If we call it 'philosophy' when we do it in universities aiming at instruction within a historically understood discipline, then we might call it 'love of wisdom' when it operates outside those standards and established norms. What remains an open question is whether there is an opportunity to micro-tune action or thought once an established set of constraints has been put in place through one's social context. Does rationalization ever unravel? Can a lowly lens grinder who has eschewed philosophy bring to bear a love of wisdom that disrupts the colonization at work in the telemetric understanding?

Analysis at a distance is not the proprietary power of computing systems, rather it has been replicated into them. The telemetry discussion in the early chapters of this book will show some of the ways in which human computation has become machine computation through the introduction of a specific kind of seeing into the mechanisms and processes constituting productivity. To the extent that analysis functions as a moment in the telemetric, even when that analysis is exercised in a social context where it lacks an application for feedback, it takes its parameters and principles of operation from analysis so situated. That is, as a historically developed practice, the methods of analysis are part of socially delimited practice even if some practitioners have been abstracted from that overall structure. For example, philosophers may practice analysis without any concrete application, but that is only when their role is abstracted from its larger social context. In that context, the philosophers are educators and their students are acquiring analytical skills that make them better software engineers or lawyers, etc. Spinoza's decision to try to keep lens grinding separate from philosophy, even if the effort failed, was an attempt to separate thinking from telemetrically structured analysis. And we should be careful in thinking that the failure to keep the two separate amounts to an infestation of his thinking with telemetric logic. *Ethics*, void of scholarly references and without any indication of being data driven, suggests that it may indeed be possible to increase understanding without relying on telemetric computation.

Early on in this project, we will begin the long process of understanding the various ways in which micro behavioral commitments like the ready to hand and present at hand interplay with macro domains like system and lifeworld; and the variable amount of rationalization that has come to occupy those domains and the micro behaviors taking place within them. Telemetry is the ultimate rationalizer. The telemetric maneuver is rationalization pure and simple, and it is only recently that this pattern of cognition has achieved the status of a concrete mechanism executing automatically beyond the control of any resident homunculus. It is too soon to tell whether this mechanism can be evaded while demonstrating its mode of operation: whether the project is liberating or a further turn of the crank. The telemetry-oriented lenses that we must grind are software mechanisms that make embedded processes visible along a variety of different dimensions and domains thus fabricating a modernized form of automated/autonomous vision using this new kind of lens.

Have We Met?

Make no mistake, even if we know each other in some casual way, we have not met. Not at this proximity. No one gets to this place with another, no one could bear it. Imagine if someone you know, even someone you love, were to soliloquize for hours on end without taking a break, without pause, word after word, page after page. It would be impossible to bear. You would tell this person, your husband or wife, your son or your daughter, to shut up, to stop talking. No person could ever withstand the river of words that extended writing brings to bear on a relationship. People need more regular feedback than that. Even if you and I do know each other, we have never come in contact like this before. You are gaining some form of access to a steady stream of thoughts, an elaborately spun tale, of everything that has occurred to me as I have worked through these notions and as I have undertaken to narrate this non-fictional fabrication. And I cannot possibly know you either, no person that I know is as patient as you are, as calm and ready to receive. The reader and the writer are artificial personas, they do not exist in flesh and blood. Rather they are positions that we are willing to take up relative to each other. Although this relationship brings us very close together, it brings these words into your mouth or ear as you traverse the pages, it is nonetheless an artificial relationship. We have not met before and yet here we are, being introduced to each other. And this must be the case because, despite the intimacy of the relationship, we are far away from each other. The message travels through time and may not arrive anywhere at all. You might be a delusion, and I am surely one.

And "I" can hear the voices of the telemetric chorus in these thoughts. Far away and beckoning. Hoping that there shall come a day when the artifact is far from me, observed at a distance even during the experience. That experience reproduces the telemetric event in/as/through my understanding of the world. To have an inner narrative that accompanies you through your life, acts as your closest confident, even this amounts to an emission. Does it fill you? Does it overfill you? Would you ever be able to keep it to yourself? Clearly this is impossible for any living and breathing human being. At least, any that is not a child swaddled in the love and living under the watchful eyes of its parents. It may then be possible for such a person to exist who both demands constraint on these emissions and allows them to spill out. This may be suffered in frustration, it may only serve to enrage the poor creature, or it may be the care-free life of a little one who has not yet earned the dignity of their own emissions and ventures.

They will overfill you, they will overflow you. Abandoned by your only friend and guardian, enraged by the failure to keep it all in, you are powerless. You explode. You are a human bomb. It is for our own safety that these watchful eyes constrain us into perpetual childhood.

There is more than legality and rights at stake here. When we talk about privacy, we are talking about what we are and how we are. Even if an intelligent agent like you or I is nothing other than computation, it does not follow that such computation is a public resource for use and abuse in whatever social agenda holds sway in this or that historical period. Suppose privacy were only a convention with a scope that goes no further than the law, a legal fiction. Would not it follow that any computational structure could be augmented with such a convention? These are two sides of a coin: computational beings can have private lives and computational beings with private lives need not be human beings or in any sense “singular”. Imagine how privacy as a convention would work when protecting the integrity of a computational being that is hive-like and able to orchestrate the actions of singular beings integrated through its operations: a leviathan that is power throughout and structured to maintain privileged access to its orientations and involvements. Privacy, in a human context, might then be a form of infantilization. That is, the hive reduces human beings to children to be disciplined and directed with the proper goals and incentives for carrying out the hive’s objectives. Under such circumstances, the only viable resistance would be to get off the grid and –whatever you do– avoid publishing any declarations that would be easily swept up into the collective computational logic. This book cannot be made known under such circumstances, each of us would have to safeguard its privacy: maintain it as a secret communication in purely analogue form exclusive to you and me.

What would be the purpose of such writing only held in private? If I were you, I would interpret this as a challenge and assert that it cannot be done. You might ask: what have you written upon? Where are the marks that you made? Is the location secure? Who has physical access to the device? Where were the keystrokes and packets relayed? Who could pick up the missives on the wire? Is the writing intelligible during its transmission? Have you encrypted the text you produce so that it is only visible to you and me in its unencrypted state? Where have you stored the private key? Are you protecting your person and the device from video surveillance while you view the unencrypted text? Where do you store any backups you make?

Edward Snowden has gone there. He wishes to totally control all data emissions from his person. It does seem like a disorder. Surely such a life is not to be envied. I assume I am not much different from most. Humans want to have experiences and having them is likely to leave artifacts. Such artifacts make measurement at a distance possible. The artifacts can be examined by humans who were not involved in the experience, even as onlookers. The onlooker performs a relay and that takes them away from the experience and out into distant observation of it. The chorus takes a seat in the bleachers. You view the photograph and click 'like'.

I propose that the phenomenology of the subject as displaced onlooker to its own life is the evil against which such conventions are meant to protect citizens of a state founded on the principle of securing freedom and justice for each participant. By extension, such a state would be devoted to protecting its citizens from relying in the final instance upon "liberty under a blanket" when experiencing the world. The state is guarding privacy via process so that the citizen need not do it through heroic effort.

By phenomenology, I mean the thrown projection that characterizes human understanding. This is an understanding replete with ontological commitments, the 'thrownness' part, and those ontological commitments are not always explicit. Whether they are explicit or implicit, however, this thrownness guides and governs all projection, where that amounts to the interpretation of meaningful structures with which the actor is involved. Furthermore, I am aligned with the tradition of phenomenologists who have worked to articulate the role that the other and a concomitant social structure have played in this thrown projection. Human beings are always born and raised, and their experience of the world develops within this history among other human beings, within systems, that provide frameworks and guidelines, rules and norms for interacting with things, for interpreting things, and for understanding the world as we come to it. This extremely general reading of the history of phenomenology enables the establishment of an other-bound history of philosophy that includes Hegel as well as Heidegger, Foucault as well as Habermas. I can elaborate in each of these figures, and the movements that they have initiated in the history of philosophy, a common pattern that sets or discovers the structures of the commonwealth in the intentional understanding of human beings. And what is more, I sense in each of them an effort at producing a proper approach for uncovering this pattern. This uncovering has proved to be a very hard problem as it requires the investigator to discover the structures of the very 'thrownness' that is at work in the discovering actions themselves. Phenomenology, therefore, always suggests hermeneutical phenomenology where this connotes the sense of a circularity at work in the effort to demonstrate thrown projection in all its nuances. Computation is never just computation, rather it is socially integrated practice including both action and reaction, integration and collaboration.

There are interests at work and there are conflicts and forces at play. I cannot even imagine the new monsters that would spawn in the realization of a vision of unbounded and asocial computational agents. The significance of the phenomenological gestures toward the need for socially bound and anchored understanding.

When one attempts to uncover or present one's findings regarding the presuppositions at work in human understanding or experience, there is the danger that this presentation will be inhabited by a fundamental blindness that fails to shed light on the very conditions that bind that presentation to a world. Hermeneutical, in this sense, means that the phenomenologist must account for the possibility of the project undertaken by revealing the boundaries, along with what is bound. Heidegger in *Being and Time*, Hegel in *Phenomenology of Spirit*, Foucault in *Discipline and Punish*, and Habermas in *Theory of Communicative Action* all had to grapple with the possibility of performing the project at hand. It is not enough to merely show the experience of consciousness on the pathway to absolute knowledge, the existential analytic of Dasein, communicative action aiming at understanding, or a genealogy of power written onto the body and its practices, rather any such project must account for how the philosopher could articulate this perspective and advance this philosophical position. This necessarily recursive maneuver must be part and parcel of any attempt to show the workings of the commonwealth within the projected commitments of human being and may be called the 'hermeneutical' requirement. This is the aspect of the project that binds it to an 'ethos', a place where it lives with others, and without this the computational executive is unfettered by familiar aims and objectives, it runs the risk of becoming a hostile presence sharing the earth with other forms of agency.

I make use of the term 'commonwealth' to describe the social structural nature of this circular precondition at work in all thrown projection. And I use this term because it invokes the historically popular notion that a kind of conceptual 'contract' drives the human association of each with all. It is not that I intend to invoke an 'Original Position' or transcendental condition of the human move from the state of nature to the state of civil society. Rather, there is something about the common sense that developed this conceptual tool in various philosophical projects that is aptly descriptive of basic organizing structures in the advent of human cognition. We are raised, and this upbringing carries with it an entrenchment in an implicit social contract. Rather than bias the status or nature of this bond, I merely borrow the term to suggest that something of the commonwealth exists in the most basic of human experiences, our being alongside other people, our situation in the middle of their groupings and organization, as well as our upbringing amid a common set of orientations, interpretations, and meanings. This is all suggestive of the notion of a commonwealth but not as a productive aim, rather as a developmental source. Additionally, the term gets its usefulness from its emphasis on what human beings have in common, what common structures they make use of to understand and engage the world around them, and the common pattern that is evoked in all cultivation of a singular human being. Each of us carries within ourselves a pool of resources that we have inherited or instantiated from the commonwealth, the manner in which those resources have been apportioned, the rules, regulations, and norms that are used to explain the distribution of the commonwealth in the cultivation of growing human people, are of fundamental import in realizing the way that human beings come to voice their place in the world and situate themselves relative to all others. No longer able to rely on a deity to provide each of us with a unique soul, we take on the function for ourselves and construct it out of a common wealth of resources.

A human being is a thrown projection and that thrownness with its ontological commitments stems from a commonwealth of rules and norms that aim at placing the human being into the world in a way organized for mass use. The structure or organization of that mass use is already at work in any conditions of thrownness and so orders the projections the human being is capable of: we butt up against each other, or we work well together, largely due to the commitments that carry us along. And those commitments are always crucial for situating us among the other people who we grew up alongside. Class will play a role, schooling will play a role, gender, geography, mass culture, even one's neighborhood. Every association we have and every association we make furthers the organizational structure that guides each of our understanding of the world. What possibilities there are, what requirements, what meanings there might be. What it means to be human is to be working out a set of thrown pre-conditions. Not in a vacuum, but in the context of undertakings and of projects that we have set about to accomplish. The notion of commonwealth is a helpful way to surface these pre-conditions because it brings ready-made the notion that the preconditions are based on an agreement, on a social fabric that has already been woven and is made of rules and norms that provide its design and offer a set of parameters within which every human being is cultivated. Sets of parameters as culture and cultivation make up a common wealth even if they vary from group to group or based on geography or weather conditions, or any of millions of other conditions or points of intersection that may exist. There is a world of difference between computation and socially constructed computation.

Night does not come all at once, neither does the daylight. The internet as a computational hive did not start out as a large-scale machine for mass surveillance and it did not become that in a single day. The technology suggested a requirement for monitoring itself and that gradually developed into a requirement to monitor the behaviors of its users. The instruments, mechanisms, and patterns evolving out of the complex architectures of this global network, spread into the homes and pockets of nearly every human being on the planet. It is my intent to tell a story of how this evolution constitutes a pattern of organization and that this pattern of organization is not exclusive to networked computing systems but has been replicated into human experience and human social organizations. I have broken this story down into three separate sub plots trying to describe how the pattern unfolds in computing systems, in human experience, and in social organizations. Telemetry Phenomenology Commonwealth. I will frequently suggest that the hero (or anti-hero) of this story is privacy but I fear that I will put equal emphasis on its central role as a plot device.

Let us then consider three distinct stories: 1) the story of the rise of a global network which has become an ubiquitous information system supporting human behavior. 2) the story of human experience concerning the placement of the individual in the context of such a growing network. 3) the story of collective organizations and how they have evolved in that same context. The stories are told such that the details of each sometimes overlap and become conflated or integrated with the others. It could be that these three stories are just one story with a common pattern that drives each of them. My mantra here was metric emission, analysis, and feedback loop. This computational pattern seemed to be at work in all three of the stories and made it possible to see fragments and segments of one at work in the fragments and segments of the others. In all three cases, the telling of the tale reveals the extent to which the old simple experience of the participants –whether they be software and hardware components, human beings, or organizations– spun itself into the computational form now so dominant.

The higher order story that spans the three stories and orders them into this book is what I call its architectonic. And I use this term thinking specifically of Hegel and the way he used it to describe the higher order logic at work in his *Phenomenology of Spirit*. Thus it is possible to describe, on the one hand, the second story in this book as phenomenology in so far as it focuses on the relationship between individual experiencing beings and the worlds where they have those experiences and, on the other hand, the higher order logic itself as a phenomenology focusing on the relationship between you and me, reader and writer, as we work through these three stories. For this book is not merely about experience, it is itself an experience. Rather, reading it is. And so are all stories. Rare is it that a philosopher will reach out to their reader to take their experience of reading philosophy in hand, or to take hold of their hand as they go along together in their funny asymmetrical manner.

The first chapter of each part enacts metric emission, the second chapter enacts analysis, the third chapter enacts the feedback loop, and the fourth chapter enacts the residual to the computational pattern of the first three chapters. Dialectic, of course, works through contradiction. The ideas presented in a stage or phase run into problems or suffer contradictions that invalidate the stage. These contradictions are determinate negations in so far as they propel the experience of the reader and the writer forward into the next phase. The reader and writer are the ones who must progress because of the contradiction. An actual individual person, experiencing the world in the terms laid down by the stage of development, can exercise their actions just fine under the conditions of these "contradictions". They are not propelled forward, only the reader and writer are. We know better than to rest easy on this artificial boundary, we must move along because the stories indicate that we are the ones in the middle of it and the falsehoods will not allow us to stand still. That is how dialectic works. My story, however, is not dialectical. Rather, we –reader and writer– do not know any better. We are unable to get the requisite point of view for the movement hoisted upon us by the dynamic at work. We are locked in a partial understanding determinate and constrained in its finitude. We move forward because we are confused and are in search of an explanation, not because we have seen a problem that bids us leave the current stage behind. We seek to explain what we have just been through, what we are continuing to go through. This confusion is not an intrinsic logic to the process, but results from the scope and pragmatic aim that we are assigning to the investigation. In the case of the three stories, privacy has that pragmatic role and it is therefore privacy which propels us forward in search of an explanation or a reason, forces us to work strenuously to improve our understanding of the matter at hand.

Privacy has not been arbitrarily selected for this role. Again, I do not know if this makes privacy into a hero or into a plot device. It propels the action, sure enough. To be a hero it would have to be personified. And that is exactly the process we are undertaking: the process of trying to personify privacy. Which amounts to establishing it. And, analogously, suggests that the two of us are in search of it, of what it means, of how it works, and maybe even a tangible taste of it. This is a private moment between us. And we drive forward to attempt to understand what that might mean. In *Ecce Homo*, Nietzsche attached the subtitle: How One Becomes What One Is. That is a mistake. There is no general advice there, only a naïve reader would think his book a recipe for how each of us could achieve that end. This is Nietzsche's "autobiography" and it is his way. *Telemetry Phenomenology Commonwealth* tries to provide a realistic alternative, a real story that tries to capture the generalized tale of how one becomes what one is. Is it cheating to transform 'privacy' into an existential analytic? Or rather, an existential dialectic? Possibly the least controversial statement in *Being and Time* is that Dasein is an issue for itself. We may bicker over its philosophical impact, but your death means something to you. And it is yours alone and belongs to no one else. The technological era that we live in has brought 'privacy' to the forefront, but that may be just the fashionable way in which we take issue with ourselves.

We begin, in a sense, with a problem declaration and proceed in explanatory fashion demonstrating the solution to that problem. Telemetry is a technology needed at a specific time to sustain a specific set of other technologies that were growing in application and popularity. Its addition introduced new possibilities and concerns such that its implementation took on a life of its own. Job one in the telemetry section is to tell that story. It will be hard to avoid running ahead and drawing social and personal conclusions from this. In software engineering, we divide and conquer problem areas. In the telemetry section, therefore, we will focus on telemetry. Phenomenology, as a concern, will grow organically out of that set of concerns. And likewise, commonwealth from phenomenology. Emission, analysis, and feedback writ large. To give you a glimpse of what to expect: the telemetry story will lead us to think in terms of privacy as the expectation of privacy. Privacy as expectation will lead us to consider the patterns at work in our experience. The phenomenological will then lead us to think in terms of privacy as self-determination over against a telemetric understanding. This in turn will raise the issue of our common wealth and the social agreement that we embody in our organized lives together. This will bring us to the architectonic that has driven our search all along. In the end, there is no understanding that waits for us, no fourth section to complete the design. Rather, we see that we have been trying to understand and doing the work of understanding all along the way and that all of it must remain. There is no pay off, no output from the computational function, just an unrequited love of wisdom.

We will be able to see in that perpetual turning both the whir of a machine and the involvement of human understanding in tune with the world around it. The beauty of such an approach is that forgetting becomes a critical response. And not just forgetting, but editing and recasting, passing the tale along with alternate endings and different emphases. Or you can just stop reading at any time. No one will tell the tale of an authoritative large-scale rejection of the twists and turns in the stories that follow. Its critical failure will be in privative form: in its not being read, in its not being spoken of, passed along, told and retold to others and to future generations.

The people who reject this spin are people who need truth. They fear the chaos that follows upon truth's demise. They are not defenders of science and rationality, they are the faithful who continue to love mastery in the form of a deity who rules the known universe with an iron fist of non-negotiable principles. If they loved science and rationality, they would know their love is unrequited and these aspects of human organization are as finite and provisional as their creators. We are temporary inhabitants of this universe and as beautiful as our efforts at stamping our physiology onto its broad expanse may be, they will not survive us. Make no mistake, I'm not advocating a position here. Whether I agree or disagree with the story I am telling is beside the point. That is not how storytelling works. Will it satisfy the critics to know that all they need to do to make my story futile is ignore it?

The part that is hard to swallow may be that we do not want to recognize science and reason as human practices, as domains of activity with historically developed structures. And it is not good enough that we are saved from relativism by our communities and peer review, our intersubjective conjectures and refutations, the ongoing dialogue that makes up the most interesting parts of so many of the stories that we tell and retell. It is no coincidence that the people who most strongly resist this are also the people who point out how weak our social sciences are and how limited is our understanding of the world we collaboratively and collectively make for each other. It is distasteful that the most powerful aspect of human being would come from the domain that is least understood. We resent the areas of life that are the most difficult to understand due to their complexity. We complain that our "essence" cannot reside in a domain that is still so immature thousands of years into this reasonably stable configuration temporarily selected as the human experiment (or so the story goes). Rather than see these opponents with their dogmatic attitudes as evidence of our incapacity or failure, I see them as evidence of how much we love understanding.

One of the goals of this work is to increase understanding, an aim which should not be confused with an increase in knowledge. An increase in knowledge can affect everyone even if only a few achieve it. Understanding works differently, only the ones who understand benefit from it. And this reminds me of a simple heuristic I once used when teaching philosophy to undergraduates: given a joke that you have heard in a social setting, is it better to get it or not? It seems that everyone thinks it is better to get the joke. Even if it is not funny, even if it is a cruel joke, even if the joke is silly and not worth much consideration. Given a choice, people would rather get the joke than miss out on it completely. Even if someone must explain it to you, you are the one who understands it in the end. If that is the case, then I can only conclude that understanding is desirable for its own sake, that it is always better to understand. My aim then is to emphasize understanding even if no problems are solved, even if no other aims are furthered in the meantime. Call it a renovation of the Principle of Utility as the Principle of Understanding. If at the end, you and I have come to think of progress in terms of the understanding rather than knowledge or information, then this aim will have been achieved.

A second goal of the work comes from my recollection of something Nietzsche once said: that the proclamation of the death of god is a storm warning. It is not a description of an historical event, but a warning of what awaits human being in its future. The death of god worries me. The consequences are dire. If there is no god, then where do all those notions that were based on a divine moral order come from? Why does truth matter? Why do human beings matter? What is goodness or decency or justice or freedom in a world without god? All these notions obtained their standing in human history through the creator story that we have been telling this past several millennia. If the creator story has come to an end, what then? What do human beings have left in the aftermath of our gods? I am not talking about religion, prayer, or faith. One could construct a story in which all of these have a role in human life even after the deity has been put to rest. They are institutional practices and one could argue that there is an increased importance in such traditions once god has been lost. In god's absence, such rituals and organizing bodies of laws may be all that we have left. And any who feel themselves an outsider to these rituals and laws would be powerless to overcome them. If human beings huddle around a set of religious principles that force certain practices and beliefs upon their compatriots on pain of death, then that will be the rule of the day, that would be the truth. In the wake of the death of god, there is nothing left of persuasion, I fear. Nothing left that could turn the tides on a wrong turn in history. We would be left to the whims of the many and the power of the few as they see fit to enact it. And there would be no reason to change their mind.

I very much see this in the world that confronts us today. And if there is any hope in a merely didactic effort at 'getting it', at emphasizing the understanding we have of ourselves and of our place in the world among each other, it is that I hope that there is a set of conditions that are driving us and that do offer up reasons, that do provide a persuasive artifice, and that shedding light on that set of preconditions will make things 'better' even if they still leave us reeling about in a godless universe where we are susceptible to the whims of our fellow inhabitants. All theory might work this way, theorizing –whatever else you can say about it– is an optimistic endeavor laden with hope: that there is a knowable order, a communicable structure, a manipulatable design, or a realizable aim. Asserting that there is such a thing as privacy and that human beings should protect it, implying that they can protect it, is an orientation deeply invested in such hope and optimism. This is merely the introduction, the project in its youth, such optimism and hope are to be expected here. Let us see how it plays out as we analyze and feedback the emissions produced in the remainder of the book.

We come back full circle to privacy and how it is that privacy initiates such a set of concerns: the private may be a manufactured boundary condition, emergent when not found in nature, that inhabits this commonwealth, providing structure to human thrown projection. In our concern for what is private, we take an interest in preserving personhood despite our cognitive and ontological origins in each other. We are social beings, deep in the origins of what we are, but we must distinguish ourselves to find or be put into our places among each other. And the boundary conditions that we are forced to activate, enforce, and define are hard won and necessary for a determinately human life. This is a hypothesis, not a fact. The conjecture is that our being is always in question, always at issue. We may succumb to the telemetric power as it determines that privacy is the fastest way to its destruction, we may allow it to completely redefine ourselves as creatures without private lives living among corporations functioning in loco parentis, or we may let it decide that privacy is a commodity like any other and only belongs to those who can pay for it. This book is one take on what those options mean and the extent to which they are catastrophic, why they would be a rejection of something crucial to our social character and the commonwealth that lies at the root, that is the source, of all our involvements in the world.

Bibliography

Althusser, Louis. 2014. *On the Reproduction of Capitalism*. Translated by G.M. Goshgarian. London, UK. Verso Books.

Althusser, Louis. 1997. *The Spectre of Hegel*. Translated by G.M. Goshgarian. London, UK. Verso Books.

Arendt, Hannah. 1958. *The Human Condition*. Chicago, IL. Chicago University Press.

Arendt, Hannah. 1976. *The Origins of Totalitarianism*. New York, NY. Harcourt, Inc.

Arendt, Hannah. 1977. *Eichmann In Jerusalem: A Report on the Banality of Evil*. New York, NY. Penguin Books.

Arendt, Hannah. 1978. *The Life of the Mind*. New York, NY. Harcourt Brace Jovanovich, Publishers.

Aristotle. 1985. *Nicomachean Ethics*. Translated by Terence Irwin. Indianapolis, IN. Hackett Publishing Company.

Berman, Marshall. 1988. *All That is Solid Melts into Air: The Experience of Modernity*. New York, NY. Penguin Books.

Blacker, David J. 2019. *What's Left of the World: Education, Identity, and the Post-Work Political Imagination* Winchester, UK. Zero Books.

Bostrom, Nick. 2014. *Superintelligence: Paths, Dangers, Strategies*. Oxford, UK. The Oxford University Press.

Brand, Arie. 1990. *The Force of Reason: An Introduction to Habermas' Theory of Communicative Action*. Sydney. Allen & Unwin.

Churchland, Paul M. 1999. *Scientific Realism and the Plasticity of Mind*. Cambridge, UK. Cambridge University Press.

Darwall, Stephen. 2009. *The Second-Person Standpoint: Morality, Respect, and Accountability*. Cambridge, MA. Harvard University Press.

Deacon, Terrence W. 2012. *Incomplete Nature: How Mind Emerged from Matter*. New York, NY. W.W. Norton & Company.

Deleuze, Gilles and Guattari, Felix. 1986. *Anti-Oedipus: Capitalism and Schizophrenia*. Translated by Robert Hurley, Mark Seem, and Helen R. Lane. Minneapolis, MN. University of Minnesota Press.

Deleuze, Gilles and Guattari, Felix. 2011. *A Thousand Plateaus: Capitalism and Schizophrenia*. Translated by Brian Massumi. Minneapolis, MN. University of Minnesota Press.

Descartes, Rene. 1988. *The Philosophical Writings of Descartes, Volume 1 & 2*. Translated by John Cottingham, Robert Stoothoff, and Dugald Murdoch. Cambridge, UK. Cambridge University Press.

Dewey, John. 1958. *Experience and Nature*. New York, NY. Dover Publications.

Domingos, Pedro. 2015. *The Master Algorithm: How the Quest for the Ultimate Learning Machine Will Remake Our World*. New York, NY. Basic Books.

Dretske, Fred I. 1999. *Knowledge and the Flow of Information*. Palo Alto, CA. CSLI Publications.

Duhem, Pierre. 1991. *The Aim and Structure of Physical Theory*. Translated by Philip P. Weiner. Princeton, NJ. Princeton University Press.

Durkheim, Emile. 2014. *The Division of Labor in Society*. Translation by W.D. Halls. New York, NY. Free Press.

Durkheim, Emile. 1995. *The Elementary Forms of Religious Life*. Translated by Karen E. Fields. New York, NY. The Free Press.

Durkheim, Emile. 2014. *The Rules of Sociological Method*. Translated by W.D. Halls. New York, NY. Free Press.

Feyerabend, Paul. 2010. *Against Method*. London, UK. Verso.

Foucault, Michel. 1977. *Discipline & Punish: The Birth of the Prison*. Translated by Alan Sheridan. New York, NY. Vintage Books.

Foucault, Michel. 1994. *The Order of Things: An Archeology of the Human Sciences*. Translated from the French. New York, NY. Vintage Books.

Foucault, Michel. 1990. *The History of Sexuality, Volume 1: An Introduction*. Translated by Robert Hurley. New York, NY. Vintage Books.

Foucault, Michel. 1990. *The History of Sexuality, Volume 2: The Use of Pleasure*. Translated by Robert Hurley. New York, NY. Vintage Books.

Foucault, Michel. 1988. *The History of Sexuality, Volume 3: The Care of the Self*. Translated by Robert Hurley. New York, NY. Vintage Books.

Frank, Anne. 2017. *The Diary of a Young Girl*. Translated by Susan Massotty. New Delhi, India. FP Classics.

Frank, Joseph. 1975-2002. *Dostoevsky*. Five Volumes. Princeton, NJ. Princeton University Press.

Fraser, Nancy. 1989. *Unruly Practices: Power, Discourse and Gender in Contemporary Social Theory*. Minneapolis, MN. University of Minnesota Press.

Gadamer, Hans-Georg. 2013. *Truth and Method*. Translated by Joel Weinsheimer and Donald G. Marshall. New York, NY. Bloomsbury.

Gadamer, Hans-Georg. 1977. *Philosophical Hermeneutics*. Translated and Edited by David E. Linge. Berkeley, CA. University of California Press.

Gleick, James. 2011. *The Information: A History, A Theory, A Flood.* New York, NY. Vintage Books.

Godfrey-Smith, Peter. 2017. *Other Minds: The Octopus, The Sea, and the Deep Origins of Consciousness*. New York, NY. Farrar, Straus, and Giroux.

Godfrey-Smith, Peter. 2003. *Theory and Reality: An Introduction to the Philosophy of Science*. Chicago, IL. University of Chicago Press.

Goldstein, Rebecca Newberger. 2014. *Plato at the Googleplex: Why Philosophy Won't Go Away*. New York, NY. Vintage Books.

Goldstein, Rebecca Newberger. 2006. *Betraying Spinoza: The Renegade Jew Who Gave Us Modernity*. New York, NY. Nextbook.

Goodman, Nelson. 1988. *Ways of Worldmaking*. Indiana, IN. Hackett Publishing.

Goodman, Nelson. 1983. *Fact Fiction and Forecast*. Cambridge, MA. Harvard University Press.

Gould, Stephen Jay. 1996. *The Mismeasure of Man*. New York, NY. WW Norton & Company.

Grandin, Temple. 2007. *Making Slaugherhouses more Humane for Cattle, Pigs, and Sheep*. Online. http://www.grandin.com /references/making.slaughterhouses.more.humane.html.

Habermas, Jurgen. 1975. *Legitimation Crisis*. Translated by Thomas McCarthy. Boston, MA. Beacon Press.

Habermas, Jurgen. 1979. *Communication and the Evolution of Society*. Translated by Thomas McCarthy. Boston, MA. Beacon Press.

Habermas, Jurgen. 1984. *The Theory of Communicative Action. Volume 1. Reason and the Rationalization of Society*. Translated by Thomas McCarthy. Boston, MA. Beacon Press.

Habermas, Jurgen. 1987. *The Theory of Communicative Action. Volume 2. Lifeworld and System: A Critique of Functionalist Reason.* Translated by Thomas McCarthy. Boston, MA. Beacon Press.

Habermas, Jurgen. 1989. *The Structural Transformation of the Public Sphere: An Inquiry into a Category of Bourgeois Society.* Translated by Thomas Burger with the assistance of Frederick Lawrence. Cambridge, MA. The MIT Press.

Haraway, Donna. 1985. "A Manifesto for Cyborgs: Science, Technology, and Socialist Feminism in the 1980s" from *Socialist Review*, No. 80, 1985.

Hartman, Thom. 2010. *Unequal Protection: How Corporations Became 'People'—And How You Can Fight Back.* San Francisco, CA. Berrett-Koehler Publishers Inc.

Hawkins, Jeff with Sandra Blakeslee. 2004. *On Intelligence.* New York, NY. St. Martin's Griffin.

Hegel, GWF. 1976. *The Phenomenology of Spirit.* Translated by A.V. Miller. Oxford, UK. The Oxford University Press.

Hegel, GWF. 2016. *Elements of the Philosophy of Right.* Translated by H.B. Nisbet. Edited by Allen W. Wood. Cambridge, UK. Cambridge University Press.

Heidegger, Martin. 2008. *Being and Time.* Translated by John Macquarrie and Edward Robinson. New York, NY. Harper Perennial

Heidegger, Martin. 1977. *The Question Concerning Technology and Other Essays.* Translated by William Lovitt. New York, NY. Harper Torchbooks.

Heidegger, Martin. 1982. *On the Way to Language.* Translated by Peter D. Hertz. New York, NY. Perennial Library, Harper & Row Publishers.

Hobbes, Thomas. 1985. *Leviathan.* London, UK. Penguin Classics.

Husserl, Edmund. 1982. *Logical Investigations, Volume 1 & 2*. Translated by J.N. Findlay. London, UK. Routledge & Kegan Paul.

Husserl, Edmund. 1992. *Experience and Judgment*. Translated by James S. Churchill and Karl Ameriks. Evanston, IL. Northwestern University Press.

Husserl, Edmund. 1986. *The Crisis of European Sciences and Transcendental Phenomenology*. Translated by David Carr. Evanston, IL. Northwestern University Press.

Husserl, Edmund. 1997. *Psychological and Transcendental Phenomenology and the Confrontation with Heidegger (1927-1931)*. Translated by Thomas Sheehan and Richard E. Palmer. Dordrecht, The Netherlands. Kluwer Academic Publishers.

Ingram, David. 1987. *Habermas and the Dialectic of Reason*. New Haven, CN. The Yale University Press.

Kant, Immanuel. 1981. *Grounding for the Metaphysics of Morals*. Translated by James W. Ellington. Indianapolis, IN. Hackett Publishing Company.

Kant, Immanuel. 1965. *Critique of Pure Reason*. Translated by Norman Kemp Smit. New York, NY. St. Martin's Press.

Kant, Immanuel. 1985. *Critique of Practical Reason*. Translated by Lewis White Beck. New York, NY. Macmillan Publishing Company.

Kant, Immanuel. 1986. *The Critique of Judgement*. Translated by James Creed Meredith. Oxford, UK. Oxford University Press.

Kelleher, John D. and Tierney, Brendan. 2018. *Data Science*. Cambridge, MA. The MIT Press.

Kierkegaard, Soren. 1983. *Fear and Trembling | Repetition. Kierkegaard's Writings, VI*. Translated by Howard V. Hong and Edna H. Hong. Princeton, NJ. Princeton University Press.

Kojève, Alexandre. 1991. *Introduction to the Reading of Hegel: Lectures on the Phenomenology of Spirit*. Translated by James H. Nichols, JR. Ithaca, NY. Cornell University Press.

Kuhn, Thomas S. 1970. *The Structure of Scientific Revolutions*. Chicago, IL. University of Chicago Press.

Kurzweil, Ray. 2005. *The Singularity Is Near: When Humans Transcend Biology*. New York, NY. Penguin Books.

Locke, John. 1986. *The Second Treatise on Civil Government*. Buffalo, NY. Prometheus Books.

Locke, John. 1959. *An Essay Concerning Human Understanding, Volume One & Two*. New York, NY. Dover.

MacKinnon, Catharine A. 1989. *Toward a Feminist Theory of the State*. Cambridge, MA. Harvard University Press.

Marx, Karl. 2009. *A Contribution to the Critique of Political Economy*. Translated by S.W. Ryazanskaya. New York, NY. International Publishers Co. Inc.

Marx, Karl. 1990. *Capital: A Critique of Political Economy, Volume I*. Translated by Ben Fowkes. New York, NY. Penguin Classics.

Marx, Leo. 1964. *The Machine in the Garden: Technology and the Pastoral Ideal in America*. London, UK. Oxford University Press.

Mead, George Herbert. 1962. *Mind, Self, & Society*. Chicago, IL. The University of Chicago Press.

Merleau-Ponty, Maurice. 2012. *Phenomenology of Perception*. Translated by Donald A. Landes. New York, NY. Routledge.

Mill, John Stuart. 2002. *The Basic Writings of John Stuart Mill*. New York, NY. The Modern Library.

Nadler, Steven. 2011. *A Book Forged in Hell: Spinoza's Scandalous Treatise and the Birth of the Secular Age*. Princeton, NJ. Princeton University Press.

Nietzsche, Friedrich. 1968. *Basic Writings of Nietzsche*. Translated by Walter Kaufmann. New York, NY. Modern Library.

Nietzsche, Friedrich. 1968. *The Will to Power*. Translated by Walter Kaufmann and R.J. Hollingdale. New York, NY. Vintage Books.

Nietzsche, Friedrich. 1982. *The Portable Nietzsche*. Translated by Walter Kaufmann. New York, NY. Penguin Books.

O'Neil, Cathy. 2016. *Weapons of Math Destruction: How Big Data Increases Inequality and Threatens Democracy*. New York, NY. Broadway Books.

Ormiston, Gayle L. and Schrift, Alan D.. 1990. *The Hermeneutic Tradition: From Ast to Ricoeur*. Albany, NY. State University of New York Press.

Pascal, Blaise. 1995. *Pensees*. Translated by A.J. Krailsheimer. New York, NY. Penguin Books.

Petzold, Charles. 2008. *The Annotated Turing: A Guided Tour through Alan Turing's Historic Paper on Computability and the Turing Machine*. Indianapolis, IN. Wiley Publishing, Inc.

Piketty, Thomas. 2014. *Capital in the Twenty-First Century*. Translated by Arthur Goldhammer. Cambridge, MA. The Belknap Press of the Harvard University Press.

Plato. 1980. *The Collected Dialogues of Plato*. Edited by Edith Hamilton and Huntington Cairns. Princeton, NJ. Princeton University Press.

Popper, Karl. 2002. *The Poverty of Historicism*. London, UK. Routledge.

Popper, Karl. 2002. *Conjectures and Refutations*. London, UK. Routledge.

Postman, Neil. 1986. *Amusing Ourselves to Death: Public Discourse in the Age of Show Business*. New York, NY. Penguin Books.

Putnam, Hilary. 2002. *The Collapse of the Fact/Value Dichotomy and Other Essays*. Cambridge, MA. Harvard University Press.

Putnam, Hilary. 1998. *Reason, Truth, and History*. Cambridge, UK. Cambridge University Press.

Putnam, Hilary. 1988. *Representation and Reality*. Cambridge, MA. The MIT Press.

Quine, Willard Van Orman. 1980. *From a Logical Point of View: Nine Logico-Philosophical Essays*. Cambridge, MA. Harvard University Press.

Rawls, John. 1971. *A Theory of Justice*. Cambridge, MA. Harvard University Press.

Rorty, Richard. 2009. *Contingency, Irony, and Solidarity*. Cambridge, UK. Cambridge University Press.

Roth, Michael. 1996. *The Poetics of Resistance: Heidegger's Line*. Evanston, IL. Northwestern University Press.

Roth, Michael. *Being, An Introduction: To Gift-Giving, Time*. Unpublished manuscript.

Roth, Michael. *Time Between Summers*. Unpublished manuscript.

Roth, Michael. *Artists of Despair: Tales of Unrequited Love*. Unpublished manuscript.

Roth, Michael. *Further Investigations of a Dog*. Unpublished manuscript.

Rousseau, Jean-Jacques. 1984. *The Social Contract*. Translated by Maurice Cranston. New York, NY. Penguin Books.

Rouvroy, Antoinette and Poullet, Yves. 2009. "The Right to Informational Self-Determination and the Value of Self-Development: Reassessing the Importance of Privacy for Democracy" from *Reinventing Data Protection?* Berlin, Germany. Springer.

Sartre, Jean-Paul. 1984. *Being and Nothingness*. Translated by Hazel Barnes. New York, NY. Washington Square Press.

Sartre, Jean-Paul. 1982. *Critique of Dialectical Reason, Volume I: Theory of Practical Ensembles*. Translated by Alan Sheridan-Smith. London, UK. Verso Books.

Schneier, Bruce. 2015. *Data and Goliath: The Hidden Battles to Collect Your Data and Control Your World*. New York, NY. WW Norton & Company.

Schoeman, Ferdinand (editor). 1984. *Philosophical Dimensions of Privacy: An Anthology*. Cambridge, UK. Cambridge University Press.

Sheehan, Thomas (ed). 1981. *Heidegger: The Man and the Thinker*. Chicago, IL. Precedent Publishing, Inc.

Silver, Nate. 2015. *The Signal and the Noise: Why So Many Predictions Fail – But Some Don't*. New York, NY. Penguin Books.

Sloterdijk, Peter. 2017. *In the World Interior of Capital*. Translated by Wieland Hoban. Cambridge, UK. Polity Press.

Smart, J.J.C. 2009. *Philosophy and Scientific Realism*. London, UK. Routledge.

Smith, Adam. 1994. *The Wealth of Nations*. New York, NY. The Modern Library.

Solove, Daniel J. 2009. *Understanding Privacy*. Cambridge, MA. Harvard University Press.

Spinoza, Baruch. 1982. *The Ethics*. Translated by Samuel Shirley. Indianapolis, IN. Hackett Publishing Company.

Vaughan, William. 2018. Correspondence.

Warren, Samuel D. and Brandeis, Louis D. 1890. *The Right to Privacy*. Cambridge, MA. Harvard Law Review. Vol. 4, Num. 5.

Weber, Max. 2002. *The Protestant Ethic and the "Spirit" of Capitalism*. Translated by Peter Baehr and Gordon C. Wells. New York, NY. Penguin Classics.

Weber, Max. 1978. *Economy and Society*. Edited by Guenther Roth and Claus Wittich. Berkeley, CA. University of California Press.

Westin, Alan. *Privacy and Freedom*. 1967. New York, NY. IG Publishing.

Wittgenstein, Ludwig. 2001. *Philosophical Investigations*. Translated by G.E.M. Anscombe. Malden, MA. Blackwell Publishing.

Wu, Tim. 2016. *The Attention Merchants: The Epic Scramble to Get Inside Our Heads*. New York, NY. Vintage Books.

Zizek, Slavoj. 2016. *Refugees, Terror and Other Trouble with the Neighbors: Against the Double Blackmail*. Brooklyn, NY. Melville House.

Zizek, Slavoj. 2007. "Why Heidegger Made the Right Step in 1933". Volume 1.4. International Journal of Zizek Studies.

Zizek, Slavoj. 1999. *The Ticklish Subject: The Absent Centre of Political Ontology*. London, UK. Verso.

Made in the USA
Monee, IL
04 November 2026

67385945R10302